Global Power Shift

Comparative Analysis and Perspectives

Ample empirical evidence points to recent power shifts in multiple areas of international relations taking place between industrialized countries and emerging powers, and between states and non-state actors. Yet there is a dearth of theoretical interpretation and synthesis of these findings, and a growing need for coherent approaches to understand and measure the transformation. This edited series aims to bring together scholars from all major world regions as well as different disciplines in order to discuss and possibly blend their different approaches and provide new frameworks for the understanding of global affairs and the governance of global power shifts.

More information about this series at
http://www.springer.com/series/10201

Maximilian Mayer • Mariana Carpes •
Ruth Knoblich
Editors

The Global Politics of Science and Technology - Vol. 1

Concepts from International Relations and Other Disciplines

Springer

Editors
Maximilian Mayer
Center for Global Studies
University of Bonn
Bonn
Germany

Mariana Carpes
GIGA German Institute of Global and Area Studies
Hamburg
Germany

Ruth Knoblich
Institute of Development Research and Development Policy (IEE)
Ruhr-University Bochum
Bochum
Germany

and

Institute of Political Science and Sociology
University of Bonn
Bonn
Germany

ISSN 2198-7343 ISSN 2198-7351 (electronic)
ISBN 978-3-642-55006-5 ISBN 978-3-642-55007-2 (eBook)
DOI 10.1007/978-3-642-55007-2
Springer Heidelberg New York Dordrecht London

Library of Congress Control Number: 2014948207

Printed on acid-free paper

Springer is part of Springer Science+Business Media (www.springer.com)

Acknowledgements

Since we have started working on this project about 2 years ago we have been receiving an enormous amount of support from institutions, colleagues and friends. Since the initial phase, the support and encouragement especially of JP Singh, Sheila Jasanoff, Johan Eriksson, and Stefan Fritsch were crucial. The discussions with many colleagues at the ISA annual conventions (2011–2013) and elsewhere were a warm welcome for us to pursue this project: gather the divers research on the global politics of science and technology as a subfield of International Relations. Many more authors than we initially had expected were willing to contribute with their ideas, insights and comments. Thus, the original plan of editing a book rapidly turned into the two volumes collection we have now the pleasure to present. We hope that this handbook will not only be helpful as an empirical and conceptual overview, a collection of theories, and a toolbox of research frameworks. It also aims at encouraging students and experts in IR and other disciplines to deepen their collaborations and jointly further this important area of study.

We want to thank all contributing authors and interview partners for their persistent engagement and dedication throughout all steps of this project. As there was no opportunity to invite all authors to a conference, the constant exchange and development of ideas between authors, interviewees, reviewers, and the editors—mostly via e-mail conversation—became very fruitful for improving the coherence, comprehensiveness, and structure of this compendium. In this journey, we are especially grateful to those researchers coming from "outside" IR. Their patience, willingness to an extended engagement and inspiring perspectives were responsible for some of the highlights within this project. Merging further the knowledge of their fields of expertise with the discipline of IR is a worthwhile task. Special thanks go, at this point, also to Peer Schouten who has contributed one interview to the book and offered to cooperate with his website: Theory Talks.

Unwavering support of our home institutions was instrumental for realizing the intellectual and practical parts of this project. We want to thank the Center for Global Studies, University of Bonn, and especially its student assistants who did a brilliant and precise job on formatting and checking chapters as well as transcribing interviews. Similarly important was the support of the Institute of Development Research and Development Policy, Ruhr-University Bochum, the Institute of Political Science and Sociology, University of Bonn, and GIGA German Institute of Global and Area Studies Hamburg. We are especially thankful to Douglas Howland, Peer Schouten, Anna Agathangelou, and Christian Bueger for their critical engagement with our two introductory texts. We furthermore want to thank Barbara Fess, Marion Kreisel, G. Ramya Prakash and the team of the section Economics and Political Science at Springer publishing for their guidance and kind assistance throughout the entire project. Finally, we are very grateful to our families and close friends for all their support and understanding.

Contents

The Global Politics of Science and Technology: An Introduction 1
Maximilian Mayer, Mariana Carpes, and Ruth Knoblich

Part I Approaches and Debates About Sciences and Technologies Within IR

From Expert Communities to Epistemic Arrangements: Situating Expertise in International Relations 39
Christian Bueger

Nuclear Science and Technology: The Race Between Weapons and Controls 55
Joseph F. Pilat

International Competition and Nanotechnology Policies: Discourse, Hegemony, and International Political Economy 75
Joscha Wullweber

Rise of a Cybered Westphalian Age: The Coming Decades 91
Chris C. Demchak and Peter J. Dombrowski

Conceptualizing the Ambivalent Role of Technology in International Relations: Between Systemic Change and Continuity 115
Stefan Fritsch

Part II Interventions from STS, History, Innovation Studies and Geography

Petroleum Supply, Marine Transportation Technology, and the Emerging International Order of the Post World War One Period 141
Peter J. Hugill

The Globalization of Intellectual Property Rights: Much Ado About Nothing? 161
Daniele Archibugi and Andrea Filippetti

Telegraph Technology and Administrative Internationalism in the Nineteenth Century 183
Douglas Howland

Biotechnology and Empire: The Global Power of Seeds and Science . . . 201
Sheila Jasanoff

Technological Collaboration and Nuclear Proliferation: A Transnational Approach 227
John Krige

Part III Interlude: Exploring a Multi-Disciplinary Field

Specialist Knowledge, Carbon Democracy, and Local Technicalities as Basis for the 'International': An Interview with Timothy Mitchell . . . 245
Peer Schouten

The Triple Helix, the Complexity of Technological Innovations, and the Decomposition of National Innovation Systems: An Interview with Loet Leydesdorff 251
Ruth Knoblich

Gaia Theory, Ecovillages, and IR Taking Material Limits Seriously: An Interview with Karen Litfin 259
Maximilian Mayer

The Role of Science and Technology in the Dynamics of Global Change and the Significance of International Knowledge Cooperation in the Post-Western World: An Interview with Dirk Messner 267
Ruth Knoblich

Nuclear Ontologies, Technopolitics in Postcolonial Spaces, and the Cold War as Transnational History: An interview with Gabrielle Hecht 275
Maximilian Mayer

Volume I: Concepts from International Relations and Other Disciplines

Daniele Archibugi is a Research Director at the Italian National Research Council in Rome, and Professor of Innovation, Governance and Public Policy at the University of London, Birkbeck College. He works on innovation and on the political theory of international relations. He has worked and taught at the Universities of Sussex, Naples, Madrid, Cambridge, Rome, London School of Economics and Political Science, Harvard, Ritsumeikan Kyoto and Swufe Chengdu. In June 2006 he was appointed Honorary Professor at the University of Sussex.

Christian Bueger is Lecturer in International Relations at Cardiff University. He holds a Ph.D. in Social and Political Sciences from the European University Institute. His core fields of research are international organizations, maritime security, contemporary piracy, international practice theory and sociology of science and expertise. Further information is available at http://bueger.info.

Mariana Carpes is a research fellow and Ph.D. candidate at the GIGA German Institute of Global and Area Studies where she is working on a historical comparative analysis of the nuclear strategies of Brazil, India, and South Africa. Her main areas of interest are history and theory of IR, nuclear politics, and foreign policy analysis. Recently Mariana has also started to work on strategic partnership between EU and Brazil. In 2012 and 2013 she held presentations on this topic at the European External Action Service (EEAS). She holds a bachelor's degree in History from the Universidade Federal Fluminense (Brazil), and a master's degree in International Relations from the Pontificia Universidade Catolica do Rio de Janeiro.

Chris C. Demchak, a professor of cybered conflict and complex socio-technical systems in the Strategic Research Department of the US Naval War College and a former Army Reserve officer. She received her Ph.D. in political science from UC Berkeley and holds an MPA in economic development from Princeton and an M.A. in energy engineering from Berkeley. She has published articles on comparative security, cyberspace, organizations, and large-scale systems surprise and three books: *Military Organizations, Complex Machines; Designing Resilience;* and

Wars of Disruption and Resilience: Cybered Conflict, Power, and National Security.

Peter J. Dombrowski is a professor of strategy at the Naval War College and chair of the Strategic Research Department. Previous positions include director of the Naval War College Press, editor of the *Naval War College Review,* co-editor of *International Studies Quarterly,* associate professor of political science at Iowa State University, and defense analyst at ANSER, Inc. He has authored over 45 articles, monographs, book chapters, and government reports and is co-editor of *Balance Sheet: The Iraq War and U.S. National Security* (Stanford University Press, 2009).

Andrea Filippetti is researcher at the National Research Council of Italy (CNR). He has graduated and taken his Ph.D. in Economic Science at the University "La Sapienza" of Rome. He has been Visiting Researcher at Birkbeck College, Harvard University and the University of Queensland, Australia. He is interested in innovation, EU innovation policy, regional development and institutions, the globalization of intellectual property rights, technological change and productivity growth. He has recently published with Daniele Archibugi a book on the effects of the economic crisis on innovation with Routledge. Together with Daniele he is also preparing the *Handbook of Global Innovation, Science, and Technology* with Wiley-Blackwell.

Stefan Fritsch is Assistant Professor of International Relations and Comparative Politics in the Department of Political Science at Bowling Green State University. He has published on topics including technology and global affairs, theories of global political economy, global trade, multinational corporations and European integration. His work has appeared in *International Studies Perspectives*, *Review of International Political Economy*, *Global Society* and the *Austrian Journal of Political Science*.

Gabrielle Hecht is Professor of History at the University of Michigan, where she also directs the Program in Science, Technology, and Society and serves as associate director of the African Studies Center. She recently published *Being Nuclear: Africans and the Global Uranium Trade* (MIT Press and Wits University Press, 2012), which has received awards from the American Historical Association and the American Sociological Association, as well as the 2013 Susanne M. Glasscock Humanities Book Prize and Honorable Mention for the African Studies Association's 2013 Herskovits Award. She is also the author of *The Radiance of France: Nuclear Power and National Identity after World War II* (MIT Press 1998 and 2009) and editor of *Entangled Geographies: Empire and Technopolitics in the Global Cold War,* editor (MIT Press, 2011).

Joseph F. Pilat is a Program Manager in the National Security Office of the Los Alamos National Laboratory and a Global Fellow at the Woodrow Wilson International Center for Scholars where he co-directs the Nonproliferation Forum. He served as Representative of the Secretary of Defense to the Fourth Review Conference of the Nuclear Non-Proliferation Treaty (NPT), and as an adviser to the US Delegation at the 1995 NPT Review and Extension Conference. Dr. Pilat also served as representative of the Secretary of Defense to the Open Skies negotiations. He has held positions in the Pentagon and the Congressional Research Service, and has taught at Cornell University, Georgetown University and the College of William and Mary. He is the editor of *Atoms for Peace: A Future after Fifty Years?* (2007).

Douglas Howland is Professor of History at the University of Wisconsin-Milwaukee (USA). He is, most recently, the author of *Personal Liberty and Public Good: The Introduction of John Stuart Mill to Japan and China* (Toronto, 2005) and co-editor (with Luise White) of *The State of Sovereignty: Territories, Laws, Populations* (Indiana, 2009).

Peter Hugill holds degrees from British, Canadian, and American universities and is currently Professor of Geography at Texas A&M University, where he has taught since 1978. His current book project is on the hegemonic struggle between Britain and America in the early twentieth century, an economic struggle between two trading states obscured by the larger, often military struggles that occupied much of the 1900s. His two major books on the relationship between technology, geography, and the world-system are published by Johns Hopkins University Press: *World Trade since 1431* in 1993, and *Global Communication since 1844*. He has also published numerous articles and book chapters.

Sheila Jasanoff is Pforzheimer Professor of Science and Technology Studies at the Harvard Kennedy School. A pioneer in her field, she has authored more than 100 articles and chapters and is author or editor of a dozen books, including *Controlling Chemicals, The Fifth Branch, Science at the Bar*, States of Knowledge, and *Designs on Nature*. Jasanoff served on the Board of Directors of the American Association for the Advancement of Science and as President of the Society for Social Studies of Science. Her grants and awards include a 2010 Guggenheim Fellowship and an *Ehrenkreuz* from the Government of Austria. She holds AB, JD, and PhD degrees from Harvard, and an honorary doctorate from the University of Twente.

Ruth Knoblich is a research fellow at the Institute of Development Research and Development Policy (IEE), Ruhr-University Bochum and at the Institute of Political Science and Sociology, University of Bonn; she teaches courses in both institutes and at the South African – German Centre for Development Research, University of the Western Cape. She studied in Bonn and Madrid and holds a master's degree in Political Science, History and Philosophy from the University of Bonn as well as two basic degrees in Medicine. Her research interests include innovation policies and innovation systems, emerging economies and South-South science and technology cooperation, global politics of intellectual property rights, science and technology in the international political economy, and national policy making in the context of globalisation and global governance.

John Krige is the Kranzberg Professor in the School of History, Technology and Society at the Georgia Institute of Technology in Atlanta. He has published extensively on the role of space and nuclear science and technology in shaping U. S.—European relations in the cold war. His current project will supplement his *American Hegemony and the Postwar Reconstruction of Science in Europe* (MIT Press, 2006) with an analysis of the role of transatlantic technological collaboration as an instrument of nonproliferation.

Loet Leydesdorff (Ph.D. Sociology, M.A. Philosophy, and M.Sc. Biochemistry) is Professor Emeritus at the Amsterdam School of Communications Research (ASCoR) of the University of Amsterdam. He is Visiting Professor of the Institute of Scientific and Technical Information of China (ISTIC) in Beijing; Visiting Professor at the School of Management, Birkbeck, University of London; and Honorary Professor of the Science and Technology Policy Research Unit (SPRU) of the University of Sussex. He has published extensively in systems theory, social network analysis, scientometrics, and the sociology of innovation (see for a list of publications at http://www.leydesdorff.net/list.htm).

Karen Litfin is an associate professor in the Department of Political Science at the University of Washington. She has a Ph.D., University of California, Los Angeles, (1992) and specializes in global environmental politics, with core interests in green theory, the science/policy interface, and what she calls "person/planet politics." Her first book, *Ozone Discourses: Science and Politics in International Environmental Cooperation* (Columbia University Press, 1994), looks at the discursive framing of science in the ozone treaties. Her second book, *The Greening of Sovereignty in World Politics* (MIT Press, 1998), explores how state sovereignty is being reconfigured as a consequence of global environmental politics. Some of the topics of her recently publications include: the politics of earth remote sensing; the political implications of Gaia Theory; the relationship between scientific and political authority in the climate change negotiations; the politics of sacrifice in an ecologically full world; and holistic thinking in the global ecovillage movement.

Maximilian Mayer is program co-chair of STAIR (Science, Technology, Arts and International Relations) a newly established section of the International Studies Association. He studied in Bochum and Shanghai and holds a master degree in East Asian politics from the Ruhr-University Bochum. He received scholarships from the German National Academic Foundation twice from 2004 to 2007 and from 2009 to 2010. He joined the Center for Global Studies in October 2009 and has worked there as managing assistant (2009–2013). He leads the CGS research group on "rising knowledge powers". His research interests include the global politics/ policy of science, innovation, and technology; Chinese foreign, and energy policy; International Relation theories.

Dirk Messner is the Director of the "German Development Institute / Deutsches Institut für Entwicklungspolitik (DIE)" since 2003 and teaches at the Institute of Political Science, University of Duisburg-Essen. He is Co-Director of the "Käte Hamburger Kolleg / Centre for Global Cooperation Research (KHK/GCR)",

University Duisburg-Essen, which was established in 2012. He furthermore is Co-Chair of the "German Advisory Council on Global Change (WBGU)", member of the "China Council on Global Cooperation on Development and Environment", member of the "Global Knowledge Advisory Commission" of the World Bank and member of the "European Commission's Scientific Advisory Board for EU development policy". Dirk Messner's research interests and work areas include globalisation and global governance, climate change, transformation towards low carbon economies, and development policy. He directed many international research programs and thus created a close international research network.

Timothy Mitchell is a political theorist and historian. His areas of research include the place of colonialism in the making of modernity, the material and technical politics of the Middle East, and the role of economics and other forms of expert knowledge in the government of collective life. Much of his current work is concerned with ways of thinking about politics that allow material and technical things more weight than they are given in conventional political theory. Educated at Queens' College, Cambridge, where he received a first-class honours degree in History, Mitchell completed his Ph.D. in Politics and Near Eastern Studies at Princeton University in 1984. He joined Columbia University in 2008 after teaching for 25 years at New York University, where he served as Director of the Center for Near Eastern Studies. At Columbia he teaches courses on the history and politics of the Middle East, colonialism, and the politics of technical things.

Joscha Wullweber is assistant professor in the department of Globalisation and Politics at the University of Kassel. He holds a Ph.D. in political science, an M.A. in global political economy and an M.Sc. in biology. He is co-representative of International Political Economy, a permanent working group of the German Political Science Association (DVPW), and founding member of the German Working Group on Theories of Hegemony and Discourse. His research interests concentrate on political theory, in particular theories of international relations and international political economy, finance and financial crises, nanotechnology and innovation policies, security studies and intellectual property rights.

The Global Politics of Science and Technology: An Introduction

Maximilian Mayer, Mariana Carpes, and Ruth Knoblich

Abstract The reality of international politics has rapidly grown in complexity. This complexity has been pressuring the discipline of International Relations (IR) to engage with new phenomena, concerns, and issue areas, and to translate them into innovative theorizations. Science and technology is one of these issues. Contemporary human life is tied to and thoroughly permeated by artifacts, technical systems and infrastructures, making it hard to imagine any international or global issue that does not have technological or scientific aspects. However, this condition remains fundamentally challenging for many approaches within IR, in which instead science and technology have been largely treated as exogenous. Although an increasing number of IR scholars is exploring the roles scientific practices and technological systems play in international affairs and global politics, the subject matter deserves much more systematic scrutiny. The following chapter articulates the conceptual, intellectual and academic contexts of this two-volume collection on the *Global Politics of Science and Technology*. After pointing out general normative challenges and briefly problematizing global technological transformations, we recapitulate the evolving IR scholarship on the topic. We argue that, although most IR theories do not grant science and technology a genuine conceptual place, there is enough research to document and reconstruct the breadth and depth of the vivid, yet unrecognized subfield of IR. While the further development of this subfield would greatly benefit from interdisciplinary conversations, we propose the notion of techno-politics to

M. Mayer (✉)
CGS, University of Bonn, Bonn, Germany
e-mail: maximilian.mayer@uni-bonn.de

M. Carpes
GIGA, Hamburg, Germany
e-mail: mariana.carpes@giga-hamburg.de

R. Knoblich
IEE, Ruhr-University Bochum, Bochum, Germany

IPWS, University of Bonn, Bonn, Germany
e-mail: ruth.knoblich@uni-bonn.de

M. Mayer et al. (eds.), *The Global Politics of Science and Technology - Vol. 1*,
Global Power Shift, DOI 10.1007/978-3-642-55007-2_1,

indicate how the discipline might rearticulate existing analytical frameworks, establish innovative conceptualizations, and advance new concerns for research.

Keywords Technology • Science • IR theory • Techno-politics • Global transformations • Technological determinism • Interdisciplinarity

1 Introduction

The reality of international affairs has rapidly grown in complexity, pressuring the discipline of International Relations (IR) to engage with new phenomena. IR scholarship thus has to address concerns and issue areas by translating them into innovative theorizing. Science and technology is the most prominent among these—it is hard to imagine any international or global issue that does not entail technological or scientific aspects. International security, statehood, global governance as well as warfare and foreign policy are thoroughly permeated by and embedded in material artifacts, technical systems and infrastructures, and scientific practices. As topic, science and technology attracts significant attention within IR; security studies are perhaps the most notable case, treating science and technology as key strategic tools in the Cold War. However, for many approaches within IR the analysis of science and technology remains fundamentally challenging. They have been largely treated as exogenous to theoretical schools and the field. Thus, while an increasing number of IR scholars are looking at the politics of science and technology, the subject matter needs to be scrutinized much more systematically. The discipline still needs to build up internal logics capable of integrally capturing the diverse meanings and dynamics of science and technologies. On the other hand, the study of (global) science and technology has become a diversified multidisciplinary effort. While a large research area exists that partly overlaps with issues in IR and often challenges central premises of IR, it lacks substantial contributions by, and sustained connections to, IR scholarship.

Having this in mind, the present two volumes explore the politics of science and technology from a variety of perspectives ranging from classic themes such as nuclear weapons to recent debates about nanotechnology, drones, peak oil, cyberspace, supercomputers and biomedical technologies. The collection pursues two central aims. First, it documents the breadth and depth of research about the *global politics of science and technology*. We argue that it already constitutes a vivid, yet largely unrecognized and still underexplored, subfield of IR. Second, these volumes present detailed empirical studies and diverse efforts of theory building by senior and junior researchers that reinforce a commitment to interdisciplinary dialogues with those research areas in which science and technology have traditionally played a central role. The first volume provides a state-of-the-art compilation of respective debates and research in IR. In addition, it introduces alternative and/or complementary contributions from geographers, theorists of science and technology, historians, and economists on the subject matter. Thereby a space for mutual learning is carved out, contributing to a better understanding of the canvass of the global politics of science and technology. The second volume features various case studies and perspectives. Based on new concepts and methods, the authors analyze how and to what extent different technical networks, artifacts, and scientific practices shape, perform, and transform global politics today. We invited authors and theorists from different theoretical perspectives and disciplinary backgrounds to articulate core issues at stake. This way we hope to clarify the historical, disciplinary and conceptual background of the global politics of science and technology.

To begin with, this introductory chapter highlights that the meanings and purposes of science and technology are deeply contested in today's social sciences. Two conflicting perspectives on technology, an optimistic and a skeptical view, raise difficult theoretical and normative challenges for IR. Historically, analyses in the discipline studied global technological transformations and military revolutions in modern world politics. In particular, nuclear strategies were a core issue of the first great wave of IR scholarship that, emerging after World War II, fully established the discipline as such. The next wave, beginning in the 1970s, shifted the focus to state-market relations, and then increasingly to global interconnectedness. Nevertheless, the overall attention of the discipline to technology and science remained remarkably narrow, particularly with regard to theorizing. While this gap would require a more comprehensive discussion, in the present chapter we can only sketch it out, by outlining clusters of interrelated intellectual contextual factors. Most importantly among them are the dominance of instrumentalist understandings of the material world and the ideational bias of leading IR schools. In part, this might explain why IR—unlike many other social science disciplines—did not develop a distinct subfield around technology and science. Finally, by employing the notion of techno-politics we explore how IR might construe a conceptual place for science and technology by reformulating existing puzzles, opening up space for new topics and synergize existing research.

2 Two Tales of Technology

Studies that focus on science and technology inescapably become embroiled in normative questions concerning meaning, purposes and consequences. IR is no exception. For the present discussion, we organize these normative issues at play in the broader social science debate around science and technology into two conflicting narratives: the first could be called a *tale of hope*. It conveys the optimist idea that advances in technology and science tend to make society better. The second is a *tale of pessimism,* assuming that new technologies and scientific advances have potentially negative or even disastrous consequences. Subsequently situating existing work in IR within these conflicting cultural narratives reveals that the subject matter hardly allows for a neutral analytical position. All politics are entwined with technologies and sciences in a great variety of ways going well beyond the conceptual apparatus of IR.

The core assumption of the first narrative is that subsequent stages of technological developments lead to significant improvements of economic progress and human conditions. It is assumed that a combination of science and technology offers better solutions to critical global challenges such as security, public health, energy, food and water supply, poverty, and climate stability. The overarching trends towards more efficiency, smartness, and artificial intelligence are understood as not only reshaping and redesigning but actually improving urban planning, reproduction, advertisement, and business models (e.g. Kurzweil 2006). From a different angle, proponents of this narrative endorse the development of smart technologies in order to mitigate the assumed limits of economic growth. Hyper-efficiency is presumed as one of the preeminent (technology-based) building blocks for fostering wealth and a decent life for nine billion people within our "planetary boundaries" (see Byrne and Glover 2005; Pielke et al. 2008). International institutions such as the World Bank and different UN initiatives call for closer international cooperation in scientific research and technological development. They argue in favor of global developmental research that makes use of open data, open access to research pools, and collaborative knowledge production. In addition, through the systematic use of "big data", the UN pushes the realization of the Millennium Development Goals and the Post-2015 Development Agenda (United Nations 2012, 2013). Furthermore, experts hope that the rapidly growing relevance of Internet-based telecommunications would render digital infrastructures and applications instrumental to modernizing agriculture (Juma 2010).

Of course, techno-optimism equally applies to the military realm, as illustrated by Anders F. Rasmussen's recent appraisal of advanced missile defense capabilities of NATO:

> To link all of these national assets together, NATO has developed, and is expanding, a technologically advanced command-and-control system, based at Ramstein Air Base in Germany. The system already can connect satellites, radars, and interceptors to defend against missile attacks, and that capability will grow more complex and agile in the years ahead. This makes NATO unique: it is the only multilateral organization that can combine the most complex systems from the world's most capable countries to create an effective whole (Rasmussen 2014).

Similarly, it is asserted that the employment of autonomous and unmanned weapon systems makes warfare more precise, allegedly less inhuman, and thus an ethical imperative (Arkin 2010; see Kaag and Kaufman 2009). The idea of the "technological fix", to put it differently, dominates decision-makers from Beijing to Kinshasa and from Washington to Brussels. In short, the optimistic tale of technology is driven by unmitigated ambitions and high expectations about continuous improvements. However, the assumption that technology and science primarily are instruments of progress is contested.

In opposition to the first narrative, many observers express skeptical or at times pessimistic views. Proponents of this view consider a fetish of modernization the belief in unending opportunities and inevitable triumph of emerging technologies and scientific knowledge. In particular, some see technological determinism as a powerful modern ideology shared across diverging cultural and historical experiences and within different political and economic systems (Adas 1989; Smith and

Marx 1994). Hence, cautious voices call for self-imposed limits in order to avoid direct links between scientific progress and weaponization, warning at the same time against the increasing militarization of academic research (Price 2011). The inherent *risks* of pollution and ultimately annihilation, paradoxically resulting from technological progress, and the objective status of related scientific knowledge itself became subject of public concerns and contestation in the late 1960s (Lear 1993; Jasanoff 1990; Nelkin 1992). According to sociologists, our societies have already reached the state of "reflexive modernity" in which unintended consequences of technologies are taken seriously in, but continue to outpace mainstream central planning and political discourse (Beck 1992; Beck et al. 1994). On the sidelines, only a few marginal voices keep insisting on "simplification", de-growth, and "appropriate technologies".[1]

From a skeptical angle, "technological progress" is thus seen as the last remaining great myth of the postmodern age. It seems resistant to an unending series of catastrophes, accidents and pending risks of annihilation related to modern weapon systems, fragile infrastructures or the proliferation of toxic particles and artifacts (see Virilio 2010). Early IR scholarship on deterrence and nonproliferation of nuclear weapons belongs, in a sense, to the pessimistic perspective as it was occupied with the nightmare of managing the "absolute weapon" in the context of strategic and ideological rivalry. However, Eric Schlosser's (2013) frightening account of the number of nuclear accidents indicates that nuclear arsenals are generative of existential security threats in addition to their strategic dimensions, on which IR puts its main emphasis (e.g. Mueller 1988). After 1945, more than 1,000 accidents have occurred in the United States alone. The consequences of nuclear catastrophes such as in Ozyorsk, Chernobyl, and most recently Fukushima are likely to last for at least ten thousands of years. They are truly global in their reach and impact—as holds true for the immense waste lands created by mining radioactive ores, testing weapons, and dumping nuclear materials (Makhijani et al. 2000; Hecht 2012).

It is in this sense that the second tale entails a dystopian register. It pictures individual and collective life emerging through oppressive technical ensembles, which have often deteriorating consequences for individual life and social coherence. From a critical view, scientific and technological progress and specific technologies as such are never neutral. They can even express irrational or inhuman ambitions (Feenberg 1999). Besides the industrially planned and organized area bombing campaigns during and after World War II, there is perhaps no better illustration of the dire consequences of technological capabilities than the crucial, yet often overseen, role of scientists and technologies in the realization of the Holocaust (Thad Allan 2001; Bauman 2013). Moreover, the extent to which modern sciences and technologies were implicated in colonial domination, racism,

[1] Precursors include Ralph Waldo Emerson, Ivan Illich's work on energy, Ernst Friedrich Schumacher's notion of "appropriate technology" and Masanobu Fukuoka's insights about farming.

and exploitative suppression, both practically and as a "white mythology",[2] has remained largely under-scrutinized in social sciences (Adas 1997; Endres 2009).

This pessimistic understanding usually refers to—correctly or incorrectly—the reasoning of influential thinkers such as Henry D. Thoreau, Max Weber, Martin Heidegger, Max Horkheimer, Lewis Mumford, Walter Benjamin, Jacques Ellul, Norbert Elias or Michel Foucault (Berman 1983; Mitcham 1994; Matthewman 2011). When the experience of the First World War shattered Europe's claim to civilizational superiority, non-Western thinkers including Mahatma Ghandi and Rabindranath Tagore also scrutinized the non-civilized and horrible side of Europe's technological advances (Mishra 2012). Public figures such as Albert Einstein and Albert Schweitzer adopted their "peripheral" outlook; a view that was in fact a Southern precursor of the anti-nuclear movement in the 1960s and 1970s. The most recent example for profound techno-pessimism is the deteriorating perception of the World Wide Web. Many are coming to see digital networks turning from an erstwhile tool of liberty, consumerism, and cultural exchange into an intrusive instrument of state oppression, censorship, and espionage inescapably bound to further erode personnel freedom and privacy (Deibert et al. 2010; MacKinnon 2013).

Of course, these two tales are over-simplifications of a much more differentiated concert of responses to the technological condition of humanity. Nevertheless, they are useful insofar as they indicate normative and conceptual challenges with which the discipline of IR has to reckon. Firstly, although early realist thinkers including Hans J. Morgenthau and John Herz occupied a fundamentally critical position towards modern technologies (Scheuerman 2009; van Munster and Sylvest 2014) the optimistic tale of technology arguably functions as the implicit default position within most IR literature. As the discipline increasingly reflects upon the significance of technology and science, it has to engage with the difficulty of retaining a seemingly "neutral" analytical position (see Richards and Ashmore 1996; Jasanoff 2003). Authors in science and technology studies (STS) as well as in feminist and postcolonial studies stress that technologies inherently *have politics*—albeit in ambivalent ways. Consequently, they reject instrumental understandings. As science and technology are deeply implicated in the fabric of life and society, they cannot be objectified, that is, treated purely as means of economic or military might.

These insights help IR approaches to cultivate a dose of sensitivity for the intimate link between normative and methodological issues.[3] Secondly, both tales—especially in their stronger deterministic versions—tend to deemphasize political deliberations and power struggles that evolve and are weaved *through* technologies and sciences. As Evgeny Morozov (2014) argues with respect to communication gadgets and networks "there are different ways to wire the world": The inherent danger of "Internet-centrism" is that our "infrastructural imagination gets atrophied to a point where we can no longer imagine how to

[2] Dinerstein (2006: 578) labels this discursive ensemble as "fusion of progress, technology, and religion" structuring Euro-American identity.

[3] See Winner (1980), Haraway (1991), Joerges (1999), Woolgar and Cooper (1999), Rose (2007).

organize our technological affairs" (see Morozov 2013). The theoretical and analytical challenges for IR, therefore, consist in neither reducing all technology to social construction, nor reducing all politics to technological determination. Additionally, a reflection on the two tales invites IR to overcome its often purely instrumental view in order to confront the hidden and often dire consequences of the pursuit of science and technologies beyond the immediate concerns of foreign policy, state power, efficient governance or national sovereignty.

3 Global Technological Transformations and the Limits of Determinism

While some narrate human history as a sequence of technological revolutions (Innis 1950; Mumford 1966; White 1966; McNeill 1982), the link between global or international transformations and technological changes became especially virulent with the onset of modernity (Harvey 1990). Yet, it remains a challenging task to analyze the nexus of technologies and changing political orders in a non-deterministic way. For example, at the peak of colonialism and imperialism, observers and colonial administrators recognized the reality of time-space compression. Technological innovations such as electronic telegraphy, new means of traffic, and other globe-spanning infrastructures rendered the geography of the globe into a single entity (Bayly 2004; Osterhammel 2009). The ability to innovate, employ, and control novel instruments of knowledge and communication also became crucial for hegemonic transition and the (de)stabilization of international dominance (Hugill 1999). Halford Mackinder famously noted that the newly materialized communication and traffic pathways constructed a "closed political system" (1904: 422). Yet, others have argued that Mackinder's claim actually had already been realized through the naval revolutions and trade networks of the fifteenth and sixteenth centuries (Cipolla 1965; Braudel 1982; Frank 1998). Technologies and scientific innovations served not only as tools for imperial projects (Headrick 1981; Zaheer 1996; Yang 2011), but also facilitated and accelerated in multiple and unforeseen ways globalization of commercial competition, statehood and modern culture (Chandler 1977; Hughes 2004).

At the turn of the twentieth century, the international political system again appeared to experience a sea change because of path-breaking information technologies. In the 1990s, theorists pinpointed the rise of an info-sphere and digitalization as redrawing fundamental parameters of society, economy, and politics (Rosenau and Singh 2002). Seen as the real engine for the latest wave of globalization, some argued that the new Information and Communication Technologies (ICTs) had given birth to a network society (Castells 1996). On the one hand, ICTs shrunk the world into a "global village" (McLuhan and Powers 1992), leading to "flattening" the international landscape of commercial competition (Friedman 2009). On the other hand, they were seen as producing a digital gap, placing those societies, firms, nations and individuals that have access to ICT networks in an advantaged socio-economic and political position. Moreover, it has been argued that a crucial side effect of ICTs is an ongoing and profound shift of power and authority towards private entities, that is, mainly economic actors in global affairs. While earlier technological changes already had ambivalent consequences for the national sovereignty, weaving an unprecedented techno-scientific web among allies and foes alike during the Cold War (Edwards 1997; Krige et al. 2013), the rapid dynamics of innovation in ICTs arguably reinforced a divergence of market actors and states.[4] In this sense, "big data" is the latest wave of ICTs' transformative power in dividing positions: those advocating the notion of cost-effective supra-governmental sources of information are contradicted by those fearing a complete loss of democratic control over data gathering, storage and privacy (Bollier 2010; Bauman et al. 2014).

Military revolutions are perceived as the most pertinent cases of novel technologies fomenting massive global changes (Delanda 1991; Hoyt 2003). The employment of new technical devices and systems recurrently transformed warfare. Among the examples are clocks, airplanes, missiles, barbed wire, diesel engines, drones, hacking software, and so forth. At the same time, a substantial share of scientific inquiry and commercial research and development came to serve the needs and desires of national armed forces (Bousquet 2009; Der Derian 2009). Aside from their effects on battlefields, advances in weapon systems have also had far-reaching political ramifications. For instance, IR regards nuclear physics most prominently for having essentially altered statecraft and international relations forever (Jervis 1989). Bernard Brodie (1946) saw the radical impact of the "Absolute Weapon" in reversing the purpose of the military establishment from winning wars to averting them. Nuclear weapons amassed more power into the hands of national armies than ever before, while apparently ending the era of wars directly fought between great powers (van Creveld 1993; Mueller 1989). However, nuclear weapons were not simply shaping history. Numerous scholars and experts saw improvements in weapon technologies causing dangerous shifts in the defense-offense balance of power, which led to uncontrollable political and military responses during the 1950s and 1960s (Freedman 2003). But linking military revolutions to determinist understandings has its pitfalls, as Francis J. Gavin points out:

> Focusing solely or even largely on nuclear weapons to the exclusion of geopolitics, ideology, and diplomacy caricatures both the Cold War and international politics today by draining them of important political and diplomatic components. (Gavin 2012: 150–151)

Moreover, the unpredictable and non-deterministic interplay between nuclear technologies, power and security is exemplary for numerous other technological and scientific harbingers of transformation. The advent of airplanes, rocket missiles, radio, telegraph, satellites, and other revolutionary technologies turned simplifying accounts of the nexus between technology and geography (or between technology and power) ultimately obsolete. The unexpected interplay between the newly

---

[4] Most prominently embodied in the massive increases of product and capital flows seemingly governed by timeless and spaceless regimes that supersede and restructure the nation state matrix (Keohane and Nye 1977; Strange 1996; Ronen 2003).

constructed undersea cables, imperial competition and governance, global and local media outlets and markets during the late nineteenth century is another vivid illustration (see Potter 2007). It is exactly this intimate, yet contingent and controversial connection between politics, technologies and scientific practices that poses difficult conceptual challenges to IR. Yet, as Morgenthau (1964) has pointed out, neither domestic politics nor geopolitics and geoeconomic patterns can be properly understood if separated from matters of technology and science.

4 From States and Markets to Complex Global Interconnectedness

Aside from security studies (Buzan and Hansen 2009), researchers who inquire into the interaction of states, markets, and enterprises have taken up the issue of science and technology. This section explores the main challenges at the forefront (and across that edge) of IR theorizing on science and technology. The idea that the state should organize education and foster sciences has a long lineage. Likely it's most prominent proponent was Wilhelm von Humboldt. At the turn of the eighteenth century, he laid the intellectual groundwork for decades of institutionalization efforts, restructuring schools, universities, and academic disciplines in Prussia and eventually in many European states (Paul 2003). Indeed, at a much broader scale and with a multi-dimensional range, experimental science, engineering, and architecture became the actual core of state formation from the eighteenth to the twentieth century—a set of practices that centralized power by manufacturing the state into landscapes, cities, epistemic infrastructures, and individual subjectivities (Shapin and Schaffer 1985; Skocpol and Rueschemeyer 1996; Carroll 2006; Guldi 2012; Scott 1998). But it was the experience of two World Wars that impelled nation-states to rigorously and persistently steer and fund scientific research. Vannevar Bush's 1945 report *Science, The Endless Frontier* to the US president, firmly cemented the notion of state involvement in basic research and development. Besides colonial expansion, security threats and Cold War enmity were the original rationales for state-financed science (Wolfe 2013). Massive public investment enabled large-scale research to produce innovations such as particle accelerators, satellite systems and supercomputers. Those "big science" projects marked a new era of science and technology policy. They raised national prestige, technological expertise (see Kevles 1977) and—as in the European case—intergovernmental collaboration that affected profoundly the international landscape of technological competition. More than any previous hegemonic power, the preeminent position of the US developed in relation to its scientific and technological leadership (Paarlberg 2004; Krige 2006; Hecht 2011); to use Adas' (2006) phrase it is *Dominance by Design*.

In the early 1960s, the Organization for Economic Co-Operation and Development (OECD) emerged as the most influential international body comparing technological progress. By measuring cross-country innovation performances by R&D

investment per GDP and the quantity and quality of "knowledge workers", OECD experts enshrined the idea of the competition-state. In addition, economists, sociologists and the OECD later promoted the concept of the "knowledge economy". As technological expertise and innovation capacities became increasingly considered decisive factors for development, economic growth, and social welfare, building up a "knowledge society" through international education and professional training programs became the key objective of international organizations (Moldaschl and Stehr 2010; UNESCO 2005). Drawing on Amartya Sen's capability approach (1999), the notion of knowledge societies reinforced the vision of development concepts that view education as crucial for sustainable development. The contemporary concern with innovation and scientific ingenuity are mirrored in the growing attention that politicians, public institutions and businesses pay to the uses, effects, and regulation of emerging technologies. Regulations of intellectual property rights turned into the most prominent battleground of rivaling interests, mainly evolving around the agreement on Trade Related Aspects of Intellectual Property Rights, the World Intellectual Property Organization and the question to what extent the standard of intellectual property protection should further become harmonized globally (May and Sell 2005; May 2010). Arguably, the focus on technological competition between states and firms was never more intensive than today.

Innovation, creativity and fast commercialization of inventions are highly valued by governments. This is nothing new as the cases of nineteenth century Germany, Japan and United States show. Today, governments of OECD countries and the BRICS are concerned with technological shortfalls and "indigenous innovation" as they aim to attract multinational companies under (commonly assumed) conditions of highly flexible production networks and a constant threat of relocating high-tech manufacturing. In OECD member countries and emerging economies, typically several ministries, state agencies, and public-private partnerships try to steer, regulate, and stimulate knowledge production, expertise, and inventions. While the controversies mirroring the typical North-South divide for decades increasingly fade, the ascent of technologically and scientifically vibrant developing economies is about to reshape alliances, interdependencies, and the rules of technology transfer and intellectual property rights (Drahos and Braithwaite 2002; Haunss and Shadlen 2009; Yu 2012). Not unlike military revolutions that have reconfigured space and speed (Hart 1946; Virilio 1986), tectonic shifts in the global political economy tend to fuel immense expectations: the possibilities and outcomes of an ever-evolving techno-science appear almost limitless. If anything, the pace with which science becomes entangled with economic competition and state rivalry is accelerating—indeed, acceleration itself may have become a major pattern of markets and politics (Der Derian 1990; Wajcman 2008).

Some IR scholars have long seen this coming. For instance, Keohane and Nye (1977) and Susan Strange (1988) pointed out that technology and knowledge production became the central battleground for state rivalry, replacing territory and population (see Singleton 2008). It became a key theme of the post-Cold War period that states face stronger competition over power and authority from transnational enterprises. Changing their institutions, procedures, and laws, states have to navigate the networked character of global information society (Luke 1998; Keohane and Nye 1998; Kahler 2009; Gilbert and Helleiner 2013). The increasing control over cyber space and the shaping of cyber security seems to bear out the argument that governments have learned to do so, despite the growing complexity and diversification of agency in global affairs (Herrera 2002; Mayer-Schönberger and Lazer 2007; Mueller 2010). In addition, realists and world system theorists stress dynamics of technological and organizational innovation as central factors behind the rise and fall of hegemonic powers (Buzan 1987; Gilpin 1981; Arrighi 1994). Dependency theorists and Neo-Marxists emphasize the significance of technological differences, capitalist property relations and the division of labor within global production chains for the reproduction of international/world order (Cox 1987; Poulantzas 1978; Rosenberg 2010; Darby 2000). But it is in the work by James N. Rosenau (1990, 2003) on non-linear dynamics that we find the most pronounced concern with techno-social changes with global impact and, arguably, the most challenging conceptual response.

The fate of the state remains the central puzzle for IR scholars in the context of continuous technological innovation and the progress of scientific expertise (Skolnikoff 1993; Drezner 2004; Eriksson and Giacomello 2009; Betz and Stevens 2011). However, the study of technologies and large technical systems, as with many other issue areas, suffers from the "territorial trap": because this common IR mindset foregrounds relations among unitary (territorial) states instead of contextualizing states within different and interlinked local, regional and transnational processes and forms of organization (Agnew 1994), technology is merely added to a prefigured theoretical puzzle. But innovation economists and sociologists generally doubt the merits of a dichotomist approach that locates technologies and technological changes in a vague conceptual region of tensions between sovereign nation-states and global forces (Etzkowitz and Leydesdorff 2000). Studies of European integration, for instance, emphasize the complex interplay of states, spaces and transnational technological systems (Misa and Schot 2005; Brenner 2004). Krige et al. (2013: 4) similarly argue that National Aeronautics and Space Administration (NASA) is typical for large technoscientific undertakings in that it needs to be situated in "transnational or global frameworks, in recognition of the interdependence and interconnectivity of the modern state". It is hardly surprising, that critical reflections within innovation studies, with regard to the analytical treatment of states and technology, run in parallel to respective IR debates:

> Nations and states are important in ways techno-nationalism does not capture, and the international and global dimension is crucial in ways which techno-globalism is ignorant of (...). In any case, politics, multinational firms, empire and race were also crucial factors in shaping the use of technology which cut across the national and global divide in complex and changing ways. The nation, the state, and the global, are central to the history of twentieth century technology, but not in the ways the relations are usually understood. We need to rethink not only nation-technology, but technology-state relations, and the place of technology at a global level too (Edgerton 2007: 1).

Arguably, an even more essential hurdle is the widely assumed ontological divide between the realms of environment and technology as well as politics and nature, that only few scholars in IR have persistantly tried to overcome (but see Camilleri 1996; Litfin 1998; Conca 2004). As the impetus to move away from the *juxtaposition* of politics and technology increasingly gains momentum in various social sciences (cf. Latour 1993; Cronon 1995; Reuss and Cutcliffe 2010), the unease of common conceptual premises becomes particularly acute when turning to some of the central concerns within contemporary global politics. Firstly, research on climate change funded by international organizations seems to challenge the neat separation of socio-political and material-technical realms. This case prompts conceptualizations of politics that do justice to the reality of socio-technical hybridization (Miller and Edwards 2001; Linklater 2009; Edwards 2010). Daniel Deudney notes:

> The human world is now 'global and planetary' due to the explosive transformation over the last several centuries of science-based technology occurring within the geophysical and biophysical features of planet Earth. The natural Earth and its relationship with humans have been massively altered by the vast amplifications in dispersed human agency produced by the emergence and spread of machine-based civilization. The overall result of these changes has been the emergence of a global- and planetary-scale material and social reality that is in some ways similar, but in other important ways radically different, from earlier times. Practices and structures inherited from the pre-global human worlds have not adequately been adjusted to take the new human planetary situation into account (Schouten 2013b).

Secondly, as the notion of the Anthropocene acknowledges that humans have turned into a force of truly planetary magnitude (Dalby 2007), it also reinforces the increasing recognition of myriads of non-state actors. These include NGOs, international organizations, megacities, and further private, commercial, and scientific agencies of all kinds. Their transnational practices, global interactions, and extended technological structures have added a significant new layer to relations between states, and governments and enterprises. Analytically, this requires not just sensitivity for the multiplicity of actors, which is well recognized by research on global governance (Rosenau 1995).[5] Following STS and Geography, it also requires exploring the *complexity of agency* itself. That is, the locus of agency that was usually assumed to lie within individuals, groups, or states moved into hybrid, networked and mediated forms of agential power.[6] Research from fields such as security, energy, environment and elsewhere illustrates that agency is increasingly enriched by *ensembles*, *cyborgs* and *non-human* actors.[7]

Thirdly, the sensitivity for the ambiguity of technology and its multipurposeness as discussed above in the case of nuclear weapons also implies that the political results always remain—despite immense efforts put into simulations and scenario building—*underdetermined*. This is not only due to the mostly unknown socio-

---

[5] See Nye and Keohane (1971), Rosenau (1990), Murphy (1994), Keck and Sikkink (1998), Sassen (2006, 2001), and Hall and Biersteker (2002).

[6] Bijker et al. (1987), Latour (1987), Haraway (2003), Whatmore (2002), Dittmer (2014).

[7] See Acuto and Curtis (2013), Agathangelou (2010), Krishna (2009), Aradau (2010), Mayer and Schouten (2011), Srnicek (2013) and Squire (2014).

material feedback mechanisms and interlinkages operating at various levels, but results from human ingenuity and creative reappropriations and repurposings of technologies (Connolly 2013; Cole 2013; Barry 2013a). Attempts of understanding and theorizing have to cope with the perpetual twists and contingencies of a full-fledged "technological drama" (Pfaffenberger 1992; Joerges 1999). Hence, carefully analyzing politics of science and technology inevitably prevents one from embracing technological determinism.

Fourthly, a multiplication of actors and the rampant proliferation of novel communication technologies may jointly produce new forms of structural and collective power. The example of big data security and digital surveillance illustrates that those are not captured by concepts such as soft, hard, or discursive power (Ansorge 2011; Singh 2013). On the one hand, common conceptual frameworks of sovereignty are challenged by the fact that governmental agencies function in a highly disaggregated manner in their daily practice within international networks (Slaughter 2004; Drake 2008). On the other hand, recent events show that *individuals, enterprises* and *public agencies* that have gained access to large data troves now possess global power to an unprecedented degree. The enormous leverage of journalists, soldiers, or system administrators becomes fully apparent from the biographies of Chelsea Manning, Julian Assange, and Edward Snowden. The technological means of gathering information and meta-data are rapidly multiplied; it is the number of people with direct access to the ever-widening realm of state secrecies and classified information. Democratic and non-democratic governments alike can do nothing but desperately hunt whistle blowers down. Perhaps, greater difficulties for states arise only from preventing algorithms of malicious or defective software from destroying data processing vital for core social systems including finance, health, energy and defense. As such, unintended consequences of novel technologies and cutting-edge scientific knowledge (see Tenner 1997; Tucker 2012), novel forms of power and temporary monopolies inherently belong to the planetary modernization project, making global politics inexorably entangled with fragile infrastructures and technical risks.

5 Science and Technology: IR's Unrecognized *Subfield*

Shortly after World War II, William F. Ogburn, the editor of the first extensive treatment of "Technology and International Relations," expected that technological change would turn into a major new field of research. Prominent scholars, including Quincy Wright, William T. R. Fox, John Herz, and Bernard Brodie, among others, investigated the impact of various modern technologies on international relations (see Ogborn 1949).[8] The handling of nuclear weapons became the central research

---

[8] Brian Schmidt's (1998) alternative hagiography of IR indicates that scholars were much earlier quite aware of the importance of technology for world affairs.

concern of a large epistemic community during the 1950s and 1960s. The research agenda turned primarily to the vital issue of nuclear deterrence, (non)proliferation, and super power relations. Other scholars such as James Rosenau, Stephen Krasner, Joseph Nye, Susan Strange, Robert Keohane, and Ernst Haas approached the emerging global cross-border communications and scientific and technological linkages through structural realism, neofunctional integration theory and regime theory. Recently, the number of studies on the political implications of information technologies and the Internet has been growing rapidly (cf. Simmons 2013), reversing a previous "silence" concerning communication technologies.

Yet, while IR scholars have persistently engaged the issue, technologies and sciences did not, become a separate research field on its own, in contrast to Ogburn's expectations. Moreover, given that Sociology, Geography, History, Anthropology, and Archeology, among others, display thriving debates, journals, and entire subfields solely dedicated to technology and science, a relative dearth of research attention in IR is remarkable. Table 1 shows the small number of IR publications explicitly dealing with the subject. Between 1990 and 2007, science and technology were covered by merely 0.7 % out of more than 21,081 published in 13 major IR journals. More importantly, an even tinier fraction genuinely deals with *theorizing* or *conceptualizing* technologies.

In short, these findings underline the assertion that IR has little explored, and much less theorized, the variety of forms of power and sites of politics related to

Table 1 Articles on science and technology published in major IR journals

Journal	Years	Overall number of articles	Articles on 'science' and 'technology'	Percent of all articles
Alternatives: Global, Local, Political	1990–2009	444	18	4,05
Foreign Affairs	1990–2009	7400	28	0,38
Foreign Policy	1990–2009	2174	13	0,60
International Affairs	1990–2007	6105	23	0,38
International Organization	1990–2009	565	6	1,06
International Political Science Review	1990–2009	447	10	2,24
International Security	1990–2007/2008	521	15	2,88
International Studies Quarterly	1990–2007	506	11	2,17
International Studies Review	1990–2007	214	2	0,93
Journal of Conflict Resolution	1990–2009	699	5	0,72
Review of International Studies	1990–2007	545	10	1,83
Third World Quarterly	1990–2007	1199	13	1,08
World Politics	1990–2009	262	2	0,76

Source: Articles were accessed via www.jstor.org database in August/September 2013 (except data on *Millennium* that were provided by SAGE journals homepage). Articles were first selected using keywords such as "science" and "technology" appearing in headlines, abstracts, or the text body. In a next step, items were carefully filtered out on a case-by-case basis concerning issue areas (applied research) and attempts of theorizing modern sciences and all kinds of technologies (Herrera 2003; Weiss 2005; Peoples 2009). Despite a degree of topical diversification of "applied research", IR is not concerned with how the actual *diversity*, *magnitude*, and *interactions* of technologies co-constitute and reshape what Tim Dant (2006) calls "material civilization". The scope of IR's academic interest neither matches the prominence of science and technology nor the extent to which they generate curiosity outside of IR circles. This seems extraordinary, in particular, when one tries to account for the plethora of emerging and constantly evolving clusters of new and old technologies and large technical systems, not to mention the amount of scientific or technology-centered controversies in contemporary global politics—only few of which we outlined above. Hence, the articulation of science and technology within IR presents a paradox of sorts for it is an "unrecognized field". The question then arises: why the issue failed to gain a central place within IR, especially in terms of theorizing?

To begin with, messiness and complexity inherent to science and technology tends to slip easily through the inter-state matrix structuring much of IR thinking and research. Because different related topics and issues have been raised in divers theoretical contexts, sciences and technologies were usually forced into a straitjacket belonging to *other* debates, concerns, and frameworks. But they have not spurred sustained attention and efforts to theorize in their own right. Another conceptual obstacle is that the study of science and technology became prey to two rivaling views: "technological determinism" and "social constructivism" (cf. Law 1991; Fritsch 2011). The former deems all politics a (by)product of technology and expertise. This kind of determinist assumptions is anathema to social scientists, who focus on social practices, institutions, and norms. Early realist thinkers, for example, refused accepting the right to dictate policies and politics that technologies of total destruction were assumed to possess. They sought to preserve progressive (enlightenment) perspectives on international politics and, above all, the possibility of democratic institutions against notions of technological determinism (see Williams 2013). Numerous IR scholars are siding with Kalevi J. Holsti, denying the idea that technological changes have *any* substantial effects on international affairs:

> (T)he foundational principles of international politics—what we term today the Westphalian system—did not undergo transformation as a result of major changes in the social and technological environment in previous centuries. Diplomatic life in 1775 was not unrecognizable from its predecessor in, let us say, 1700 despite that major intellectual upheaval of the eighteenth century, the Enlightenment. Similarly, the Industrial Revolution, surely a change as momentous as globalization is today, did not reorder major international institutions, except perhaps in the domain of war (Holsti 2004: 19).

As a result, technologies are deprived of their historical significance. At the same time, most approaches to IR are actually dominated by instrumentalist views on technologies (Singh 2002). Social constructivism, liberalism and realism alike tend to picture technologies and sciences as neutral tools. They figure merely in the service of exercising power, realizing interests or carrying meaning. Technology and science, consequently, only possess secondary importance for theoretical explanations and systemic models, encouraging a systematic externalization of infrastructures, technical networks and other material artifacts. In turn, treating technologies as *deus ex machina* became a recurring practice among IR scholars (Herrera 2003: 569) and has to be understood as flipside of instrumentalism.

Instrumentalism and externalism are both reinforced by an ideational bias at the ontological level of IR theories. For instance, in Kenneth Waltz's classical work Man, the State and War, we encounter a theoretical model that eschews the material world. The perennial problem of IR as such, namely the danger of warfare due to anarchy, is "weightless" to the extent that collective action—that is more or less rational calculation—is played out primarily in human minds (Waltz 1959: 169; Jervis 1988).[9] In Waltz's systemic theory, the material world features merely in the form of "capabilities", which bear on rational calculations by great powers as computable variables (1979).[10] Hedley Bull's *Anarchical Society* similarly pictures the political world exclusively in terms of social practices. Bull sees the international society as a functioning rule-based order. "International order", he proposes, "is order among states; but states are simply groupings of men" (Bull 1977: 20). These examples are not meant to indicate that Waltz's or Bull's approaches display logical inconsistencies but to highlight that these influential works epitomize a mindset within IR, which externalizes science and technology. While representing an essentially social world, material artifacts and infrastructures are not deemed conceptually constitutive to the world assembled in these classical works.

A few IR scholars went to considerable length conceptually in order to acknowledge the existence of material agencies. Regime theory, for instance, was advanced in response to the mutual constraints imposed by nuclear weapons, increased technological interconnections, and trans-boundary flows between nations (Keohane and Nye 1977). Because interdependence "provides opportunities for actors to externalize costs of their actions onto others", Keohane notes that "institutions for global governance will need to limit the negative externalities of decentralized action." (2001: 2) So the intriguing complexity of the technological world, by implication, is reduced to an issue of "beggar thy neighbor". Because -they treat technical systems as "apolitical" and "irrelevant", institutionalist approaches see the design of functioning institutions of governance merely as a *social* problematique (Porter 2003: 524, Pinch 2008). Advancing a neorealist-structural approach, Barry Buzan is among the few who have tried granting technologies a critical role. He suggests that technological progress changes the "interaction capacity", and therefore, the properties of international systems. This explains the historical diversity of system types—an empirical fact that is omitted by Waltz and others who foreground "like-units" neglecting the possibility of differentiation (Buzan et al. 1993, cap. 4).

[9] Although Waltz initially argued that even nuclear weapons are not able to override the constant threat of war between independent sovereign states (Waltz 1959: 235–237), he later claimed that nuclear weapons, aside from bipolarity, secure world peace (Waltz 1990).

[10] Lebow (1994) and Schmidt (2007) detail the difficulties of realist theories, and particularly Waltz's structural realism, to make sense of nuclear weapons and other technologies coherently.

Alexander Wendt also places technology exogenous to a purely social system. Wendt's *Social Theory* rests on a realist meta-theoretical foundation that explicitly acknowledges the existence of an "objective reality". Nevertheless, *intersubjectively* enacted identities are the central domain to order and explain international relations (Zehfuss 2001). Although Went draws on the notion of "interdependence" and "interaction capabilities" to make sense of technological change (Wendt 1999: 243–249), the latter remains a "master variable" off-limits for any theoretical gaze. Poststructuralist scholars also tend to externalize technology by analyzing international relations in terms of "discourses", "speech acts", "subjectivities" or "ideas".[11] Material artifacts are elements of performative processes, however, the very existence of objects (and subjects) is conditioned by discourse (Walker 1993; Campbell 2007).[12] Constructivism and poststructuralism advance a principally similar emphasis: apart from some inescapable constraints of "rump matter", what counts as important for IR is eventually "ideas all the way down" (Wendt 1999: 110). Ultimately it is people who, by attaching meaning to materials—not vice versa—make certain infrastructures, innovative products, or new military items, for instance, more or less relevant (Adler 1997; see Pouliot 2010). Hence, conceptual omission of technologies and sciences can be partly explained by the fact that "both postmodernism and constructivism have been marked by a strong tendency to go too far in their emphasis of the ideational and, by going too far in the search for a pure or nearly pure social ontology."[13]

To summarize, notwithstanding the differences or even incommensurable theoretical positions of Realism, English School, Liberalism, Constructivism and Poststructuralism, these schools found common ground by conceptually externalizing the technological world, while presuming it in their projects. Ideational bias precludes them from substantially contributing to this emerging field. But to sublimate or externalize material agencies is no less common to studies of globalization and world order.[14] The history of technological permeation, composition,

[11] Securitization theory, for instance, eclipses technological or material elements of reality (McDonald 2008).

[12] Poststructuralism nevertheless can involve a subtler reading of materiality than other IR theories would allow for (see Jasanoff 2004b).

[13] Deudney in Schouten (2013b). See also Patomäki and Wight (2000), Buzan (2004).

[14] For instance, David Held et al. (1999) mention in the last pages of their monumental volume *Global Transformations: Politics, Economics and Culture* that "the immense increase in global and regional interactions of all kinds has been supported by a series of transformations in the infrastructures of global interaction." However, they also claim that "the invention of these technologies is not sufficient by itself to account for their deployment, use and growth; but their contribution to both the increased volume and transformed character of contemporary globalization is undeniable." (1999: 428) Apart from a few comments on military infrastructures, *Global Transformations* does not dedicate a single entire page to the technological world. Samuel Huntington's *Clash of Civilizations* is another case in point. While Huntington critically notes that the expansion of the western civilization was owed to military superiority (1997: 51), he prefers to tie his line of arguments into culture, religion, and identity rather than relating it to a technological world.

and remaking of international affairs is widely missing, particularly within systemic theories of IR (Buzan and Lawson 2013; Schouten 2013a; Mayer 2014). Hence, a broader understanding of the global politics of science and technology not just requires the integration of traditional notions and approaches into more comprehensive analytical frameworks. New concepts, approaches and perspectives need to be developed, perhaps under the notion of "integrative pluralism" (Dunne et al. 2013). A more comprehensive analytical toolbox enables incorporating an enormous panorama of empirical materials, cases, and puzzles into the subfield. Thus, far from approaching its endpoint, IR theory is challenged and inspired.

6 Approaching the Global Politics of Science and Technology

The discipline of IR is an outlier. In comparison to other social sciences, it seems much less aware of the extraordinary extent to which (global) politics evolve through sciences and technologies.[15] Nonetheless, decades of innovative research have accumulated a large body of knowledge. Given this condition, it is crucial to incorporate empirical insights and theoretical notions from other disciplines. Making the existing disciplinary boundaries more flexible will make a major contribution to the subfield and to IR in general (see Buzan and Little 2001). Interdisciplinary conversations help to deepen and differentiate IR's understanding of science and technology, both theoretically and analytically (see Bray 2012; Barry 2013b). In this sense, both practitioners and students of the global politics of science and technology might greatly benefit from mutual learning and by reflectively absorbing a rich tradition of debates, puzzles and methods of neighboring disciplines.

The notion of "techno-politics" (Mitchell 2002; Hecht 1998) provides a promising vantage point for this purpose by enhancing IR's vocabulary and perspectives on science and technology. For one, the term can be employed as a boundary concept—a relay between dissimilar disciplinary terminologies as well as confusingly overlapping research agendas and methods. For another, techno-politics functions as an umbrella. Instead of conveying a single definition, it suggests a range of different conceptualizations of technology and science. Their common denominator is a focus on the "middle zone" that covers to deserted area between technological determinism and social constructivism.[16] The occupation and widening of this zone is achieved through the integration of IR and non-IR approaches.

[15] For respective compilations in STS see Jasanoff et al. (1995); for History see Krige and Barth (2006); for political theory see Braun and Whatmore (2010); for Geography see Brunn et al. (2004); for Anthropology see Star (1999) and Stroeken (2013); for Philosophy see Scharff et al. (2013).

[16] See Herrera (2003) and Fritsch (2014, in this volume) for an elaboration of this argument.

It represents, in the view of the editors, the best choice to stimulate a productive theoretical discourse and innovative empirical research. The following conceptually structured set of examples indicates how studying the global politics of science and technology might shed fresh light on existing puzzles within IR and contributes to ongoing debates about ontology and methodology that currently structure the discipline.

- *Constructivist studies of technology* are perhaps the most obvious entry point for IR approaches (cf. Bijker 1993; Ruggie 1993). Here, the notion of technopolitics challenges the instrumentalist idea that social actors can simply attach "meaning" to technological artifacts that, in reverse, turn into powerful carriers of identity or simply express social norms. For instance, Herrera's work on large technological systems (2006), Der Derian's studies of global war and media (2003) and Biggs' (1999) and Branch's (2011) exploration of mapping technologies in the context of early modern state formation show an intricate historical interplay of meaning, representation, and emerging technical practices. Recently, the surge of recording devices, online content, and communication data related to the Internet significantly complicates processes of signification, perceptions and the fixation of intersubjective meaning. In addition, technopolitics imply that scientific practices, epistemic communities and technical designs are not just objective and neutral phenomena but deeply interwoven with the fabric of power (Litfin 1994, Adler and Bernstein 2005). While generations of IR scholars have dealt with the role of scientific knowledge for international politics,[17] it is literature inspired by the "practice turn" in IR,[18] which speaks most immediately to core issues raised in constructivist studies of technology and science. Objective knowledge and technical standards are enmeshed in cultural traditions, ideological views and partly products of political struggles as MacKenzie's famous study on nuclear missile guidance (1993) and Susan Greenhalgh's analysis of China's one-child policy (2008) stress. In the same vein, Sheila Jasanoff's work reinforces symmetrical understandings. Suggesting the idiom of co-production, she "calls attention to the social dimensions of cognitive commitments and understandings, while at the same time underscoring the epistemic and materials correlates of social formations" (Jasanoff 2004a: 3). Christian Reus-Smit's (1996) critical analysis of technological and economic progress as the modern "state purpose", functioning as master narrative of International Society, could open up a symmetrical-constructivist perspective on global order. Similarly the constitution of markets for novel technologies such as nanotechnologies also depends on a fragile, yet powerful co-production of technical expertise on risks, hegemonic practices and public controversies.[19] In this sense, even oceans are subject to enormous construction efforts and, hence, historically evolving hybrids (Steinberg 2001).

[17] See Bueger (2014, in this volume) for an overview.

[18] See Adler and Pouliot (2011), Bigo (2011), Bueger (2013).

[19] See Wullweber (2014, in this volume) for an overview.

- *Assemblage approaches* emphasize complexity, turning global infrastructures, networks, and often overlapping socio-technical ensembles into primary objects of inquiry. Vantage point is the empirical observation that "politics" and "the economy" *emerge* from a messy density. As a consequence, observers cannot priori distinguish between "social" and "material" (Latour 2005). Following Michel Foucault seminal insights, (global) governance is seen as a process of stabilization and ordering though logistical devices, data and various technologies of inscriptions (Foucault 1991; Rose and Miller 1992; Larner and Walters 2004). It is in this sense that "emergence" is key to rearticulating both the agent-structure problem and the ideal-material divide of international affairs beset with techno-politics:

> techno-politics is always a technical body, an alloy that must emerge from a process of manufacture whose ingredients are both human and nonhuman, both intentional and not, and in which the intentional or the human is always somewhat overrun by the unintended. But it is a particular form of manufacturing, a certain way of organizing the amalgam of human and nonhuman, things and ideas, so that the human, the intellectual, the realm of intentions and ideas seems to come first and to control and organize the nonhuman (Mitchell 2002: 42–43).

Although intimate links exist between modern state formation and infrastructure (Mann 1984), large technical systems, often stretching across the planet, as well as the technological zones that facilitate production, trade, finance, communication, surveillance, and weapon systems are far more complex, multi-sited, and interconnected than any state-centric framework of social collective action allows for (Porter 2003; Sassen 2006; Barry 2006).[20] Assemblage theories, in addition, point out different ways how infrastructures and material inscriptions can wield power. Latour (1987), Mann (2008) and Agnew (2005) suggest that they allow for "acting at a distance" and are co-generative of uneven political space and asymmetric power distribution.[21] For example, the successful European integration is increasingly understood through the lens of transnational infrastructures, including railways, river management, electricity grids, roads, border protection, common currency and so fourth that have immensely contributed to the *materializing* of unification (Barry 2001; Badenoch and Fickers 2010). From a similar perspective, Ruth Oldenziel (2011) and Bélanger and Arroyo (2012) show the crucial importance of hundreds of tiny islands and strategic sites for the logistical maintenance of global surveillance and defense installations underpinning the US empire's use of military force. Stuart Elden's (2013) notion of "vertical geopolitics" indicates how the "depth" of power and influence can be structured through infrastructures such as sewage tunnels, roads, digital networks and so fourth. Even power grids can be deliberately constructed and utilized as instrument of infrastructural oppression (Shamir 2013). In short, this group of conceptual approaches translates into IR

[20] In particular, studies of global air traffic, airports and border technologies exemplify the fruitfulness of this line of inquiry within IR (cf. Salter 2007; Bellanova and Duez 2012; Schouten 2014; Bigo 2014).

[21] An insight, that speaks to early *realist explorations* of spatiality and power (e.g. Carr 1942; Ogborn 1949).

approaches, which do not just emphasize "things", but highlight the variety of non-human agencies without becoming determinist (Latour 1992; Steinberg 2013; Acuto and Curtis 2013; Salter and Mark 2014).

- *Critical and subaltern approaches* stress scientific and technological dimensions of core IR themes such as security, warfare, anarchy, and capitalism (Agathangelou and Ling 2009; Grovogui 1996; Sabaratnam 2011); for instance, the intimate link between Cold War strategic priorities and the transfer of scientific and technical know-how in agriculture to developing countries (Perkins 1997; McNeill and Unger 2010). Critical studies of capitalist reconfigurations of scientific research and progress center around the translation and valorization of knowledge and human bodies in global biomedical markets (Sunder Rajan 2012; Cooper 2008). Gabrielle Hecht (2010) illustrates the shifting ontology of nuclearity and its consequences for post-colonial spaces. James Der Derian and others critically document a fundamental remaking, that is, virtualization of warfare and terror at the nexus of images, entertainment and communication technologies (Der Derian 2009; see also Dillon 2003; Bratton 2009). William Walters (2009) demonstrates the crucial, yet deliberately hidden, role that complex layers of surveillance and border control technologies play for prohibiting (and co-producing) the flow of illegal migration. Critical liberal approaches to drones technology scrutinize the connection between asymmetric warfare and democratic institutions. This latter strand links IR theorizing to the concerns of pragmatist philosophy by John Dewey (1927), who critically examined the interplay of democratic mass societies and emerging media technology (Marres 2007). Furthermore, critical studies converge with constructivist approaches in deciphering and challenging the ways in which technology (and science) is turned into (and misused as) an ideology in the service of powerful commercial, political, or military elites (see Aronowitz 1988; Price 2011).
- Approaches to *technology-based power* raise the diversity and paradoxes of techno-political power shifts, influence and control. Reviel Netz's (2004) account of barbed wire as the quintessential modern power instrument, prohibiting human movements and creating new spaces of control, exemplifies the surprising global effects of a seemingly mundane and local technology. With respect to the Internet, Laura DeNardis' (2014) exploration of its complex material dimensions has, for example, uncovered the infrastructural layers subjected to power struggles. Explorations of time and speed as dimensions of power and governance also show the intricacies of metropolitan and transnational realities technologies help to generate (Thrift 1996; Kern 2003; Nanni 2012). As has already been noted, not even military technologies can be simply reduced to their instrumental functions. The technical design employed by governments does not necessarily lend itself to the intended tactical or strategic goals (Adas 2006: 281ff.). The pervasiveness of multi-layered path-dependencies and unintended consequences of technological systems require, if at all, a highly sophisticated understanding of "causation" (Heilbroner 1994; Hutchby 2001; Feenberg 2010). Conceptualizations, in turn, have to capture more subtle

differences than a broad correlation between major weapon systems and specific sets of security practices (Deudney 2000). Numerous vehicles—from specific technologies to the philosophy of science—are in *differing ways* highly significant for the exercise of (state) power (e.g. Reisch 2005; Innis 2008). Holistic versions of medium theory may offer a way to study how evolving communication technologies led to a digitally structured world order (Deibert 1997; Poe 2011). The ramifications of the ubiquity of recording devices are perhaps the best illustration of the paradoxes of "technological power": intelligence services of five aligned states ("five eyes") and several telecommunication enterprises now possess historically unparalleled surveillance capabilities, while whistle blowing individuals have equally unparalleled leverage in world affairs. The advent of a massive global "surveillance industrial complex" (Gray 2014) necessitates a sustained engagement between IR and scholarship in surveillance studies (see Lyon 2007; Green 1999).

- Infusing conceptualizations of techno-politics into *International Political Economy* leads to approaches that zoom in at technological macro-processes and micro practices that create and stabilize "the world economy". Çalışkan's and Callon's (2009) notion of "economization", Gavin Bridge's (2010) analysis of technological shifts in the carbon economy, and the fusion of sociology and STS to study financial markets and the economy in general (Knorr Cetina and Preda 2004; Pinch and Swedberg 2008) are indicative of the crucial importance of local agency and connected infrastructures that "perform" resources, markets, growth, and prizes. Craig N. Murphy and JoAnne Yates (2009) capture the significance that the rapid evolution of global container infrastructures, technical standards and related business models had in establishing the explosion of world trade after the 1960s. Peter Drahos (2010) illustrates the enormous amount of intimate technocratic cooperation between national patent bureaucracies and multinational enterprises that is required to construct a working global system for the protection of intellectual property. Sensitivity for contextual nuance, meticulous detail, and thick empirical description refers back to Karl Marx's work. As a careful anthropologist of technological change, Marx saw machines and infrastructures not only as tools of capitalist exploitation and oppression, but also as generative for the liberation and empowerment of the masses (see Matthewman 2011: 29–49). Analytical perspectives such as the technology club approach (Castellacci and Archibugi 2008), creative destruction (Acemoglu and Robinson 2012), and the digital divide (Norris 2001; Youngs 2007) capture the amplification of techno-political asymmetries in a globalized world. In addition, transnational cultural industries, arts, design and virtual worlds such as video games embody new politico-technical practices and institutions that underlay (and arise from) the commercialization of imaginations of values, identity, and warfare (Dyer-Witheford and De Peuter 2009; Singh 2010). Another major theme is the contingent dynamics of innovations processes and their impact on the formulation and success of industrial and innovation policies. National innovation systems have to operate within a basically global system (Etzkowitz and Leydesdorff 2000). Standard setting, technological monopolies, trade barriers and migration dynamics exert enormous influence on catching up of emerging economies.[22]

These examples are not meant to present a comprehensive and ultimate mapping of the subfield of the global politics of sciences and technologies. Instead, they exemplify ways in which the notion of techno-politics can function as umbrella of various IR approaches that do justice to the highly complex, hybrid, and dynamic character of the subject matter. Widening and exploring the conceptual zone that transcends technological determinism and social constructivism is most conducive for further theorizing as well as interdisciplinary exchanges. In this sense, the double volume *The Global Politics of Science and Technology* raises multiple related perspectives, concepts, approaches, issue areas, and methods.

The first book summarizes time-tested approaches for studying global politics of science and technology from an IR perspective. The structured overviews in the first part include: three generations of research on experts and scientific expertise in international relations (*Christian Bueger*); postsructuralist, Gramscian, and Marxian studies of hegemony, discourse, and political economy (*Joscha Wullweber*); a conceptual analysis of the place of technology within the main theoretical schools of IR (*Stefan Fritsch*); the evolution and management of nuclear weapon technologies (*Joe Pilat*); the interplay of cyberspace, states, and international anarchy (*Chris C. Demchack* and *Peter Dombrowski*). The second part provides empirical, theoretical, and conceptual interventions from Geography (*Peter Hugill*), History (*Douglas Howland, John Krige*), Innovation Studies (*Daniele Archibugi and Andrea Filippetti*), and Science and Technology Studies (*Sheila Jasanoff*) in order to rearticulate and reframe IR approaches. The final part consists of five interviews. *Loet Leydesdorff, Gabrielle Hecht, Dirk Messner, Timothy Mitchell* and *Karen Litfin* address various empirical and theoretical aspects and possibilities of cross- and multi-disciplinary collaboration. Thereby, a space for mutual learning is carved out to work towards understanding the canvass of the global politics of science and technology.

The second book features a collection of issue areas, actors, and cases, advancing IR research on science and technology. Besides detailed empirical studies, it aims at offering a toolbox that entails theoretical perspectives and analytical frameworks transcending both technical determinism and social constructivism.[23] The first part contains *interactional* approaches that inquire into the relevance and consequences of science and technology for the pursuit of foreign policy, regionalization, and international relations. The chapters of the second part explore *co-production* processes through which sciences and technologies become generative for the *emergence* of collective action and new entities, processes and actors.

Acknowledgment The authors are thankful to Douglas Howland, Anna Agathangelou, Peer Schouten, and Christian Bueger for their insightful comments and helpful suggestions. They were instrumental for writing this introductory chapter.

[22] E.g. Stopford et al. (1991), Chang (2002), Hugill and Bachmann (2005), Breznitz (2007).

[23] The introduction of the companion volume provides detailed chapter summaries.

References

Acemoglu, D., & Robinson, J. A. (2012). *Why nations fail: The origins of power, prosperity, and poverty*. New York: Crown.

Acuto, M., & Curtis, S. (Eds.). (2013). *Reassembling international theory: Assemblage thinking and international relations*. London: Palgrave Macmillan.

Adas, M. (1989). *Machines as the measure of men: Science, technology, and ideologies of western dominance*. Ithaca, NY: Cornell University Press.

Adas, M. (1997). On science technology and colonial expansion and dominance of European states: A field matures: Technology, science, and western colonialism. *Technology and Culture, 38*(2), 478–487.

Adas, M. (2006). *Dominance by design: Technological imperatives and America's civilizing mission*. Cambridge, MA: Harvard University Press.

Adler, E. (1997). Seizing the middle ground: constructivism in world politics. *European Journal of International Relations, 3*(3), 319–363.

Adler, E., & Pouliot, V. (2011). International practices. *International Theory, 3*(1), 1–36.

Adler, E., & Bernstein, S. (2005). Knowledge in power: The epistemic construction of global governance. In M. Barnett & R. Duvall (Eds.), *Power and global governance* (pp. 294–318). Cambridge, MA: Cambridge University Press.

Agathangelou, A. M. (2010). Bodies of desire, terror and the war in Eurasia: Impolite disruptions of (neo) liberal internationalism, neoconservatism and the 'new' Imperium. *Millennium Journal of International Studies, 38*(3), 693–722.

Agathangelou, A. M., & Ling, L. H. M. (2009). *Transforming world politics: From empire to multiple worlds*. London: Routledge.

Agnew, J. (1994). The territorial trap: The geographical assumptions of international relations theory. *Review of International Political Economy, 1*(1), 53–80.

Agnew, J. (2005). Sovereignty regimes: Territoriality and state authority in contemporary world politics. *Annals of the Association of American Geographers, 95*(2), 437–461.

Ansorge, J. T. (2011). Digital power in world politics: Databases, panopticons. *Millennium Journal of International Studies, 40*(1), 65–83.

Aradau, C. (2010). Security that matters: Critical infrastructure and objects of protection. *Security Dialogue, 41*(5), 491–514.

Arkin, R. C. (2010). The case for ethical autonomy in unmanned systems. *Journal of Military Ethics, 9*(4), 332–341.

Aronowitz, S. (1988). *Science as power: Discourse and ideology in modern society*. Minneapolis: University of Minnesota Press.

Arrighi, G. (1994). *The long twentieth century: Money, power, and the origins of our times*. New York: Verso.

Badenoch, A., & Fickers, A. (Eds.). (2010). *Materializing Europe: Transnational infrastructures and the project of Europe*. London: Palgrave Macmillan.

Barry, A. (2001). *Political machines: Governing a technological society*. London: Athlone.

Barry, A. (2006). Technological zones. *European Journal of Social Theory, 9*(2), 239–253.

Barry, A. (2013a). *Material politics: Disputes along the pipeline*. New York: Wiley.

Barry, A. (2013b). The translation zone: Between actor-network theory and international relations. *Millennium: Journal of International Studies, 41*(3), 413–429.

Bauman, Z. (2013). *Modernity and the holocaust*. Hoboken, NJ: Wiley.

Bauman, Z., Bigo, D., Esteves, P., Guild, E., Jabri, V., Lyon, D., et al. (2014). After snowden: Rethinking the impact of surveillance. *International Political Sociology, 8*(2), 121–144.

Bayly, C. A. (2004). *The birth of the modern world, 1780-1914: Global connections and comparisons*. Oxford: Blackwell.

Bélanger, P., & Arroyo, A. S. (2012). Logistics Islands: The global supply archipelago and the topologics of defense. *Prism, 3*(4), 54–75.

Bellanova, R., & Duez, D. (2012). A different view on the 'making' of European security: The EU Passenger Name Record System as a socio-technical assemblage. *European Foreign Affairs Review, 17*(Special Issue), 109–124.

Berman, M. (1983). *All that is solid melts into air: The experience of modernity*. London: Verso.

Biggs, M. (1999). Putting the state on the map: Cartography, territory, and European state formation. *Comparative Studies in Society and History, 41*(2), 374–405.

Bigo, D. (2011). Pierre Bourdieu and international relations: Power of practices, practices of power. *International Political Sociology, 5*(3), 225–258.

Bigo, D. (2014). The (in)securitization practices of the three universes of EU border control: Military/Navy-border guards/police-database analysts. *Security Dialogue, 45*(3), 209–225.

Bijker, W. E., Hughes, T. P., & Trevor Pinch, J. (Eds.). (1987). *The social construction of technological systems: New directions in the sociology and history of technology*. Cambridge, MA: MIT Press.

Bijker, W. E. (1993). Do not despair: There is life after constructivism. *Science, Technology and Human Values, 18*(1), 113–138.

Beck, U. (1992). *Risk society: Towards a new modernity*. New Delhi: Sage.

Beck, U., Giddens, A., & Lash, S. (1994). *Reflexive modernization: Politics, tradition and aesthetics in the modern social order*. Stanford: Stanford University Press.

Betz, D., & Stevens, T. (2011). *Cyberspace and the state: Toward a strategy for cyber-power*. Abingdon; New York: Routledge.

Bollier, D. (2010). *The promise and peril of big data*. Washington, DC: The Aspen Institute.

Bousquet, A. J. (2009). *The scientific way of warfare: Order and chaos on the battlefields of modernity*. New York: Columbia University Press.

Branch, J. (2011). Mapping the sovereign state: Technology, authority, and systemic change. *International Organization, 65*(1), 1–36.

Bratton, B. H. (2009). On geoscapes and the Google caliphate reflections on the Mumbai Attacks. *Theory, Culture and Society, 26*(7–8), 329–342.

Braudel, F. (1982). *Civilization and capitalism, 15th-18th century: The perspective of the world* (Vol. 3). Berkeley: University of California Press.

Braun, B., & Whatmore, S. (Eds.). (2010). *Political matter: Technoscience, democracy and public life*. Minneapolis: University of Minnesota Press.

Bray, F. (2012). Only connect: Comparative, national, and global history as frameworks for the history of science and technology in Asia. *East Asian Science, Technology and Society: An International Journal, 6*(2), 233–241.

Brenner, N. (2004). Urban governance and the production of new state spaces in Western Europe, 1960–2000. *Review of International Political Economy, 11*(3), 447–488.

Breznitz, D. (2007). *Innovation and the state*. New Haven, CN: Yale University Press.

Bridge, G. (2010). Geographies of peak oil: The other carbon problem. *Geoforum, 41*(4), 523–530.

Brodie, B. (1946). *The absolute weapon: Atomic power and world order*. New York: Harcourt.

Brunn, S. D., Cutter, S. L., & Harrington, J. W., Jr. (Eds.). (2004). *Geography and technology*. Berlin: Springer.

Bueger, C. (2013). Pathways to practice: Praxiography and international politics. *European Political Science Review, 6*(3), 383–406.

Bull, H. (1977). *The anarchical society: A study of order in world politics*. New York: Columbia University Press.

Buzan, B. (1987). *An introduction to strategic studies: Military technology and international relations*. London: Macmillan.

Buzan, B. (2004). *From international to world society?: English school theory and the social structure of globalization*. Cambridge: Cambridge University Press.

Buzan, B., & Lawson, G. (2013). The global transformation: The nineteenth century and the making of modern international relations. *International Studies Quarterly, 57*(3), 620–634.

Buzan, B., & Hansen, L. (2009). *The evolution of international security studies*. Cambridge, MA: Cambridge University Press.

Buzan, B., & Little, R. (2001). Why international relations has failed as an intellectual project and what to do about it. *Millennium Journal of International Studies, 30*(1), 19–40.
Buzan, B., Jones, C., & Little, R. (1993). *The logic of anarchy*. New York: Columbia University Press.
Byrne, J., & Glover, L. (2005). Ellul and the weather. *Bulletin of Science, Technology and Society, 25*(1), 4–16.
Çalışkan, K., & Callon, M. (2009). Economization, part 1: Shifting attention from the economy towards processes of economization. *Economy and Society, 38*(3), 369–398.
Camilleri, J. (1996). Impoverishment and the national state. In F. Osler & J. Reppy (Eds.), *Earthly goods. Environmental change and social justice* (pp. 122–153). Ithaca: Cornell Univeristy Press.
Campbell, D. (2007). Geopolitics and visuality: Sighting the Darfur conflict. *Political Geography, 26*(4), 357–382.
Carroll, P. (2006). *Science, culture, and modern state formation*. Berkeley: University of California Press.
Castellacci, F., & Archibugi, D. (2008). The technology clubs: The distribution of knowledge across nations. *Research Policy, 37*, 1659–1673.
Castells, M. (1996). *The rise of the network society* (The information age: Economy, society and culture, Vol. 1). Cambridge: Blackwell.
Chandler, A. D. (1977). *The visible hand: The managerial revolution in American business*. Cambridge, MA: Harvard University Press.
Chang, H.-J. (2002). *Kicking away the ladder: Development strategy in historical perspective*. London: Anthem.
Cipolla, C. M. (1965). *Guns, sails and empires: Technological innovation and the early phases of European expansion, 1400-1700*. New York: Pantheon Books.
Cole, D. (2013). Agentic capacities and capacious historical materialism: Thinking with new materialisms in the political sciences. *Millennium Journal of International Studies, 41*(3), 451–469.
Conca, K. (2004). Ecology in an age of empire: A reply to (and extension of) Dalby's imperial thesis. *Global Environmental Politics, 4*(2), 12–19.
Connolly, W. E. (2013). The 'new materialism' and the fragility of things. *Millennium Journal of International Studies, 41*(3), 399–412.
Cooper, M. (2008). *Life as surplus: Biotechnology & capitalism in the Neoliberal era*. Seattle: University of Washington Press.
Cox, R. W. (1987). *Production, power, and world order: Social forces in the making of history*. New York: Columbia University Press.
Creveld, V., & Martin, L. (1993). *Nuclear proliferation and the future of conflict*. New York: Free Press.
Cronon, W. (Ed.). (1995). *Uncommon ground: Rethinking the human place in nature*. New York: W.W. Norton.
Dalby, S. (2007). Anthropocene geopolitics: Globalisation, empire, environment and critique. *Geography Compass, 1*(1), 103–118.
Dant, T. (2006). Material civilization: Things and society. *The British Journal of Sociology, 57*(2), 289–308.
Darby, P. (Ed.). (2000). *At the edge of international relations: Postcolonialism, gender and dependency*. London/New York: Continuum.
Deibert, R. J. (1997). *Parchment, printing, and hypermedia: Communication in world order transformation*. New York: Columbia University Press.
Deibert, R., Palfrey, J., Rohozinski, R., & Zittrain, J. (Eds.). (2010). *Access controlled: The shaping of power, rights, and rule in cyberspace*. Cambridge, MA: MIT Press.
Delanda, M. (1991). *War in the age of intelligent machines*. New York: Zone Books.
De Nardis, L. (2014). *The global war for internet governance*. New Haven: Yale University Press.

Der Derian, J. (1990). The (S)pace of international relations: Simulation, surveillance, and speed. *International Studies Quarterly, 34*(3), 295–310.
Der Derian, J. (2003). The question of information technology in international relations. *Millennium Journal of International Studies, 32*(3), 441–456.
Der Derian, J. (2009). *Virtuous war: Mapping the military-industrial-media-entertainment-network*. London: Routledge.
Deudney, D. (2000). Geopolitics as theory: Historical security materialism. *European Journal of International Relations, 6*(1), 77–107.
Dewey, J. (1927). *The public and its problems*. Ohio: Ohio University Press.
Dillon, M. (2003). Virtual security: A life science of (dis) order. *Millennium Journal of International Studies, 32*(3), 531–558.
Dinerstein, J. (2006). Technology and its discontents: On the verge of the posthuman. *American Quarterly, 58*(3), 569–595.
Dittmer, J. (2014). Geopolitical assemblages and complexity. *Progress in Human Geography, 38* (3), 385–401.
Drahos, P. (2010). *The global governance of knowledge: Patent offices and their clients*. Cambridge, MA: Cambridge University Press.
Drahos, P., & Braithwaite, J. (2002). *Information feudalism: Who owns the knowledge economy?* London: Earthscan.
Drake, W. J., & Wilson, E. J. (Eds.). (2008). *Governing global electronic networks. International perspectives on policy and power*. Cambridge, MA: The MIT Press.
Drezner, D. W. (2004). The global governance of the internet: Bringing the state back in. *Political Science Quarterly, 119*(3), 477–498.
Dyer-Witheford, N., & De Peuter, G. (2009). *Games of empire: Global capitalism and video games*. Minneapolis, MN: University of Minnesota Press.
Dunne, T., Hansen, L., & Wight, C. (2013). The end of International Relations theory? *European Journal of International Relations, 19*(03), 405–425.
Edgerton, D. E. H. (2007). The contradictions of techno-nationalism and techno-globalism: A historical perspective. *New Global Studies, 1*(1), 1–32.
Edwards, P. N. (1997). *The closed world: Computers and the politics of discourse in Cold War America*. Cambridge: MIT Press.
Edwards, P. N. (2010). *A vast machine: Computer models, climate data, and the politics of global warming*. Cambridge, MA: MIT Press.
Eriksson, J., & Giacomello, G. (2009). Who controls the Internet? Beyond the obstinacy or obsolescence of the state. *International Studies Review, 11*(1), 205–230.
Elden, S. (2013). Secure the volume: Vertical geopolitics and the depth of power. *Political Geography, 34*, 35–51.
Endres, D. (2009). The rhetoric of nuclear colonialism: Rhetorical exclusion of American Indian arguments in the Yucca Mountain nuclear waste siting decision. *Communication and Critical/Cultural Studies, 6*(1), 39–60.
Etzkowitz, H., & Leydesdorff, L. (2000). The dynamics of innovation: From National Systems and 'Mode 2' to a Triple Helix of university–industry–government relations. *Research Policy, 29* (2), 109–123.
Feenberg, A. (1999). *Questioning technology*. East Sussex: Psychology Press.
Feenberg, A. (2010). *Between reason and experience: Essays in technology and modernity*. Cambridge, MA: MIT Press.
Foucault, M. (1991). *Discipline and punish. The birth of the prison*. London: Penguin.
Frank, A. G. (1998). *ReOrient: Global economy in the Asian age*. Berkeley: University of California Press.
Freedman, L. (2003). *The evolution of nuclear strategy*. London: Palgrave Macmillan.
Friedman, T. (2009). *The world is flat. The globalized world in the twenty first century*. London: Penguin.

Fritsch, S. (2011). Technology and global affairs. *International Studies Perspectives, 12*(1), 27–45.
Gavin, F. J. (2012). *Nuclear statecraft: History and strategy in America's atomic age*. Ithaca, NY: Cornell University Press.
Gilbert, E., & Helleiner, E. (Eds.). (2013). *Nation-states and money: The past, present and future of national currencies*. London: Routledge.
Gilpin, R. (1981). *War and change in world politics*. Cambridge, MA: Cambridge University Press.
Gray, J. (2014). The Snowden files: So much more than state surveillance. *Open Democracy*. February 6. http://www.opendemocracy.net/ourkingdom/jonathan-gray/snowden-files-so-much-more-than-state-surveillance. Accessed 19 Feb 2014.
Green, S. (1999). A plague on the panopticon: Surveillance and power in the global information economy. *Information, Communication and Society, 2*(1), 26–44.
Greenhalgh, S. (2008). *Just one child: Science and policy in Deng's China*. Berkeley: University of California Press.
Grovogui, S. N.' Z. (1996). *Sovereigns, quasi sovereigns and Africans: Race and self-determination in international law*. Minneapolis: University of Minnesota Press.
Guldi, J. (2012). *Roads to power: Britain invents the infrastructure state*. Cambridge: Harvard University Press.
Hall, R. B., & Biersteker, T. J. (Eds.). (2002). *The emergence of private authority in global governance*. New York: Cambridge University Press.
Haraway, D. (1991). *Simians, cyborgs, and women: The reinvention of women*. New York: Routledge.
Haraway, D. (2003). Cyborgs to companion species. Reconfiguring Kinship in technoscience. In D. Ihde & E. Selinger (Eds.), *Chasing technoscience: Matrix for materiality* (pp. 58–82). Bloomington: Indiana University Press.
Hart, H. (1946). Technological acceleration and the atomic bomb. *American Sociological Review, 11*(3), 277–293.
Harvey, D. (1990). *The condition of postmodernity: An enquiry into the conditions of cultural change*. Oxford: Blackwell.
Haunss, S., & Shadlen, K. C. (Eds.). (2009). *Politics of intellectual property: Contestation over the ownership, use, and control of knowledge and information*. Northhampton, MA: Edward Elgar.
Headrick, D. (1981). *The tools of empire. Technology and European imperialism in the nineteenth century*. New York: Oxford University Press.
Hecht, G. (1998). *The radiance of France: Nuclear power and national identity after world war II*. Cambridge, MA: MIT Press.
Hecht, G. (2010). The power of nuclear things. *Technology and Culture, 51*(1), 1–30.
Hecht, G. (Ed.). (2011). *Entangled geographies: Empire and technopolitics in the global Cold War*. Cambridge, MA: MIT Press.
Hecht, G. (2012). *Being nuclear: Africans and the global uranium trade*. Cambridge, MA: MIT Press.
Heilbroner, R. (1994). Technological determinism revisited. In M. R. Smith & L. Marx (Eds.), *Does technology drive history? The dilemma of technological determinism* (pp. 67–78). Cambridge, MA: MIT Press.
Held, D., Goldblatt, D., McGrew, A., & Perraton, J. (1999). *Global transformations: Politics, economics and culture*. Stanford: Stanford University Press.
Herrera, G. L. (2002). The politics of bandwidth: International political implications of a global digital information network. *Review of International Studies, 28*(1), 93–122.
Herrera, G. L. (2003). Technology and international systems. *Millennium Journal of International Studies, 32*(3), 559–593.
Herrera, G. (2006). *Technology and international transformation: The railroad, the atom bomb, and the politics of technological change*. Albany: State University of New York Press.
Hoyt, T. D. (2003). Technology and security. In M. E. Brown (Ed.), *New world: Security challenges in the 21st century* (pp. 17–37). Grave, Georgetown: Georgetown University Press.

Holsti, K. J. (2004). *Taming the sovereigns: Institutional change in international politics*. Cambridge, MA: Cambridge University Press.
Hutchby, I. (2001). Technologies, texts and affordances. *Sociology, 35*(2), 441–456.
Hughes, T. P. (2004). *Human-built world: How to think about technology and culture*. Chicago: University of Chicago Press.
Hugill, P. J. (1999). *Global communications since 1844: Geopolitics and technology*. Ann Arbor: University of Michigan.
Hugill, P. J., & Bachmann, V. (2005). The route to the techno-industrial world economy and the transfer of German organic chemistry to America before, during, and immediately after world war I. *Comparative Technology Transfer and Society, 3*(2), 158–186.
Huntington, S. P. (1997). *The clash of civilizations and the remaking of world order*. London: Simon & Schuster.
Innis, H. (1950). *Communication and empire*. Oxford: Oxford University Press.
Innis, H. (2008). *The bias of communication. Second Edition with a new introduction by Alexander J.* Watson, Toronto: University of Toronto Press.
Jasanoff, S. (1990). *The fifth branch: Science advisers as policymakers*. Cambridge, MA: Harvard University Press.
Jasanoff, S. (2003). Technologies of humility: Citizen participation in governing science. *Minerva, 41*(3), 223–244.
Jasanoff, S. (2004a). The idiom of co-production. In J. Sheila (Ed.), *States of knowledge: The co-production of science and social order* (pp. 1–12). London: Routledge.
Jasanoff, S. (2004b). Ordering knowledge, ordering society. In S. Jasanoff (Ed.), *States of knowledge. The co-production of science and social order* (pp. 13–45). London: Routledge.
Jasanoff, S., Markle, G. E., Peterson, J. C., & Pinch, T. (Eds.). (1995). *Handbook of science and technology studies*. London: Sage.
Joerges, B. (1999). Do politics have artefacts? *Social Studies of Science, 29*(3), 411–431.
Jervis, R. (1988). The political effects of nuclear weapons: A comment. *International Security, 13* (2), 80–90.
Jervis, R. (1989). *The meaning of the nuclear revolution: Statecraft and the prospect of Armageddon*. Ithaca, NY: Cornell University Press.
Juma, C. (2010). *The new harvest: Agricultural innovation in Africa*. Oxford: Oxford University Press.
Kaag, J., & Kaufman, W. (2009). Military frameworks: Technological know-how and the legitimization of warfare. *Cambridge Review of International Affairs, 22*(4), 585–606.
Kahler, M. (Ed.). (2009). *Networked politics: Agency, power, and governance*. Ithaca, NY: Cornell University Press.
Keck, M. E., & Sikkink, K. (1998). *Activists beyond borders: Advocacy networks in international politics*. Ithaca, NY: Cornell University Press.
Keohane, R. O. (2001). Governance in a partially globalized world. *American Political Science Review, 95*(1), 1–13.
Keohane, R. O., & Nye, J. S. (1977). *Power and interdependence: World politics in transition*. Boston: Little, Brown.
Keohane, R. O., & Nye, J. S. (1998). Power and interdependence in the information age. *Foreign Affairs, 77*(5), 81–94.
Kern, S. (2003). *The culture of time and space 1880-1918*. Cambridge, MA: Harvard University Press.
Kevles, D. J. (1977). The national science foundation and the debate over postwar research policy, 1942–1945: A political interpretation of science–the endless frontier. *Isis, 68*(1), 4–26.
Knorr Cetina, K., & Preda, A. (2004). *The sociology of financial markets*. New York: Oxford University Press.
Krige, J. (2006). *American hegemony and the postwar reconstruction of science in Europe*. Cambridge, MA: MIT Press.

Krige, J., Callahan, A. L., & Maharaj, A. (2013). *NASA in the world: Fifty years of international collaboration in space*. New York: Macmillan.
Krige, J., & Barth, K.-H. (Eds.). (2006). *Global power knowledge: Science and technology in international affairs*. Chicago: University of Chicago Press.
Krishna, S. (2009). The social life of a bomb: India and the ontology of an 'overpopulated' society. In I. Abraham (Ed.), *South Asian cultures of the bomb: Atomic publics and state in India and Pakistan* (pp. 68–88). Bloomington: Indiana University Press.
Kurzweil, R. (2006). *The singularity is near: When humans transcend biology*. New York: Viking.
Larner, W., & Walters, W. (Eds.). (2004). *Global governmentality: Governing international spaces*. Oxon: Routledge.
Latour, B. (1987). *Science in action: How to follow scientists and engineers through society*. Cambridge, MA: Harvard University Press.
Latour, B. (1992). Where are the missing masses? The sociology of a few mundane artifacts. In W. E. Bijker & J. Law (Eds.), *Shaping technology/building society: Studies in sociotechnical change* (pp. 225–58). Cambridge, MA: MIT.
Latour, B. (1993). *We have never been modern*. Cambridge, MA: Harvard University Press.
Latour, B. (2005). *Reassembling the social-an introduction to actor-network-theory*. New York: Oxford University Press.
Law, J. (1991). *A sociology of monsters: Essays on power, technology, and domination*. London: Routledge.
Lebow, R. N. (1994). The long peace, the end of the cold war, and the failure of realism. *International Organization, 48*(2), 249–277.
Lear, L. J. (1993). Rachel Carson's "silent spring". *Environmental History Review, 17*(2), 23–48.
Linklater, A. (2009). Human interconnectedness. *International Relations, 23*(3), 481–497.
Litfin, K. T. (1994). *Ozon discourses. Science and politics in global environmental cooperation*. New York: Columbia University Press.
Litfin, K. T. (1998). *The greening of sovereignty in world politics*. Cambridge, MA: The MIT Press.
Luke, T. W. (1998). Running flat out on the road ahead: nationality, sovereignty, and territoriality in the world of the information superhighway. In G. Ó. Tuathail & S. Dalby (Eds.), *Rethinking geopolitics* (pp. 175–212). London: Routledge.
Lyon, D. (2007). *Surveillance studies: An overview*. New York: Polity.
Mackinder, H. J. (1904). The geographical pivot of history. *The Geographical Journal, 23*(4), 421–437.
MacKenzie, D. A. (1993). *Inventing accuracy: An historical sociology of nuclear missile guidance*. Cambridge, MA: MIT Press.
MacKinnon, R. (2013). *Consent of the networked: The worldwide struggle for internet freedom*. New York: Perseus.
Makhijani, A., Howard, H., & Yih, K. (Eds.). (2000). *Nuclear wastelands: A global guide to nuclear weapons production and its health and environmental effects*. Cambridge, MA: MIT Press.
Mann, M. (1984). The autonomous power of the state: Its origins, mechanisms and results. *European Journal of Sociology, 25*(2), 185–213.
Mann, M. (2008). Infrastructural power revisited. *Studies in Comparative International Development, 43*(3), 355–365.
Matthewman, S. (2011). *Technology and social theory*. Basingstoke: Macmillan.
Marres, N. (2007). The issues deserve more credit pragmatist contributions to the study of public involvement in controversy. *Social Studies of Science, 37*(5), 759–780.
May, C. (2010). *The global political economy of intellectual property rights. The new enclosures* (2nd ed.). London: Routledge.
May, C., & Sell, S. K. (2005). *Intellectual property rights. A critical history*. London: Boulder.
Mayer, M. (2014). *The unbearable lightness of international relations – technological innovations, creative destruction, and explorative realism*. Berlin: Springer.

Mayer, M., & Schouten, P. (2011). Energy security and climate security under conditions of the Anthropocene. In L. Anceschi & J. Symons (Eds.), *Energy security in the era of the climate change. The Asia-Pacific experience* (pp. 13–35). London: Palgrave Macmillan.
Mayer-Schönberger, V., & Lazer, D. (Eds.). (2007). *Governance and information technology: From electronic government to information government*. Cambridge, MA: MIT Press.
McDonald, M. (2008). Securitization and the construction of security. *European Journal of International Relations, 14*(4), 563–587.
McLuhan, M., & Powers, B. R. (1992). *The global village: Transformations in world life and media in the 21st century*. New York: Oxford University Press.
McNeill, W. H. (1982). *The pursuit of power. Technology, armed force, and society since A.D. 1000*. Chicago, IL: Chicago University Press.
McNeill, J. R., & Unger, C. R. (Eds.). (2010). *Environmental histories of the cold war*. Cambridge, MA: Cambridge University Press.
Miller, C., & Edwards, P. (Eds.). (2001). *Changing the atmosphere: Expert knowledge and environmental governance*. Cambridge, MA: MIT Press.
Misa, T. J., & Schot, J. (2005). Introduction: Inventing Europe: 1. Technology and the hidden integration of Europe. *History and Technology, 21*(1), 1–19.
Mishra, P. (2012). *From the ruins of empire. Intellectuals who remade Asia*. New York: Farrar Straus & Giroux.
Mitcham, C. (1994). *Thinking through technology: The path between engineering and philosophy*. Chicago, IL: University of Chicago Press.
Mitchell, T. (2002). *Rule of experts: Egypt, techno-politics, modernity*. Berkeley: University of California Press.
Moldaschl, M., & Stehr, N. (Eds.). (2010). *Wissensökonomie und Innovation. Beiträge zur Ökonomie der Wissensgesellschaft*. Marburg: Metropolis.
Morgenthau, H. J. (1964). Modern science and political power. *Columbia Law Review, 64*, 1386–1409.
Morozov, E. (2013). *To save everything, click here: Technology, solutionism, and the urge to fix problems that don't exist*. London: Penguin.
Morozov, E (2014). More political interference! *Frankfurter Allgemeine Zeitung*, 15 Jan 2014. http://www.faz.net/aktuell/feuilleton/medien/evgeny-morozov-s-response-to-sascha-lobo-more-political-interference-12752837.html?printPagedArticle=true. Accessed 19 Feb 2014
Mueller, J. (1988). The essential irrelevance of nuclear weapons: Stability in the postwar World. *International Security, 13*(2), 55–79.
Mueller, J. (1989). *Retreat from doomsday: The obsolescence of major war*. New York: Basic Books.
Mueller, M. L. (2010). *Networks and states. The global politics of Internet governance*. Cambridge, MA: MIT.
Mumford, L. (1966). *The myth of the machine: Technics and human development*. New York: Harcourt Brace.
Murphy, C. N., & Yates, J. A. (2009). *The International Organization for Standardization (ISO): Global governance through voluntary consensus*. London: Routledge.
Murphy, C. N. (1994). *International organization and industrial change: Global governance since 1850. Europe and the international order*. New York: Oxford University Press.
Nanni, G. (2012). *The colonisation of time: Ritual, routine and resistance in the British empire*. Manchester: Manchester University Press.
Nelkin, D. (Ed.). (1992). *Controversy: Politics of technical decisions* (3dth ed.). Newbury Park, CA: Sage.
Netz, R. (2004). *Barbed wire: An the ecology of modernity*. Middletown: Wesleyan University Press.
Norris, P. (2001). *Digital divide: Civic engagement, information poverty, and the Internet worldwide*. Cambridge, MA: Cambridge University Press.

Nye, J. S., & Keohane, R. O. (1971). Transnational relations and world politics: An introduction. *International Organization, 25*(3), 329–349.
Oldenziel, R. (2011). Islands: The United States as a networked empire. In G. Hecht (Ed.), *Entangled geographies: empire and technologies in the global Cold War* (pp. 13–42). Cambridge, MA: MIT.
Ogborn, W. F. (1949). *Technology and international relations.* Chicago, IL: University of Chicago Press.
Osterhammel, J. (2009). *The transformation of the world. A history of the 19th century*. Princeton: Princeton University Press.
Paarlberg, R. L. (2004). Knowledge as power: Science, military dominance, and US security. *International Security, 29*(1), 122–151.
Patomäki, H., & Wight, C. (2000). After postpositivism? The promises of critical realism. *International Studies Quarterly, 44*, 213–237.
Paul, H. W. (2003). *From knowledge to power: The rise of the science empire in France, 1860-1939*. Cambridge, MA: Cambridge University Press.
Peoples, C. (2009). Technology, philosophy and international relations. *Cambridge Review of International Affairs, 22*(4), 559–561.
Perkins, J. H. (1997). *Geopolitics and the green revolution*. Oxford, NY: Oxford University Press.
Pfaffenberger, B. (1992). Social anthropology of technology. *Annual Review of Anthropology, 21*, 491–516.
Pielke, R., Wigley, T., & Green, C. (2008). Dangerous assumptions. *Nature, 452*(7187), 531–532.
Pinch, T., & Swedberg, R. (2008). *Living in a material world: Economic sociology meets science and technology studies* (Vol. 1). Cambridge, MA: MIT.
Pinch, T. (2008). Technology and institutions: Living in a material world. *Theory and Society, 37* (5), 461–483.
Poe, M. T. (2011). *A history of communications: Media and society from the evolution of speech to the internet*. Cambridge: Cambridge University Press.
Porter, T. (2003). Technical collaboration and political conflict in the emerging regime for international financial regulation. *Review of International Political Economy, 10*(3), 520–551.
Potter, S. J. (2007). Webs, networks, and systems: Globalization and the mass media in the nineteenth-and twentieth-century British empire. *The Journal of British Studies, 46*(3), 621–646.
Pouliot, V. (2010). The materials of practice: Nuclear warheads, rhetorical commonplaces and committee meetings in Russian–Atlantic relations. *Cooperation and Conflict, 45*(3), 294–311.
Poulantzas, N. (1978). *State power and socialism*. London: NLB.
Price, D. H. (2011). *Weaponizing anthropology: Social science in service of the militarized state*. Petrolia and Oakland, CA: CounterPunch and AK.
Rasmussen, A. (2014). NATO's new frontier. *World Affairs,* February 11. http://www.project-syndicate.org/commentary/anders-fogh-rasmussen-presents-the-benefits-of-creating-a-european-missile-defense-system. Accessed 02 Feb 2014.
Reisch, G. A. (2005). *How the cold war transformed philosophy of science: to the icy slopes of logic*. Cambridge, MA: Cambridge University Press.
Reus-Smit, C. (1996). The normative structure of international society. In F. Osler & J. Reppy (Eds.), *Earthly goods. Environmental change and social justice* (pp. 196–221). Ithaca: Cornell Univeristy Press.
Reuss, M., & Cutcliffe, S. H. (Eds.). (2010). *The illusory boundary: Environment and technology in history*. Charlottesville: University of Virginia Press.
Richards, E., & Ashmore, M. (1996). More sauce please! The politics of SSK: Neutrality, commitment and beyond. *Social Studies of Science, 26*(2), 219–228.
Ronen, P. (2003). *The offshore world: Sovereign markets, virtual places, and nomad millionaires*. Ithaca: Cornell University Press.
Rose, N. (2007). *The politics of life itself: Biomedicine, power, and subjectivity in the twenty-first century*. Princeton: Princeton University Press.

Rose, N., & Miller, P. (1992). Political power beyond the state: Problematics of government. *British Journal of Sociology, 43*(2), 173–205.
Rosenau, J. N. (1990). *Turbulence in world politics: A theory of change and continuity*. Princeton: Princeton University Press.
Rosenau, J. N. (1995). Governance in the twenty-first century. *Global Governance, 1*(1), 13–43.
Rosenau, J. (2003). *Distant proximities: Dynamics beyond globalization*. Princeton: Princeton University Press.
Rosenau, J. N., & Singh, J. P. (Eds.). (2002). *Information technologies and global politics. The changing scope of power and governance*. Albany: State University of New York Press.
Rosenberg, J. (2010). Basic problems in the theory of uneven and combined development. Part II: unevenness and political multiplicity. *Cambridge Review of International Affairs, 23*(1), 165–189.
Ruggie, J. G. (1993). Territoriality and beyond: Problematizing modernity in international relations. *International Organization, 47*(1), 139–174.
Sabaratnam, M. (2011). IR in dialogue... but can we change the subjects? A typology of decolonising strategies for the study of world politics. *Millennium Journal of International Studies, 39*(3), 781–803.
Sassen, S. (2001). *Global networks, linked cities*. London: Routledge.
Sassen, S. (2006). *Territory, authority, rights: From medieval to global assemblages*. Princeton: Princeton University Press.
Salter, M. B. (2007). Governmentalities of an airport: Heterotopia and confession. *International Political Sociology, 1*(1), 49–66.
Salter, M. B. (Ed.). (2014). *Making things international I: Circulation*. Minneapolis: University of Minnesota Press.
Scharff, R. C., & Dusek, V. (Eds.). (2013). *Philosophy of technology: The technological condition: An anthology*. Hoboken, NJ: Wiley.
Schmidt, B. (1998). *The political discourse of anarchy: A disciplinary history of international relations*. Albany, NY: State University of New York Press.
Schmidt, B. C. (2007). Realist conceptions of power. In F. Berenskoetter & M. J. Williams (Eds.), *Power in world politics* (pp. 43–64). London: Routledge.
Scheuerman, W. E. (2009). Realism and the critique of technology. *Cambridge Review of International Affairs, 22*(4), 563–584.
Schlosser, E. (2013). *Command and control: Nuclear weapons, the Damascus accident, and the illusion of safety*. New York: Penguin.
Schouten, P. (2013a). The materiality of state failure: Social contract theory, infrastructure and governmental power in Congo. *Millennium Journal of International Studies, 41*(3), 553–574.
Schouten, P. (2013b). Theory talk No. 60: Daniel Deudney on mixed ontology, planetary geopolitics, and Republican Greenpeace. http://www.theorytalks.org/2013/11/theory-talk-60_285.html. Accessed 20 Feb 2014
Schouten, P. (2014). Security as controversy: Reassembling security at Amsterdam airport. *Security Dialogue, 45*(1), 23–42.
Scott, J. C. (1998). *Seeing like a state: How certain schemes to improve the human condition have failed*. New Haven: Yale University Press.
Sen, A. K. (1999). *Development as freedom*. Oxford: Oxford University Press.
Shamir, R. (2013). *Current flow: The electrification of palestine*. Stanford, CA: Stanford University Press.
Shapin, S., & Schaffer, S. (1985). *Leviathan and the air-pump: Hobbes, boyle, and the experimental life*. Princeton, NJ: Princeton University Press.
Simmons, B. A. (2013). Preface: International relationships in the information age. *International Studies Review, 15*(1), 1–4.
Singh, J. P. (2002). Introduction: Information technologies and the changing scope of global power and governance. In J. N. Rosenau & J. P. Singh (Eds.), *Information technologies and global politics. The changing scope of power and governance* (pp. 1–38). Albany: State University of New York Press.

Singh, J. P. (Ed.). (2010). *International cultural policies and power*. Basingstoke: Macmillan.

Singh, J. P. (2013). Information technologies, meta-power, and transformations in global politics. *International Studies Review, 15*(1), 5–29.

Singleton, R. (2008). Knowledge and technology: The basis of wealth and power. In D. N. Balaam & M. Veseth (Eds.), *Introduction to international political economy* (4th ed., pp. 196–214). Upper Saddle River, NJ: Prentice Hall.

Skocpol, T., & Rueschemeyer, D. (Eds.). (1996). *States, social knowledge, and the origins of modern social policies*. Princeton, NJ: Princeton University Press.

Skolnikoff, E. B. (1993). *The elusive transformation: Science, technology and the evolution of international politics*. Princeton: Princeton University Press.

Slaughter, A.-M. (2004). *A new world order*. Princeton: Princeton University Press.

Smith, M. R., & Marx, L. (Eds.). (1994). *Does technology drive history? The dilemma of technological determinism*. Cambridge, MA: MIT.

Srnicek, N. (2013): *Representing complexity: The material construction of world politics* (PhD thesis). The London School of Economics and Political Science (LSE).

Squire, V. (2014). Reshaping critical geopolitics? The materialist challenge. *Review of International Studies*. doi:10.1017/S0260210514000102.

Star, S. L. (1999). The ethnography of infrastructure. *American Behavioral Scientist, 43*, 377–391.

Steinberg, P. E. (2001). *The social construction of the ocean*. Cambridge, MA: Cambridge University Press.

Steinberg, P. (2013). Of other seas: Metaphors and materialities in maritime regions. *Atlantic Studies, 10*(2), 156–169.

Stopford, J. M., Strange, S., & Henley, J. S. (1991). *Rival states, rival firms: Competition for world market shares*. Cambridge, MA: Cambridge University Press.

Strange, S. (1988). *State and markets*. London: Bloomsbury Academic.

Strange, S. (1996). *The retreat of the state*. Cambridge, MA: Cambridge University Press.

Stroeken, K. (Ed.). (2013). *War, technology, anthropology*. New York/Oxford: Berghahn Books.

Sunder Rajan, K. (Ed.). (2012). *Lively capital: biotechnologies ethics and governance in global markets*. Durham, NC: Duke University Press.

Tenner, E. (1997). *Why things bite back: Technology and the revenge of unintended consequences*. New York: Vintage.

Thad Allan, M. (2001). Modernity, the holocaust, and machines without history. In A. C. Hughes, M. Thad Allan, & G. Hecht (Eds.), *Technologies of power: Essays in honor of Thomas Parke Hughes and Agatha Chipley Hughes* (pp. 175–214). Cambridge, MA: MIT.

Thrift, N. (1996). *Spatial formations*. London: Sage.

Tucker, J. B. (Ed.). (2012). *Innovation, dual use, and security: Managing the risks of emerging biological and chemical technologies*. Cambridge: MIT Press.

United Nations. (2012). *Big data for development. Challenges and opportunities*. New York: UN Publishing.

United Nations (2013). Statistics and Indicators for the post-2015 development agenda. New York: UN publishing. http://www.un.org/en/development/desa/policy/untaskteam_undf/UNTT_MonitoringReport_WEB.pdf. Accessed 19 Feb 2014

UNESCO. (2005). *Towards knowledge societies*. Paris: UNESCO publishing. http://unesdoc.unesco.org/images/0014/001418/141843e.pdf. Accessed 19 Feb 2014.

Van Munster, R., & Sylvest, C. (2014). Reclaiming nuclear politics? Nuclear realism, the H-bomb and globality. *Security Dialogue* (in press).

Virilio, P. (1986). *Speed and politics: an essay on dromology*. New York: Columbia University.

Virilio, P. (2010). *Grey ecology*. New York: Atropos Press. Translated by Drew Burk.

Wajcman, J. (2008). Life in the fast lane? Towards a sociology of technology and time. *The British Journal of Sociology, 59*(1), 59–77.

Walker, R. B. J. (1993). *Inside/outside: International relations as political theory*. Cambridge, MA: Cambridge University Press.

Walters, W. (2009). Anti-political economy: Cartographies of "illegal immigration" and the displacement of the economy. In J. Best & M. Paterson (Eds.), *Cultural political economy* (pp. 113–138). London: Routledge.
Waltz, K. N. (1959). *Man, the state, and war: A theoretical analysis*. New York: Columbia University Press.
Waltz, K. N. (1979). *Theory of international politics*. Boston: MacGraw-Hill.
Waltz, K. N. (1990). Nuclear myths and political realities. *American Political Science Review, 84* (3), 731–745.
Weiss, C. (2005). Science, technology and international relations. *Technology in Society, 27*(3), 295–313.
Wendt, A. (1999). *Social theory of international politics*. Cambridge, MA: Cambridge University Press.
Whatmore, S. (2002). *Hybrid geographies: Natures cultures spaces*. London: Sage.
Williams, M. C. (2013). In the beginning: The international relations enlightenment and the ends of international relations theory. *European Journal of International Relations, 19*(3), 647–665.
Winner, L. (1980). Do artifacts have politics? *Daedalus, 109*(1), 121–136.
White, L. J. (1966). *Medieval technology and social change*. New York: Oxford University Press.
Wolfe, A. (2013). *Competing with the Soviets: Science, technology, and the state in Cold War America*. Baltimore: John Hopkins University Press.
Woolgar, S., & Cooper, G. (1999). Do artifacts have ambivalence? Moses' bridges, Winner's bridges and other urban legends in S&TS. *Social Studies of Science, 29*(3), 433–449.
Yang, D. (2011). *Technology of empire: Telecommunications and Japanese expansion in Asia, 1883–1945*. Cambridge, MA: Harvard University Asia Center.
Youngs, G. (2007). *Global political economy in the information age: Power and inequality*. London: Routledge.
Yu, P. K. (2012). The rise and decline of intellectual property powers. *Campbell Law Review, 34* (3), 525–577.
Zaheer, B. (1996). *The science of empire: Scientific knowledge, civilization, and colonial rule in India*. Albany: SUNY Press.
Zehfuss, M. (2001). Constructivism and identity: A dangerous liaison. *European Journal of International Relations, 7*(3), 315–348.

Part I
Approaches and Debates About Sciences and Technologies Within IR

From Expert Communities to Epistemic Arrangements: Situating Expertise in International Relations

Christian Bueger

Abstract The role and functions of expertise in international politics is, since decades, a core research theme. This chapter outlines a history of how the relation between science and international politics has been approached through the lenses of expertise. My intention is to offer a heuristic device. I argue that the debate can be structured in three generations. A first generation is interested in experts as actors that have a causal influence on international politics. The second generation scrutinizes discourses of expertise and their constitutional role in making the international. And the third generation concentrates on practices of expertise and the way these perform the epistemic arrangements of the international. To think about the study of expertise in the frame of three generations each offering different insights and carrying advantages and problems provides not only a practical tool for sorting ideas, but clarifies what one 'buys in' by following a specific generation.

Keywords Expertise • Epistemic communities • Discourse theory • Practice theory • Sociology of the discipline of international relations

1 Science, Technology and International Relations

The status of science and technology in international politics has been a recurrent theme in the discipline of International Relations (IR). All major theoretical developments within the discipline have implicitly or explicitly come along with assertions of the status of science and technologies. Post-World War II thought was pre-occupied with weapon technology, the use of science and technology in national security strategies and its link to state capacity and power. The rise of technocratic thinking as well as the emerging knowledge and post-industrial society

C. Bueger (✉)
Cardiff University, Cardiff, UK
e-mail: cbueger@gmx.de, http://bueger.info

M. Mayer et al. (eds.), *The Global Politics of Science and Technology - Vol. 1*, Global Power Shift, DOI 10.1007/978-3-642-55007-2_2,

discourse in the 1960s opened a new agenda. Academic interest turned towards the question of how developments in science and technology have an impact on the working of the international system. Research started to focus on how scientific inventions and new research developments spur new patterns of state interaction, increase the likelihood of cooperation between states and the foundation of international organizations and regimes. This 'regime' research in turn opened a new agenda that attempted to introduce scientists as an influential (transnational) actor type in the international system. Together with the rise of the constructivist agenda that highlighted the importance of knowledge in shaping the international system a growing number of studies became concerned with the various communities of scientists and experts and how they influence global politics. This extension of thinking about science and technology has last but not least also challenged the assumption of an essential differentiation between the natural sciences and the social sciences. Not only natural scientists, but the broader set of scientific experts became part of the agenda. Hence, increasing attention was paid to the various experts and knowledge-based professions—stretching from molecular biologists and climate scientists to international lawyers, and indeed also the professors of the discipline of IR themselves.

The result of these multiple extensions of thinking about science, technology and international relations is that today we are facing a quite scattered, fragmented system of various discourses mainly concerned with either distinct technologies (e.g. information technologies), or policy fields and their experts (e.g. international environmental or development policy). Such fragmentation and specialization suggests that that the importance of science and technology is increasingly recognized and mainstreamed across the disciplinary subfields. It points, moreover, to the growing openness of IR to the concepts and results of other scientific disciplines interested in the global, including anthropology and science and technology studies.

Irrespective of the proliferation of studies on science and technology and the diversification of agendas, I argue in this chapter that if this debate is read through the lenses of the concept of "expertise"—a main way of thinking the relation between science, technology and international politics—the debate can be ordered in three 'generations' of research. A first generation is interested in *experts as actors* that have a *causal influence* on international politics. The second generation scrutinizes *discourses* of expertise and their *constitutional* role in making the international. And the third generation concentrates on *practices of expertise* and the way these *perform* the epistemic arrangements of the international.

In proposing such a structuring of the debate, my intent is not to impose order on the discourse, but merely to suggest a practical tool for sorting thoughts that is useful in at least three ways: to identify different ways of situating techno-science in international relations and locating the epistemic, to clarify advantages and disadvantages of different accounts of the epistemic, and to elaborate on the spectrum of choice for empirical study. Creating such a heuristic tool, however, implies to create boxes, simplify and gloss over many details of the debate. The metaphor of generations is useful here as it emphasizes what bonds together is not shared belief or a coherent Lakatosian research program, but Wittgensteinian 'family

Table 1 Three generations

	Actors: 1st	Discourses: 2nd	Practice: 3rd
Ontology	Divergent	Convergent	Multiplicity
Mechanism	Causation	Constitution	Performation

resemblance', that is, a number of inter-related themes and assumptions. With employing the metaphor of generations it is, moreover, not my intention to argue for a linear, progressive or evolutionary narrative and suggest that one generation is superior to the others. Although discussing contributions temporally makes some sense, historicity should not be over emphasized, as work in the line of each generation tend to persist. Generations develop different discursive spaces for their arguments. These spaces overlap and also hybrids exist. Hence, the generations should not be seen as clearly demarcated from each other. This concerns especially the difference between the 2nd and the 3rd generation, which is rather one of emphasis than principle.

My reading of these generations is based on the observation that accounts differ over at least two core dimensions. Firstly, the ontological question whether there are specific logics to science and politics, and whether these converge or diverge. Secondly, the question of mechanisms, that is, how expertise relates science and politics to each other. Table 1 summarizes the answers to these question each generation provides.

In the following I elaborate on these three generations. While much of this presentation is abstract, my intent is to review the core texts and research approaches in each of the three generations. I introduce each generation and reflect on their advantages and challenges. The conclusion briefly reflects on how we can use the heuristic of three generations.

2 Expert Communities

Researchers that became known as IR 'functionalists' were perhaps the first ones calling for close scrutiny on the role of scientific experts in international relations. Functionalist studies of international cooperation—especially by Ernst Haas and his disciples—became translated into what is today the most well-known approach to study experts as actors of international politics: The epistemic community framework. This 1st generation relies on an understanding of science and politics as following divergent logics and aims at understanding the role of science by conceptualizing expertise as a form of causal mechanism.

In 1943 David Mitrany outlined core functionalist ideas by introducing his vision of what he called a "working peace system" (Mitrany 1943).[1] He argued

[1] For earlier scholarly debates about the role of experts and technical expertise within nineteenth century international legal organizations see Howland (2014).

for distinguishing between two forms of international cooperation, political and functional, and understood by the latter cooperation by experts on issues of technical nature (drawing on examples such as the regulation of post and telecommunications). For Mitrany, functionalist cooperation provided the mean to overcome the conflicts of political cooperation. The basis of his argument was a core characteristic of what I coin here the first generation, the assumption that science and politics have fundamentally divergent logics. As Haas (1964: 9) put it, "the distinction between "politicians" (concerned with doctrine and ideology) and "technicians or "experts" (concerned with doing practical tasks) is implicit or explicit in most Functional writing."

Mitrany (1943) understood technical problems as in need of practical solutions—solutions that 'work' or 'function' and which are carried out by committees and regulatory agencies. He hence spoke of 'practical arrangements' constituted by the need to solve a concrete practical coordination problem by other means than political struggle. His argument was that these arrangements follow a logic of "technical self-determination"—an argument that sees the solution of the problem as inserted in the problem. For Mitrany a working peace system was based on this functional cooperation from where it can spill over from one issue area into the other. Mitrany's argument became problematic, as the IOs directed at technical problems, set up in the 1950s and 1960s seemed not to respond to his analysis: even technical regimes were overshadowed by political conflict and struggles (Haas 1992b: 12). The functionalist argument was hence in need for revision.

Haas advanced the functionalist agenda through detailed empirical investigations of international regimes and international organizations (Haas 1964; Ruggie and Haas 1975; Haas et al. 1977; Haas 1990). Against Mitrany he argued that technical cooperation processes involve both scientists and politicians, and that in consequence close scrutiny has to be paid to the question of how scientists and politicians interact in technical cooperation. Experts, which included for Haas "physicists, epidemologists, metereologists, agronomists, management specialists, and lawyers" (Haas 1964: 455), have to be analysed in relation to "their position in the process of international decisionmaking" (Haas 1964: 455). In Haas' (1990: 46) view, "the language of science is becoming a world view that penetrates politics everywhere". Throughout his works he advanced three influential arguments. Firstly, the importance of uncertainty: As he suggested in his outline of a functionalist theory (Haas 1964), the most important expert in cooperation is what he called the "expert as crisis counsellor" (Haas 1964: 455). This expert is able to "analyse the past and projects aspects of the future without falling victim to highly specialized and subordinate objectives" (Haas 1964: 456). Secondly, Haas demonstrated that there is a substantial variation of expert-politician interactions across issue domains which are largely related to the degree to which expert knowledge is consensual or contested (Haas et al. 1977). Hence, he claimed that experts mattered only in special fields: "The more dependent an issue area becomes on technical information, the greater the likelihood that epistemic communities gain in influence" (Haas and Haas 2002: 592). In his later works he, thirdly, elaborated on the concept of "learning" as the core mechanism for change in international organizations (Haas 1990). For Haas the knowledge provided by scientists can inform the construction of regimes and the adjustment of state expectations, if policymakers "learn" from it. The relationship between science and politics becomes, therefore, a question of how politicians are able to learn from the scientists. These three core themes of uncertainty, diversity across issue domains, as well as learning as the core causal mechanism, opened an innovative agenda. These also become the major themes when a new generation of functionalists formulated the epistemic community framework.

The concept of epistemic communities initially articulated by Haas (1990: 41) was developed in collaboration of a group of scholars led by Emanuel Adler and Peter Haas and received attention when a special issue on the concept was published in IR's leading journal International Organization (Haas 1992a). The core contribution of the concept was to further develop functionalist insights, provide a heuristic device, and liberate the study of experts from the focus on the question of how authority is transferred to international institutions. In offering a unifying concept of the collective agency of experts it opened up the study of expertise to the sub-state level, wider questions of transformations of the international system, and also broadened the agenda towards other experts than techno-scientists. Drawing on earlier concepts of scientific communities including Thomas Kuhn's notion of 'paradigm' and Ludwik Fleck's concept of 'thought collective', Haas defined an 'epistemic community' as a "network of knowledge-based experts" (Haas 1992b: 2) "with recognized expertise and competence in a particular domain and an authoritative claim to policy-relevant knowledge within that domain or issue-area" (Haas 1992b: 3). The signifying features, which also allowed distinguishing an epistemic community from other actor types,[2] are a "(1) a shared set of normative and principled beliefs [...]; (2) shared causal beliefs, which are derived from their analysis of practices [...]; (3) shared notions of validity—that is, intersubjective, internally defined criteria for weighing and validating knowledge [...]; and (4) a common policy enterprise—that is, a set of common practices associated with a set of problems to which their professional competence is directed" (Haas 1992b: 3).

The influence of these expert communities was related to the rise of situations of uncertainty, in which there is a lack of information, interests are undefined or the available courses of action are indeterminate. These situations create a growing demand for the advice of experts. Policymakers learn from this advice, and the knowledge provided by epistemic communities becomes part of a process of diffusion on national, transnational or international scales, as well as eventually institutionalized in bureaucracies, regimes and organizations. The case studies of the volume showed a varying influence of epistemic communities. As summarized in Adler and Haas (1992), epistemic communities create "shared interpretations" (Adler and Haas 1992: 371), promote "values and practices" (Adler and Haas 1992: 372), produce

[2] Including scientific disciplines, professions, interest groups, social movements or bureaucracies (Haas 1992b: 16–20).

"intellectual innovations" (Adler and Haas 1992: 373) "alter perceptions and frame the context for collective responses to international problems" (Adler and Haas 1992: 376), might have an "influence on the setting of standards and the development of regulations" (Adler and Haas 1992: 378), narrow "the range within which political bargains could be struck" (Adler and Haas 1992: 378), or provide a channel for international diffusion (Adler and Haas 1992: 374). The conceptual framework spurred a quite substantial range of studies on different forms of epistemic communities and their influence in national and international contexts.[3] As Cross (2012) suggests, the majority of these studies, however, narrow the concept down to refer to natural scientists or technicians.

Haas' functionalism and the epistemic community framework are two of the major perspectives paradigmatic of a first generation of research on expertise in world politics. If paradigmatic, not all research representing this generation is directly related to or sees itself in this tradition. A substantial amount of other studies have been produced. This firstly concerns the field of foreign policy analysis, in which the role of experts and advisors in foreign policy processes has become a substantial research question. Drawing on survey data, Jacobs and Benjamin (2005: 117), for instance, come to the conclusion that "experts appear to be a potent force affecting the views of [US foreign] policymakers". Kaufmann (2004: 45) investigates the threat construction process prior to the US's Iraq invasion and argues that independent experts were largely unsuccessful to shape the debate on the Iraqi threat. Secondly, in the field of the sociology of the discipline of IR a number of studies investigate the role of the discipline's experts in policy processes.[4] While much of this debate is prescriptive in character—laying out how IR experts can or should relate to the policy process (e.g. Nincic and Lepgold 2000)—Mallin and Latham (2001), for instance, provide an empirical discussion that demonstrates that scholars shape the intellectual environment, provide practical innovation, engage in issue advocacy and contribute to agenda setting. Thirdly, in the study of transnational governance there is substantial further amount of studies of collective expert agencies or 'professions', of which Slaughter (2004) study of transnational expert networks is certainly the most well-known.

In what way do these studies converge so it becomes meaningful to speak about a first generation? Firstly, these studies conceptualize experts as an actor type that can be differentiated from others (e.g. lobbyists, advocates, or social movements). Secondly, the core tenet is to ask for the influence of this actor type on policy making and to identify the causal mechanisms by which this influence is exerted. The process of uncertainty, learning, diffusion and institutionalization described in the epistemic community model is one instance of such a mechanism. Thirdly, what underlies these studies of experts is a more or less strong divergence claim about

---

[3] Besides the contributions to original special issue, see the reviews of the literature in Antoniades (2003) and Cross (2012).

[4] Overviews of the literature are provided in Eriksson and Sundelius (2005) and Bueger and Gadinger (2007).

scientific knowledge. Scientific knowledge is considered to have specific qualities—e.g. for Haas it represented 'consensual knowledge'—and hence fundamentally differs from political knowledge. The core challenge for 1st generation research lies hence in how to 'isolate' experts as distinct group from other actors. Research is faced with what can be described as a 'boundary problem', that is, by which criteria can it be established who is an expert and who is not? Having convincing criteria is the precondition for isolating experts as an independent causal factor. As the discussion in the sociology of science[5] as well as around the epistemic community concept[6] documents, finding a convincing set of internal criteria which specifies who should count as expert is, however, an intricate problem.

3 Expertise as Discourse

The second generation breaks with the first in several ways. It criticizes the restriction of the term episteme to the policy relevant knowledge held by expert communities. The intention is to re-align the notion of episteme to the way it was developed in post-structuralist thought, centrally the work of Michel Foucault. This induces a shift in perspective from studying the influence of experts to the investigation of *expertise* as an expression of epistemic structures and regimes of truth. Extending the term in such a way to refer to the discourses that constitute the reality of international politics, has the intention of relating knowledge to power (going beyond questions of causal influence), and pointing to the contingency of all forms of knowledge. Apprehending the contingency and political character of all forms of knowledge, leads to a claim that scientific knowledge (production) is not superior or different from other forms. If Haas was committed to an understanding of scientific knowledge as a form of different, even superior knowledge, since it represented consensual knowledge—which is in more narrow form also present in the epistemic community framework's criterion of shared notions of validity—the 2nd generation fundamentally departs from this claim. The second generation hence elaborates a strong notion of the convergence of science and other forms of knowledge. Following a constitutive logic, expertise is studied as productive of social reality. If the first generation congregates around clearly established research frameworks (Haas' functionalism and the epistemic community framework), the second generation is less organized in such a way and linked to the increasing engagement with critical social theory and post-structuralism in IR starting in the late 1980s (Ashley and Walker 1990). In consequence, the style of reasoning becomes more philosophical and historical, and studies tend to operate on a higher degree of tempo-spatial aggregation, often referring to centuries rather than decades.

---

[5] See for instance Collins and Evans (2002) and the related debate on how to demark experts by internal criteria.

[6] See notably Toke (1999) and the summary of the controversy in Antoniades (2003: 25–26).

Contrary to an understanding of episteme as a set of validated beliefs held by groups of actors, the second generation elaborates an understanding of episteme as "the deepest layer of social knowledge, which, productive of what social reality is, helps constitute the order of global things (Adler and Bernstein 2005: 295). In this sense episteme refers to the background knowledge, discourses and practices in which people organize and create their world.[7] Decentering their investigations from the knowing subjects (the expert) and how they relate science and politics to each other, the core question becomes how expertise is a relational process that re-enforce discourses, that is, structures of meaning, in which international politics is organized.

For the advocates of such an approach in IR the initial object of study became the discipline of international relations—notably in the expression of realism—and how it is part of processes of expertise. As Jim George and David Campbell (1990: 285) paradigmatically summarized this line of research, the perspective shifted to "the practices that give rise to the subjects of international relations and constitute the domain to which International Relations theory is purportedly only responding. In this more far-reaching conception neither theorist nor theory escapes attention. For the conception invoked by the rationalists and realists, theory remains a tool for analysis, something that might be more sensitively honed but that nonetheless can continue to serve us (the theorists and, of course, the practitioners). For the conception associated with poststructuralism, theory is as much the object of analysis as the tool for analysis. The concern, although no less practical in its implications, is how analytic approaches privilege certain understandings of global politics and marginalize or exclude others." Authors such as Ashley (1987), Klein (1994) or Walker (1993) hence conducted analyses of how IR stabilizes certain understandings of the world and marginalizes others. While the main focus in these early analyses was realism and its understandings of strategy, balance of power and sovereignty, more recent arguments have generalized this argument. Steve Smith, for instance, argued that the discipline of IR has helped "to reinforce Western, predominately U.S., practices of statecraft"(Smith 2004: 499) and has been complicit in producing "the world that made September 11 possible" (Smith 2004: 500). This type of 2nd generation research hence primarily takes an introspective perspective and elaborates an understanding of expertise as a form of self-reflection and internal disciplinary critique. This debate however also shifted the emphasis from the 1st generation focus on techno-science to the social sciences and humanities. It opened productive avenues for studying expertise across domains not conventionally presumed technical in character.

Although the discursive understanding of expertise has become increasingly widespread across IR's several subdomains, it is notably in the study of security, international law and environmental politics that such a perspective is developed.

[7] For an elaboration of such an understanding of episteme that contrasts it with the understanding of epistemic communities see, besides Adler and Bernstein (2005), Antoniades (2003) and Litfin (1995).

Within critical security studies it was particularly David Campbell's *Writing Security* (1992), Bradley S. Klein's *Strategic Studies and World Order* (1994) as well as, a decade later, Lene Hansen's *Security as Practice* (2006) which have popularized not only the study of security as a textual structure, but also to consider expertise as one of the core regimes that produce and stabilize the meaning of security. Studies such as these hence include expert statements and documents in the body of texts they reconstruct. In studies of global environmental politics, a field in which the epistemic community approach became very prevalent, Karen Litfin's (1995) landmark study of the discourse on the ozone problem revealed how the discursive understanding can illuminate processes of expertise. For Litfin (1995: 254) "environmental problems [...] are discursive phenomena that can be studied as struggles among contested knowledge claims, which become incorporated into divergent narratives about risk and responsibility." To study such struggles she suggested understanding expertise as a set of 'discursive strategies' and 'linguistic practices' which frame knowledge in a certain way. Based on such an understanding she demonstrated how a certain type of discourse, precautionary discourse, came to structure the ozone problem and the regime that responds to it. In the study of international law David Kennedy has become one of the core researchers that shows how international cooperation and international law is a discourse fundamentally shaped by the vocabulary and language of expertise. For Kennedy (2001) the discourse of 'global governance' is the product of what he calls "invisible colleges" which above all render international policy making in technical terms which disables more pragmatic and progressive forms of policy.

The 2nd generation hence is structured by a broader understanding of knowledge as a shared textual structure in which language is a primary mean of producing the international and its problems. Researchers undermine the idea of science as a site of privileged knowledge and document the importance of knowledge generation across all social domains. Science is seen as part of the processes of expertise which produce and stabilize the orders of knowledge of world politics. In contrast to the first generation the analysis is decentered from subjects and there is a strong convergence claim that rejects ideas of a foundational boundary between science and politics. The focus shifts from identifying causal mechanisms to constitutive research and understanding the production of discourse. Given the conflation of science and political knowledge the boundary problem dissolves. Yet, this comes at a price: The main challenge of the second generation is how to produce meaningful statements of the relation of science, expertise and politics in the face of the strong conflation thesis (Bueger and Villumsen 2007). The study of discourse, moreover, has been primarily equated with the investigation of language, vocabulary and linguistic practices. Such a perspective, if it claims broader applicability, raises the question of what the place of non-linguistic activities as well as technologies and things are.

4 Epistemic Practice

A third generation takes up core ideas from both the first and second generation. To some degree the third generation can be understood as an *empirical* version of the second generation, as it aims at delivering detailed accounts of how epistemes are produced. Like the 2nd generation it operates under a broader understanding of episteme. Yet, paying attention to these details involves to some degree taking up ideas from the first generation, that is, to explore the difference between the scientific and political production of the episteme as well as to re-emphasize the importance of agency and the creative capacities of actors. In contrast to the 1st generation, the assumption is, however, not that there is an *ontological* difference between science and politics, but a practical one. Engaging in a dialogue with Post-Kuhnian Sociology of Science and Cultural Studies of Science (Rouse 1996), the 3rd generation, moreover, shifts the unit of analysis from actors and discourses to *practices*. This crucial move justifies distinguishing such work as a generation in its own right. The concept of practices occupies an ontological middle ground between actors and discourses. It brings back agency in that it centers on the practical activities of producing the epistemic, without falling back to a form of research primarily concerned with the (causal) influence of actors. Moreover, the concept of practice returns the material dimension of the production of knowledge, that is, the importance of bodily movements, of doings, as well as things and artifacts, which are major devices in producing knowledge, or in which knowledge is inscribed. The core question hence becomes by which practices (authoritative) expertise is produced.

Practices have been defined as configurations of material and bodily activities (doings and sayings), forms of knowledge (rules, habits, projects or affections), and objects and artefacts (technologies and things).[8] To study expertise as practice is hence the attempt to reconstruct such configurations. In IR it is mainly three approaches which use such a perspective, the study of 'fields of expertise' primarily drawing on interpretations of Pierre Bourdieu, the study of expertise as governmentality dwelling on Michel Foucault's practice theoretical work, and the study of expertise as arrangements, drawing on Actor-Network Theory or Assemblage Theory.

The study of fields of expertise has mainly been formulated in security studies. Scholars such as Bigo (2008), Huysmans (2006), Michael Williams or Berling (2011, 2012) develop such a perspective to understand how issues become securitized and policy proposals are formed. Huysmans (2006), for instance, outlines an understanding of expertise in arguing that the politics of (in)security is bifurcated in a politics of the political spectacle and technocratic politics. The political spectacle refers to "public visible contests that are mediated by political institutions such as parliament, political rituals such as elections and addresses to the nation, and

[8] See in IR, Adler and Pouliot (2011) as well as Bueger and Gadinger (2014) for outlines of such definitions.

formats of publicly visible dispersing ideas and symbols such as the news media, pamphlets and public opinion polling" (Huysmans 2006: 153), in sum, a publicized process of seeking and contesting legitimacy. Technocratic politics, then, refers to the rather less visible processes of "lobbying, instituting routine, struggles over expertise, and the development of forms, databases and other technologies" (Huysmans 2006: 154).

The expertise as governmentality perspective has mainly been developed in the study of transnational governance. Scholars such as Leander and van Munster (2007), Fougner (2008), Jaeger (2010), Sokhi-Bulley (2011a, b) or Hagmann and Cavelty (2012) argue that expertise is not a form of knowledge that precedes or informs politics, but is a form of governing in its own right, it is governance *through* expertise. Governmentality scholars emphasize the importance of expert technologies, such as statistics, indices and benchmarking or risk registers. Studying the Millennium Development Goals, Jaeger (2010) for instance argues that "the logic of benchmarking" displaces UN authority by "highlighting that the latter is no inherently a function of state power or institutional legitimacy, but follows a self-validating and self-propagating technical logic". Fougner (2008: 319), also discussing benchmarking, suggests that this expertise technology "constitutes states as technocratic agencies acting in accordance with expert determination of what works best". Sokhi-Bulley studies human rights indicators and statistics and shows that these not only "translate reality into documentary form", but also "allow for the normalization of the identities of both the 'victim' of rights violations and the 'guardian' of rights (Sokhi-Bulley 2011a: 241).

A third line of reasoning develops ideas of expertise as arrangements. Drawing on Actor-Network theory or the Deleuzian notion of assemblage the focus is more straightforwardly on the practices that produce the relations which carefully arrange elements to each other.[9] Such a perspective enables an understanding of the hybridity between science and politics as much as between social (human) and material (non-human). In IR, Rolf Lidskog and Göran Sunqvist (2002) premiered such an understanding. Investigating an international environmental regime they showed how knowledge is always embedded and carried by social arrangements in which the value of scientific knowledge is negotiated. Porter (2012) has demonstrated how numerical indices and peer review processes organize and maintain enormous networks. These devices are part of "successive translations" reducing "the complexity of the world" within which they "operate to a set of smaller, more visible and more manipulable relationships. Once these relationships are modified and reintroduced into the world, and once they enrol actants and networks, the indices and peer review that identified them are empowered, as are the individuals involved most directly with the production" of devices (Porter 2012: 553). Developing a similar account based on Actor-Network theory, Mayer (2012) demonstrated empirically that the securitization of climate is best understood as "the manufacturing of knowledge and risks through multidimensional, mutually

[9] See Best and Walters (2013) and Acuto and Curtis (2014).

responsive, and often seamless relationships between scientists, technical experts, and politicians".

In my own work I have argued for an understanding of arrangements as epistemic infrastructures characterized by practices and relations which maintain international issues. Drawing on the case of maritime piracy I argued that such a perspective allows for investigating how international problems are produced by a heterogeneous set of technical, scientific, administrative and political practices (Bueger 2014). In Bueger and Bethke (2014) we intend to show how an actor-network of academics, international administrations and foreign policy makers created and maintained the concept of the failed state. Bueger and Gadinger (2007) brought such a perspective to the sociology of the discipline and argued for also understanding IR as an arrangement primarily characterized by the practices of establishing and maintaining relations between scholarly knowledge and a broad range of actors including politicians or publics.

The third generation then is characterized by empirical inquiries on the practices of relating science, politics and other forms of knowledge, actors and practices to each other. Combining insights from the first and second generation the interest is primarily empirical and on the details of fields, governmentality or arrangements. No clear ontological position concerning the nature of scientific and political knowledge is taken here, other than the claim that both are effects of relations and practices. Multiplicity is foregrounded, that is, the claim that there are not only various potential constellations across, but also within, situations. The mechanism described in such studies can be perhaps best coined as "performation" to pick up a concept suggested by Michel Callon (2007). The notion highlights that expertise brings about certain worlds (performativity) as well maintain and stabilize them (performances). But the third generation also faces a range of challenges. As for other practice-theoretical works (Bueger and Gadinger 2014), the challenge arises that given the primacy of the empirical how to formulate statements which are relevant beyond the concrete situations studied and how to cope with the complexity that a practice perspective unleashes.

5 Three Generations of Expertise Studies: Summary and Challenges

This chapter outlined a history of how the relation between science and international politics has been approached in IR through the lenses of expertise. My intention was to offer a heuristic device. To think about the study of expertise in the frame of three generations each offering different insights and carrying advantages and problems provides not only a practical tool for sorting ideas, but clarifies what one 'buys in' by following a specific generation. Firstly, understanding expertise through a focus on communities, discourses or practices, each raises major challenges. A community focus leads to the boundary challenge. The discourse perspective to the quest of how one can account for the empirical, notably, material intricacies and the fragility of processes of knowing. The practice focus leads to the challenge of how to produce statements relevant outside the immediate situation one studies. Secondly, studying expertise involves ontological positioning, in the sense of what Patrick T. Jackson (2008) has called 'philosophical ontology'. Studies of expertise rely on claims about the ontological character of knowledge, that is, assertions about the characteristics of scientific knowledge and how it differs from political knowledge. Various positions of convergence and divergence can be justified and, as we have seen, they transform quite fundamentally how we approach expertise.

If the three generations differ in important regards, they however also face a set of common challenges. This concerns mainly the 'normative evaluation' of expertise. Across the generations there are mainly two normative interpretations: A *liberal-progressive narrative* foregrounds that expertise leads to better futures and contributes to the common good, enables cooperation, rights, peace or development. A *critical narrative* emphasizes the 'dark side of expertise' and reveals how expertise narrows down available policy options, stabilizes power relations, hinders emancipation or marginalizes actors. The majority of studies on expertise in IR embrace one of these narratives in paying little attention to the other. One of the challenges of research is hence how to relate both narratives to each other, and to study the situations in which expertise has what kind of negative or positive consequences against the backdrop of what kind of evaluative criteria. In engaging in such a debate, which is not only about how politics should be demarcated from science, and who should have the authority to govern in international politics, it is important not to fall back to a pejorative debate, in which participants are accused of arguing for expertocracy or technocracy. Instead of denunciation the debate has to recognize that the Golem of expertise is neither the enemy of legitimate political orders, nor by necessity its friend. What are required are hence studies which combine empirical insights with explicit normative considerations and explore which political situations require what forms of expertise .[10]

As this discussion has shown, IR has not only taken expertise seriously and produced a quite substantial amount of research on the issue—which is much more extended than what could be reviewed in the limits of this chapter—, but also outlined a number of distinct frameworks useful for the future study of international politics. This pertains to the various subfields of IR and as we have seen in the research I sampled in this chapter, expertise is a core concern in the debates on international organizations, security policy, environmental politics or international law. The study of expertise and the relation of science, technology and politics it entails have been and continue to be a major field of research in IR. This field is moreover increasingly trans- or at least inter-disciplinary in nature and—if sometimes with a considerable time lag—takes the dialogue and engagement with

[10] See notably Turner (2003), Pielke (2007) and Brown (2009) for important primers in this regards.

insights from other disciplines interested in knowledge (production) seriously. If I see certain advantages in third generation research and my own interest is to contribute to such a perspective, this does not imply that the other generations are in any way inferior or do not provide important insights. Inter-generational dialogue will be as much important as is inter-disciplinary. A core challenge for this dialogue will be, as I have suggested, how to connect detailed empirical studies to normative considerations in order to explore which situations require which forms of expertise. This is also a reflexive challenge as it concerns how IR as a discipline should partake in the games of expertise and how the 'meta-experts' (the experts studying experts) should feed into the debate.

Acknowledgements For discussions and comments which have informed and improved this chapter I am grateful to Maximilian Mayer and Trine Villumsen Berling. Research for this article has been supported by the Economic and Social Research Council [ES/K008358/1].

References

Acuto, M., & Curtis, S. (Eds.). (2014). *Re-assembling international theory. Assemblage thinking and international relations*. Basingstroke: Palgrave MacMillan.

Adler, E., & Bernstein, S. (2005). Knowledge in power: The epistemic construction of global governance. In M. Barnett & R. Duvall (Eds.), *Power in global governance* (pp. 294–318). Cambridge: Cambridge University Press.

Adler, E., & Haas, P. M. (1992). Epistemic communities, world order, and the creation of a reflective research program. *International Organization, 46*(1), 367–390.

Adler, E., & Pouliot, V. (2011). International practices. *International Theory, 3*(1), 1–36.

Antoniades, A. (2003). Epistemic communities, epistemes and the construction of (world)politics. *Global Society, 17*(1), 21–38.

Ashley, R. K. (1987). The geopolitics of geopolitical space: Toward a critical social theory of international politics. *Alternatives: Global, Local, Political, 12*(4), 403–434.

Ashley, R. K., & Walker, R. B. J. (Eds.) (1990). Speaking the language of exile. Dissidence in International studies. Special Issue of *International Studies Quarterly* 34 (3).

Berling, T. V. (2011). Science and securitization: Objectivation, the authority of the speaker and mobilization of scientific facts. *Security Dialogue, 42*(4–5), 385–397.

Berling, T. V. (2012). Bourdieu, international relations, and European security. *Theory and Society, 41*(5), 451–478.

Best, J., & Walters E, (Eds.) (2013). Forum on "Actor-Network Theory" and international relationality: lost (and found) in Translation, *International Political Sociology* 7(3).

Bigo, D. (2008). The emergence of a consensus: Global terrorism, global insecurity, and global security. In A. C. D'Appollonia & S. Reich (Eds.), *Immigration, integration and security: America and Europe in comparative perspective* (pp. 67–94). Pittsburgh, PA: University of Pittsburgh Press.

Brown, M. B. (2009). *Science in democracy: Expertise, institutions, and representation*. Cambridge: MIT Press.

Bueger, C. (2014). Making Things Known: Epistemic Practice, the United Nations and the Translation of Piracy, *mimeo*, Cardiff: Cardiff University.

Bueger, C., & Bethke, F. (2014, January 18). Actor-networking the failed state: An enquiry into the life of concepts. *Journal of International Relations and Development*, advance online publication. 17(1): 30–60 doi:10.1057/jird.2012.30.

Bueger, C., & Gadinger, F. (2007). Reassembling and dissecting: International relations practice from a science studies perspective. *International Studies Perspectives, 8*(1), 90–110.

Bueger, C., & Gadinger, F. (2014). The play of international practice. Minimalism, pragmatism and critical theory. *International Studies Quarterly*, forthcoming.

Bueger, C., & Villumsen, T. (2007). Beyond the gap: Relevance, fields of practice and the securitizing consequences of (democratic peace) research. *Journal of International Relations and Development, 10*(4), 417–448.

Callon, M. (2007). What does it mean to say that economics is performative? In D. MacKenzie, F. Muniesa, & L. Siu (Eds.), *Do economists make markets? On the performativity of economics* (pp. 311–357). Princeton, NJ: Princeton University Press.

Campbell, D. (1992). *Writing security: United States foreign policy and the politics of identity*. Minneapolis, MN: University of Minnesota Press.

Collins, H. M., & Evans, R. (2002). The third wave of science studies: Studies of expertise and experience. *Social Studies of Science, 32*(2), 235–295.

Cross, M. K. D. (2012). Rethinking epistemic communities twenty years later. *Review of International Studies, 39*(1), 137–160.

Eriksson, J., & Sundelius, B. (2005). Molding minds that form policy: How to make research useful. *International Studies Perspectives, 6*(1), 51–72.

Fougner, T. (2008). Neoliberal governance of states: The role of competitiveness indexing and country benchmarking. *Millennium: Journal of International Studies, 37*(2), 303–326.

George, J., & Campbell, D. (1990). Patterns of dissent and the celebration of difference: Critical social theory and international relations. *International Studies Quarterly, 34*(3), 269–293.

Haas, E. B. (1964). *Beyond the nation-state: Functionalism and international organization*. Stanford: Stanford University Press.

Haas, E. B. (1990). *When knowledge is power. Three models of change in International Organizations*. Berkeley, CA: University of California Press.

Haas, P. M. (1992a). Introduction: Epistemic communities and international policy coordination. *International Organization, 46*(1), 1–35.

Haas, P. M. (Ed.) (1992a). *Knowledge, power, and international policy coordination*. Special Issue of *International Organization* 46 (1).

Haas, P. M., & Haas, E. B. (2002). Pragmatic constructivism and the study of international institutions. *Millennium: Journal of International Studies, 31*(3), 573–601.

Haas, E. B., Williams, M. P., & Babai, D. (1977). *Scientists and world order: The uses of technical knowledge in international organizations*. Berkeley, CA: University of California Press.

Hagmann, J., & Cavelty, M. D. (2012). National risk registers: Security scientism and the propagation of permanent insecurity. *Security Dialogue, 43*(1), 79–96.

Hansen, L. (2006). *Security as practice. Discourse analysis and the Bosnian War*. Milton Park and New York: Routledge.

Howland, D. (2014). Telegraph technology and administrative internationalism in the 19th century. In M. Mayer, M. Carpes, & R. Knoblich (Eds.), *The global politics of science and technology*. Heidelberg: Springer.

Huysmans, J. (2006). *The politics of insecurity. Fear, migration and asylum in the EU*. Milton Park: Routledge.

Jackson, P. T. (2008). Foregrounding ontology: Dualism, monism, and IR theory. *Review of International Studies, 34*(1), 129–153.

Jacobs, L. R., & Benjamin, I. P. (2005). Who influences U.S. foreign policy? *American Political Science Review, 99*(1), 107–123.

Jaeger, H.-M. (2010). UN reform, biopolitics, and global governmentality. *International Theory, 2*(1), 50–86.

Kaufmann, C. (2004). Threat inflation and the failure of the marketplace of ideas. *International Security, 29*(1), 5–48.

Kennedy, D. (2001). The politics of invisible college: International Governance and the politics of expertise. *European Human Rights Law Review, 5*, 463–497.

Klein, B. S. (1994). *Strategic studies and world order: The global politics of deterrence*. Cambridge: Cambridge University Press.
Leander, A., & van Munster, R. (2007). Private security contractors in the debate about darfur: Reflecting and reinforcing neo-liberal governmentality. *International Relations, 21*(2), 201–216.
Lidskog, R., & Sundqvist, G. R. (2002). The role of science in environmental regimes: The case of LRTAP. *European Journal of International Relations, 8*(1), 77–100.
Litfin, K. (1995). Framing science: Precautionary discourse and the ozone treaties. *Millennium: Journal of International Studies, 24*(2), 251–277.
Mallin, M., & Latham, R. (2001). The public relevance of international security research in an era of globalism. *International Studies Perspectives, 2*(2), 221–230.
Mayer, M. (2012). Chaotic climate change and security. *International Political Sociology, 6*, 165–185.
Mitrany, D. (1943). *A working peace system*. London: Royal Institute of International Affairs.
Nincic, M., & Lepgold, J. (2000). *Being useful. Policy relevance and international relations theory*. Ann Arbor: University of Michigan Press.
Pielke, R. A., Jr. (2007). *The honest broker. Making sense of science in politics*. Cambridge: Cambridge University Press.
Porter, T. (2012). Making serious measures: Numerical indices, peer review, and transnational actor-networks. *Journal of International Relations and Development, 15*, 532–557.
Rouse, J. (1996). *Engaging science. How to understand its practices philosophically*. Ithaca: Cornell University Press.
Ruggie, J. G. & Haas, E. B. (Eds.) (1975). International responses to technology. Special Issue of *International Organization* 29 (3).
Slaughter, A.-M. (2004). *A new world order*. Princeton, NJ: Princeton University Press.
Smith, S. (2004). Singing our world into existence: International relations theory and september 11. Presidential address to the international studies association, February 27, 2003, Portland, OR. *International Studies Quarterly* 48: 499–515.
Sokhi-Bulley, B. (2011a). Governing (through) rights: Statistics as technologies of governmentality. *Social & Legal Studies, 20*(2), 139–155.
Sokhi-Bulley, B. (2011b). Government(ality) by experts: Human rights as governance. *Law and Critique, 22*(3), 251–271.
Toke, D. (1999). Epistemic communities and environmental groups. *Politics, 19*(2), 97–102.
Turner, S.P. (2003). *Liberal democracy 3.0 civil society in an age of experts*. London: Sage Publications.
Walker, R. B. J. (1993). *Inside/outside: International relations as political theory. Cambridge studies in international relations*. Cambridge: Cambridge University Press.

Nuclear Science and Technology: The Race Between Weapons and Controls

Joseph F. Pilat

Abstract The revolution in physics in the early twentieth century was utilized before its mid-point to develop revolutionary weapons of extraordinary power, following a path that has been repeated again and again in history with other scientific advances. The peaceful and military uses of the atom were seen as "Siamese twins" or as "Janus headed," which meant that not only the proliferation threat emanating from dedicated weapon programs in states, but also that of the misuse of civil programs by states, needed to be addressed. Nuclear science and technology was developed in wartime for military purposes, but since the dawn of the nuclear age, its civil possibilities were foreseen, and its dual nature was present in policy makers' and academics' minds. The dual-use nature of the technology, and the high technological hurdles to its development at the outset of the nuclear age led to predominantly supply-side approaches such as export controls to deal with the risks of proliferation. While this focus has persisted, it has evolved over the decades in response to experience and today is complemented by growing attention to the demand side. From this perspective, the history of nuclear energy, the past history and current and future prospects of nuclear proliferation and the development of the nonproliferation regime, can be seen as elements of a race between nuclear technology diffusion on the one hand and technological and institutional efforts to combat proliferation on the other. This paper explores the military influences that shaped the earliest applications of nuclear science and technology, the promise of nuclear energy for peaceful purposes that highlighted the dual nature of the atom and the long standing efforts and debates over the control of this revolutionary science and technology.

The views expressed are the author's own and not those of the Los Alamos National Laboratory, the National Nuclear Security Administration, the Department of Energy or any other agency.

J.F. Pilat (✉)
Los Alamos National Laboratory, Los Alamos, NM, USA
e-mail: jpilat@lanl.gov

M. Mayer et al. (eds.), *The Global Politics of Science and Technology - Vol. 1*, Global Power Shift, DOI 10.1007/978-3-642-55007-2_3,

Keywords Nuclear science and technology • Nuclear energy • History • US

1 Introduction

The revolution in physics in the early twentieth century was utilized before its mid-point to develop revolutionary weapons of extraordinary power, following a path that has been repeated again and again in history with other scientific advances. The peaceful and military uses of the atom were seen as "Siamese twins" or as "Janus headed," which meant that not only the proliferation threat emanating from dedicated weapon programs in states, but also that of the misuse of civil programs by states, needed to be addressed.[1]

Any use of nuclear energy to produce electricity or for medical or industrial purposes would, in fact, create a military capability. This dual nature of the atom meant that most applications of nuclear science created nuclear latency, which can be viewed as the possession of most or all of the technologies, facilities, materials, expertise (including tacit knowledge), resources and other capabilities necessary for the development of nuclear weapons, without full operational weaponization. While latency is largely tied technologically to the dual-use nature of the atom, it has to be seen historically. It involves the full range of capability possessed by aspiring, existing and former nuclear-weapon states, and the possible diffusion of nuclear-weapon relevant equipment and information via a number of outlets, including non-state nuclear supply networks, the Internet, etc.

Nuclear science and technology was developed in wartime for military purposes, but since the dawn of the nuclear age, its civil possibilities were foreseen, and its dual nature was present in policy makers' and academics' minds from the beginning. Nuclear shares some issues with other dual-use technologies, especially the creation of latency that poses an over-the-horizon threat and could result in strategic surprise. It has unique elements as well. Unlike most other dual-use technologies, which pose a serious strategic military or other security challenge only when linked to other technologies or to new operational concepts, nuclear technology was seen at the outset to be in itself revolutionary. Although originally conceived and used as a more powerful strategic bomb, it was rapidly understood to be the "absolute weapon," so powerful that it could deter other states from an attack using these

[1] In the 1940s, the threat was seen to emanate from states. Over the intervening decades, the threat has evolved and today covers a wide range of possibilities. Proliferation and terrorism could involve state or nonstate actors, or state support of nonstate actors. The threat could take a variety of forms. Nuclear proliferation can involve everything from the establishment of a virtual capability to produce weapon-usable materials to a full-blown weapon program. Nuclear and radiological terrorism can range from the threat or use of a nuclear weapon to the dispersal of radiological material and attacks or sabotage against nuclear facilities or transport. Not only are there different forms of proliferation and terrorism, but each differs in terms of enabling and other factors such as the materials and facilities involved, the technological sophistication required of the state or nonstate actors, their strategies, motivations and intentions, etc., and, most notably, the probability and the consequence of successful execution.

weapons or conventional forces. Moreover, nuclear is unusual to the extent that the technologies that were essential to creating the bomb have been critical to efforts to control the bomb.

The dual-use nature of the technology, and the high technological hurdles to its development at the outset of the nuclear age led to predominantly to supply-side approaches such as export controls to deal with the risks. While this focus has persisted, it has evolved over the decades in response to experience and today is complemented by growing attention to the demand side. The history of nuclear energy, the past history and current and future prospects of nuclear proliferation and the development of the nonproliferation regime, can be seen from this perspective as elements of a race between nuclear technology diffusion on the one hand and technological and institutional efforts to combat proliferation on the other.[2]

This paper explores the military influences that shaped the earliest applications of nuclear science and technology, the promise of nuclear energy for peaceful purposes that highlighted the dual nature of the atom and the long standing efforts and debates over the control of this revolutionary science and technology.

2 The Possibility of Nuclear Weapons

The path from science to weaponry in the nuclear arena was recognized early and was realized in practice in a remarkably short period. In *The World Set Free* (2014), H. G. Wells imagined nuclear energy being used for peaceful and military uses, from electrical power generation and transportation to bombs. He wrote of the impact of atomic energy on industry, arguing it would first be used in electrical generation. Other applications would include use in automobiles and airplanes. Soon, he anticipated,

> a gigantic replacement of industrial methods and machinery was in progress all about the habitable globe. . . .And with an equal speed atomic engines of various types invaded industrialism. The railways paid enormous premiums for priority in the delivery of atomic traction engines, atomic smelting was embarked upon so eagerly as to lead to a number of disastrous explosions due to inexperienced handling of the new power, and the revolutionary cheapening of both materials and electricity made the entire reconstruction of domestic buildings a matter merely dependent upon a reorganization of the methods of the builder and the house-furnisher. Viewed from the side of the new power and from the point of view of those who financed and manufactured the new engines and material it required the age of the Leap into the Air was one of astonishing prosperity" (Wells 2014: 15).

Yet Wells believed this technology would be used for military purposes. The same technology that transformed industry world transforms the military. Wells clearly understood the revolutionary nature of the atomic bomb:

[2] For further discussion of issues addressed here, see Schroeer (1984); Bundy (1990); Newhouse (1989); Hewlett (1962); Hewlett and Duncan (1969); Hewlett and Holl (1989); Rhodes (1986); Holloway (1994); Gowing and Arnold (1974); Scheinman (1965); Lewis and Xue (1988); Cohen (1998); Perkovich (2001).

> Never before in the history of warfare had there been a continuing explosive; indeed, up to the middle of the twentieth century the only explosives known were combustibles whose explosiveness was due entirely to their instantaneousness; and these atomic bombs which science burst upon the world that night were strange even to the men who used them. Those used [by one side in the war]... were lumps of pure Carolinum, painted on the outside with unoxidised cydonator inducive enclosed hermetically in a case of membranium. A little celluloid stud between the handles by which the bomb was lifted was arranged so as to be easily torn off and admit air to the inducive, which at once became active and set up radio-activity in the outer layer of the Carolinum sphere. This liberated fresh inducive, and so in a few minutes the whole bomb was a blazing continual explosion. ... [The other side's bombs were] the same, except that they were larger and had a more complicated arrangement for animating the inducive.
>
> Always before in the development of warfare the shells and rockets fired had been but momentarily explosive, they had gone off in an instant once for all, and if there was nothing living or valuable within reach of the concussion and the flying fragments then they were spent and over. But Carolinum...once its degenerative process had been induced, continued a furious radiation of energy and nothing could arrest it (Wells 2014: 31).

Wells described the "degenerative process" as akin to a "miniature active volcano" that could erupt from weeks to years depending on the size of the bomb, and once launched was "absolutely unapproachable and uncontrollable until its forces were nearly exhausted." (Wells 2014: 31). He also predicted the persistent radioactive consequences of the use of atomic bombs. After the war, in every country where bombs had struck, he wrote that there were areas,

> ...a score of miles in diameter, [that] mark the position of the dying atomic bombs and the death areas that men have been forced to abandon around them. Within these areas perished museums, cathedrals, palaces, libraries, galleries of masterpieces, and a vast accumulation of human achievement, whose charred remains lie buried, a legacy of curious material that only future generations may hope to examine... (Wells 2014: 58–59).

Wells' vision was prophetic, and foreshadowed the views of many in 1945 after the bombing of Hiroshima and Nagasaki. His science was wildly inaccurate—in Leo Szilard's view, "moonshine"—but Szilard believed Wells understood the possible practical applications of the science better than the physicists (Szilard 1978: 38–40 citing Hirst 1934).

Two decades later, a breakthrough in physics opened up the prospects of Wells' predictions. At a time when the first clouds of the coming global war were discernible, the path foreseen by Wells in which peaceful development of nuclear technology led to its military uses was reversed. The fears of war focused attention on the military possibilities.

In experiments in 1934, Enrico Fermi bombarded various materials with neutrons to see if he could induce radioactivity. His experiment bombarding uranium and thorium created new elements. He reported his results, but he did not identify what had occurred as fission (Fermi 1934: 283; Fermi et al. 1934: 48–500).[3] It was only in December 1938 that the German chemists Otto Hahn and Fritz

[3] Sidney Drell has wondered what might have occurred had fission been identified in 1934 rather than 1938.

Strassmann detected barium after bombarding uranium with neutrons. Lise Meitner and Otto Frisch understood this was result of nuclear fission. Frisch confirmed this the following year.

Upon learning of fission, Albert Einstein wrote to President Franklin Delano Roosevelt, about the possibility of the bomb. He stated:

> In the course of the last four months it has been probable through the work of Joliot in France as well as Fermi and Szilard in America—that it may become possible to set up a nuclear chain reaction in a large mass of uranium, by which vast amounts of power and large quantities of new radium-like elements would be generated. Now it appears almost certain that this could be achieved in the immediate future.
>
> This new phenomenon would also lead to the construction of bombs, and it is conceivable—though much less certain—that extremely powerful bombs of a new type may thus be constructed. A single bomb of this type, carried by board and exploded in a port, might very well destroy the whole port together with some of the surrounding territory. However, such bombs might very well prove to be too heavy for transportation by air (Einstein 1939).

The Frisch-Perls "Memorandum on the Properties of a Radioactive Super-Bomb" of March 1940 confirmed this possibility, and how it might be achieved, stating:

> The possible construction of "super-bombs" based on a nuclear chain reaction in uranium has been discussed a great deal and arguments have been brought forward which seemed to exclude this possibility. We wish here to point out and discuss a possibility which seems to have been overlooked in these earlier discussions (Frisch-Perls Memorandum 1940).

Elaborating, the memorandum states:

> Bohr has put forward strong arguments for the suggestion that the fission that this isotope has, on the whole, a much greater fission probability than the common isotope ^{238}U. Effective methods for the separation of isotopes have been developed recently, of which the method of thermal diffusion is simple enough to permit separation of a fairly large scale.
>
> This permits, in principle, the use of nearly pure ^{235}U in such a bomb, a possibility which apparently has not so far been seriously considered. We have discussed this possibility and come to the conclusion that a moderate amount of would ^{235}U indeed constitute an extremely efficient explosive (Frisch-Perls Memorarandum 1940).

Nuclear weapons were a reality 31 years after Wells' vision and only 6 years after Einstein's warning. Nuclear technology was forged during wartime; it can be used for peaceful purposes. The destructive power of nuclear weapons developed in the Manhattan Project and demonstrated in Hiroshima and Nagasaki defined the nuclear landscape in the aftermath of World War II, along with unrealistic expectations of the peaceful benefits of the new power source. Many believed this revolutionary weapon would spread widely as new weapons had always spread in the past. In a world of states, the risk of a nuclear arms race and nuclear war was believed to be unacceptably high. There were, of course, those who saw something different in these weapons and believed they should be controlled internationally. The prospect of a world without nuclear weapons may never have been a real option at the time of the Cold War was imminent, but the expected allure of nuclear power for civil, peaceful purposes complicated the post-war situation dramatically.

3 The International Control of Nuclear Energy

Written in the aftermath of Hiroshima and Nagasaki, the seminal Acheson-Lilienthal Report of 1946 reflects the challenges arising from the dual nature of the atom, and the race between nuclear technology diffusion on the one hand and international efforts to combat proliferation. Concerns about nuclear latency, primarily in the form of the possible misuse of civilian nuclear power programs for military purposes was on the minds of the authors of the report and the Baruch Plan that was based on the report. The report "recognized that the industry required and the technology developed for the realization of atomic weapons are the same industry and the same technology which play so essential a part in man's almost universal striving to improve his standard of living and his control of nature." (International Control of Atomic Energy 1946: 2) More specifically, it declared: "The development of atomic energy for peaceful purposes and the development of atomic energy for bombs are in much of their course interchangeable and interdependent" (International Control of Atomic Energy 1946: 4).

While recognizing the inherent dual-use nature of the nuclear fuel cycle, the report drew distinctions between "dangerous" and "safe" activities. In its reasoning, a nuclear activity is dangerous if it offers a solution to any of three major problems of proliferation: obtaining raw material, producing fissionable materials and fabricating a weapon. Dangerous activities were identified as uranium and thorium mining, operation of plutonium production reactors and reprocessing facilities and weaponization. Safe activities were those involving relatively small amounts of fissile material, fissile material production facilities of low capacity and certain types of power reactors. The report proposed that "dangerous" nuclear activities be reserved for an international "Atomic Development Authority" (ADA) that would own and license the use of all fissile material. "Safe" activities would be allowed to be carried out by individual nations.

The line between dangerous and safe activities was not set in stone and could be affected by technological action. This issue was raised in the Acheson-Lilienthal report's discussion of "denaturing." According to the report:

> ...U 235 and plutonium can be denatured; such denatured materials do not readily lend themselves to the making of atomic explosives, but they can still be used with no essential loss of effectiveness for the peaceful applications of atomic energy. They can be used in reactors for the generation of power or in reactors useful in research and in the production of radioactive tracers. It is important to understand the sense in which denaturing renders material safer. In the first place, it will make the material unusable by any methods we now know for effective atomic explosives unless steps are taken to remove the denaturants. In the second place, the development of more ingenious methods in the field of atomic explosives which make this material effectively useable is not only dubious, but is certainly

> not possible without a very major scientific and technical effort (International Control of Atomic Energy 1946: 26–27).

The report recognized that denaturing could be reversed, but held that "doing so calls for rather complex installations which, though not of the scale of those at Oak Ridge or Hanford, nevertheless will require a large effort and, above all, scientific and engineering skill of an appreciable order for their development" (International Control of Atomic Energy 1946: 27).

The authors of the report were overly hopeful of denaturing, even though they did not give full support to the dream of dividing the civil from the military atom. Inspections were also soundly dismissed in Acheson–Lilienthal, where it was unanimously concluded that "there is no prospect of security ... [for a system] which relies on inspection and similar police-like methods (International Control of Atomic Energy 1946: 4)."

Given these limits, and the dual nature of the atom, it was held that in order to best ensure against technological surprise and breakout, the proposed Atomic Development Authority would have to be at the cutting edge of all nuclear science and technology, including weapon R&D:

> ...one of the important things that the Authority will have to do is research in atomic explosives. We are by no means sure that important new discoveries in this field do not lie ahead. Possibly the study of atomic explosives may yield byproducts useful in peaceful activities. But this will not be the main purpose of the Authority's research. Only by preserving its position as the best informed agency will the Authority be able to tell where the line between the intrinsically dangerous and the non-dangerous should be drawn. If it turns out at some time in the future, as a result of new discoveries, that other materials lend themselves to dangerous atomic developments, it is important that the Authority should be the first to know. At that time measures would have to be taken to extend the boundaries of safeguards (International Control of Atomic Energy 1946: 36).

Even this would not, it was recognized, be sufficient in a world without nuclear weapons. The report understood that there was a need inter alia "to provide security while allowing states to maintain a relatively secure position if there are problems; to be able to deal with unanticipated new threats; to provide unambiguous and reliable danger signals of noncompliance; [and] to deal with the problem of enforcement (International Control of Atomic Energy 1946: 9–10). The report stated:

> In strengthening security, one of the primary considerations will relate to the geographical location of the operations of the Authority and its property. For it can never be forgotten that it is a primary purpose of the Atomic Development Authority to guard against the danger that our hopes for peace may fail, and that adventures of aggression may again be attempted. It will probably be necessary to write into the charter itself a systematic plan governing the location of the operations and property of the Authority so that a strategic balance may be maintained among nations. In this way, protection will be afforded against such eventualities as the complete or partial collapse of the United Nations or the Atomic Development Authority, protection will be afforded against the eventuality of sudden seizure by any one nation of the stockpiles, reduction, refining, and separation plants, and reactors of all types belonging to the Authority (International Control of Atomic Energy 1946: 47).

The authors of the report argued that the United States, with Hanford, Oak Ridge, and Los Alamos, may evoke fear in other states that can only be mitigated to the extent that

> the Atomic Development Authority locates similar dangerous operations within their borders. Once such operations and facilities have been established by the Atomic Development Authority and are being operated by that agency within other nations as well as within our own, a balance will have been established. It is not thought that the Atomic Development Authority could protect its plants by military force from the overwhelming power of the nation in which they are situated. Some United Nations military guard may be desirable. But at most, it could be little more than a token. The real protection will lie in the fact that if any nation seizes the plants or the stockpiles that are situated in its territory, other nations will have similar facilities and materials situated within their own borders so that the act of seizure need not place them at a disadvantage (International Control of Atomic Energy 1946: 47).

More would be needed, in their view:

> The design of primary production plants should make them as little dangerous as possible. The stockpiles of materials suitable for the production of bombs should be kept as small as possible consistent with sensible economics and engineering. So far as practicable, stocks should be denatured or kept in low concentrations unsuitable for the production of bombs. In other words, the design and operating procedures should definitely prevent the accumulation of substantial amounts of material quickly convertible into important quantities of explosives (International Control of Atomic Energy 1946: 47–48).

In effect, the response to noncompliance and the threat of breakout would be the prospect that other advanced states could arm or rearm in reaction. The distribution of latent capability across the globe was to enable this response. Although this would provide a measure of security for those states that feared the United States rearming, the very idea would be highly unstable as states rushed to arm at the first sign of a crisis. Nonetheless, it was regarded as a deterrent by the authors of the report.

The Acheson-Lilienthal report was not a proposal for the control of atomic energy per se, but it was the basis for the Baruch plan, which was presented to the United Nations Atomic Energy Commission on June 14, 1946. The Baruch plan recognized the issues laid out in the report, and proposed "...the creation of an International Atomic Development Authority, to which should be entrusted all phases of the development and use of atomic energy, starting with the raw material" (Baruch 1946). The Authority would have, *inter alia*, managerial control or ownership of all dangerous atomic energy activities; the power to control, inspect and license all other atomic activities; the promotion of the beneficial uses of atomic energy; and research and development intended to ensure the Authority in the forefront of atomic knowledge and able to detect the misuse of atomic energy (Baruch 1946).

The Baruch plan, however, it did not provide the same solution for addressing latency-driven breakout scenarios as the report. Rather than ensuring nuclear latency was distributed across the globe so that a state's efforts to break out could be countered by all other states, Baruch called for "condign punishment" of violators. He stated that: "We must provide the mechanism to assure that atomic energy is used for peaceful purposes and preclude its use in war. To that end, we must provide immediate, swift, and sure punishment of those who violate the agreements that are reached by the nations. Penalization is essential if peace is to be more than a feverish interlude between wars. And, too, the United Nations can prescribe individual responsibility and punishment on the principles applied at Nuremberg by the Union of Soviet Socialist Republics, the United Kingdom, France and the United States–a formula certain to benefit the world's future" (Baruch 1946).

The heart of his message was the need to deal with any violations effectively, including the elimination of the Security Council veto for the five permanent members. This appears to have reflected concern about the ideas on this topic in the original report. According to Baruch:

> It would be a deception, to which I am unwilling to lend myself, were I not to say to you and to our peoples that the matter of punishment lays at the very heart of our present security system. It might as well be admitted, here and now, that the subject goes straight to the veto power contained in the Charter of the United Nations so far as it relates to the field of atomic energy. The Charter permits penalization only by concurrence of each of the five great powers - the Union of Soviet Socialist Republics, the United Kingdom, China, France, and the United States.
>
> I want to make very plain that I am concerned here with the veto power only as it affects this particular problem. There must be no veto to protect those who violate their solemn agreements not to develop or use atomic energy for destructive purposes (Baruch 1946).

4 The Cold War, Atoms for Peace and the Nonproliferation Regime

Amid emerging Cold War rivalries, and Soviet suspicion of US motives, the proposal languished in the UNAEC. In the Third Report of the Atomic Energy Commission to the Security Council, of May 17, 1948, the Commission reported that it cannot prepare a treaty draft. After Baruch, the idea of an international control regime to reduce the dangers of nuclear arms and energy was no longer a focus of policy makers. Disarmament was largely off the table during the Cold War as the "arms race" resulted in larger stockpiles of nuclear weapons and weapons became more powerful with the development of thermonuclear weapons. As nuclear weapons spread from a US monopoly to be possessed by the Soviet Union and the United Kingdom, the nuclear threat was seen to be spiraling out of control. By this time, it was recognized that these weapons were fundamentally different from other weapons. The threat was largely dealt with through deterrence, which depended ultimately on the destructive power of these weapons. At the same time, halting, modest arms control efforts were also first being discussed as more realistic alternative to the failed proposals for international controls. The Atoms-for-Peace speech of President Eisenhower laid the foundations for a far less ambitious international nuclear nonproliferation regime.

In the early 1950s, President Eisenhower believed something must be done as the race between technology diffusion and control efforts was being lost. The US believed that the use of nuclear technology for peaceful programs would be inevitable, providing the great majority of, perhaps *all*, states, with nuclear weapons capabilities. As a result, President Eisenhower delivered the Atoms for Peace speech at the UN General Assembly on December 8, 1953. He stated:

> . . .the dread secret, and the fearful engines of atomic might, are not ours alone
>
> In the first place, the secret is possessed by our friends and allies, Great Britain and Canada, whose scientific genius made a tremendous contribution to our original discoveries, and the designs of atomic bombs.
>
> The secret is also known by the Soviet Union....
>
> If at one time the United States possessed what might have been called a monopoly of atomic power, that monopoly ceased to exist several years ago. Therefore, although our earlier start has permitted us to accumulate what is today a great quantitative advantage, the atomic realities of today comprehend two facts of even greater significance.
>
> First, the knowledge now possessed by several nations will eventually be shared by others–possibly all others.
>
> Second, even a vast superiority in numbers of weapons, and a consequent capability of devastating retaliation, is no preventive, of itself, against the fearful material damage and toll of human lives that would be inflicted by surprise aggression (Eisenhower 1953).

President Eisenhower called for an international atomic energy agency that would both provide assistance to states on progress for peaceful acts of nuclear energy, and ensure that the assistance provided was not measured in, or diverted to, military programs, declaring:

> The governments principally involved, to the extent permitted by elementary prudence, should begin now and continue to make joint contributions from their [military] stockpiles of normal uranium and fissionable materials to an international atomic energy agency. We would expect that such an agency would be set up under the aegis of the United Nations. . . .
>
> The atomic energy agency could be made responsible for the impounding, storage and protection of the contributed fissionable and other materials. The ingenuity of our scientists will provide special safe conditions under which such a bank of fissionable material can be made essentially immune to surprise seizure (Eisenhower 1953).

The Atoms-for-Peace proposal and the international nuclear nonproliferation regime that emerged from the Atoms-for-Peace bargain—centered on the International Atomic Energy Agency (IAEA) and the Treaty on the Nonproliferation of Nuclear Weapons (NPT),[4] and supported by safeguards, export controls, national laws and regulations, economic and security assurances, sanctions, and regional arrangements of various types, including nuclear-weapon-free zones—were based on the atom's dual nature.

Disarmament was effectively off the table at this time. The Cold War was fully engaged, and an arms race had begun. As stockpiles of nuclear weapons expanded rapidly, and the prospect of even larger arsenals appeared likely, Atoms for Peace was an arms control as well as a nonproliferation initiative. States were to provide fissile material from weapons programs to an international fuel bank set up by the

---

[4] For details on these programs, See Pilat (2007).

agency. The agency through the bank would then distribute these materials to states for peaceful nuclear energy programs. Because fissile material production was seen as a choke point in the production of nuclear arms, it was believed this provision would significantly draw down military stockpiles. This assumption was to an extent true, but it was definitely overstated. In any case, only limited, symbolic contributions to the bank were made.

This swords-to-plowshares vision of the nuclear future has from the beginning been surrounded by controversy and its wide acceptance and influence did not end concerns. Critics were concerned about the very dual nature of the atom that made the proposal possible, and pointed out the dangers of a program that was spreading nuclear technology—declassification of information, provision of research reactors and other capabilities, training nuclear technologists, etc. —that could be used for military purposes. However, the dominant view was that the diffusion would have occurred any way, and that the Atoms for Peace framework and the IAEA established as a result of the proposal was a practicable means of addressing the risks.

5 Debating Nuclear Energy and Proliferation

For decades after this landmark speech, those who believed we could manage the risks of nuclear power have recognized its inherent dangers, but pointed to the dedicated military programs of most proliferants as the real source of concern. Those who were less sanguine about the prospect of harnessing a promising but dangerous technology in a world rife with conflict often argued against the use of nuclear power altogether, or at least opposed closing the nuclear fuel cycle.

The debate is now over 60 years old. It has waxed and waned, depending upon real-world developments such as the concerns derived from extrapolations of rapid, even exponential, growth in nuclear power and by the actual emergence of proliferation threats, notably the Indian program in the 1970s, the Pakistani program in the 1980s and the Iraqi, Iranian, North Korean and Syrian programs since the 1990s. Although the prospects of nuclear accidents had not been at the forefront of the debate from the 1940s to the 1970s, the accidents at Three Mile Island and Chernobyl greatly affected the debate over civil uses of nuclear power, as has the accident at Fukushima Daiichi.

In recent years, the debate is beginning to be reengaged on a level not seen since the 1970s. At that time, the Indian test of a so-called "peaceful nuclear explosion" and the prospect of dramatic, even exponential, nuclear power growth raised fears of uncontrolled proliferation—of losing the race between technology diffusion and controls—and spurred further debate over the threat of proliferation and the adequacy of the nonproliferation regime. The report of the Ford-MITRE Nuclear Energy Policy Subgroup *Nuclear Power: Issues and Choices*, reflected the concerns of the times, arguing on the basis of the longstanding recognition of the link between nuclear power and weapon capability that the "growth and diffusion of nuclear power . . .inevitably enhance the potential for the proliferation of nuclear weapons" (Nuclear Power: Issues and Choices: 1977: 22). The result was the creation of the Nuclear Suppliers Group and a tightening of export controls, as well as US domestic policies opposing reprocessing and civil use of plutonium. This response was a means to regain advantage in the race. It was not well received and led to a rift between North and South that continues today, and between the US and key allies in Europe and East Asia as well that were only mitigated by the International Nuclear Fuel Cycle Evaluation (INFCE) and by the Reagan Administration's decision to reassure allies on the policy's scope and application.

There are similarities between the debate now and 40 years ago, for example:

- Expectations of dramatic growth in nuclear power;
- Concerns about reprocessing and plutonium use; and
- Perceptions of rising proliferation and terrorism threats.

But there are major differences. On the one hand, while US and Russian stockpiles have been dramatically reduced, proliferation dangers appear more real or concrete, if not necessarily greater than they did 40 years ago when attention focused on plutonium. The risks from highly enriched uranium (HEU) are now seen as greater.[5] The risks are also increasingly seen to be emerging from unanticipated sources, including non-state actors. The prospect of nuclear terrorism is receiving unprecedented attention (although it was a factor in the debate during the 1970s). After 9/11, some concluded the danger of any use of nuclear power was too great to accept (see Nader 2007).

On the other hand, the desire for energy independence has led to increased interest in nuclear energy. And global warming concerns have convinced many, including some staunch environmentalists, of the need to pursue nuclear power aggressively (See Moore 2008; Lovelock 2004). Moreover, to address rising concerns about proliferation and terrorism, strong efforts to reduce nuclear power's risks and vulnerabilities are being proposed and undertaken, including efforts to avoid separation of plutonium in the future.

Threats posed to the nuclear enterprise are changing. In the context of rising regional instability and conflict, along with increased incidents of global terrorism, in a dynamic, uncertain security environment, emerging nuclear and other weapons of mass destruction (WMD) threats—both proliferation and terrorism—are seen as growing dangers giving rise to increasing global insecurity. Many observers believe today that additional states as well as non-state actors will obtain a nuclear-weapon capability or nuclear weapons, and that these weapons are more likely be used than in the past. Although the notion of nuclear anarchy or a "nuclear armed crowd" (Wohlstetter et al. 1976) may represent the worst-case scenario—which is by no

[5] This perception is largely based on the spread of centrifuge enrichment technology by the A.Q. Khan network and, to the extent it may represent an overreaction, may be reconsidered in the future.

means the inevitable or even the most likely future—it is clear that the threat situation today covers a wide range of possibilities.

6 Technological and Institutional Responses to Nuclear Threats

Among the risks of proliferation and terrorism we confront today, there are those that were understood and anticipated from the beginning, along with those that have surprised us. Efforts to reduce, mitigate or eliminate these risks—at least those we anticipated—are decades old. From a US perspective, proliferation has been seen from the beginning as a global problem. Responses to the threat in the nuclear realm have been primarily global in nature, from the Baruch Plan to the Atoms-for-Peace initiative to the IAEA and the NPT.

The global treaty approach has been important for setting norms concerning nuclear and other weapons of mass destruction and missiles, and the treaties have been influential in redefining thinking about the problem. However, the NPT and the international nuclear non-proliferation regime were created in a different time to deal with different threats. If all of the problems with, and stresses on, the regime portended the regime's collapse or increasing irrelevance, the spread of nuclear energy would be very dangerous. The efforts to address proliferation are complicated by dissatisfaction with the progress on disarmament, which is called for in Article VI of the NPT.

Will the regime be able to address the challenges of today, along with those that will emerge with an expansion of nuclear energy around the world? It remains to be seen whether the regime will meet the challenges ahead.[6] However, in the face of these challenges, the regime is being reformed. As it has in the past, the regime is evolving as threats have changed, as is evident in the case of safeguards.

In the last two decades, the International Atomic Energy Agency has been transforming its safeguards system to address, in part, the limits of its verification mandate and the burden of noncompliance issues, which have raised questions about the value and effectiveness of international safeguards. In this context, the IAEA is adopting a fundamentally new approach to implementing safeguards based on the strengthening measures developed in the 1990s and the lessons learned from Iraq, North Korea, Libya and Iran. It is recognized that an effective, strengthened international safeguards system, with a strong focus on searching for undeclared nuclear materials and activities, is essential to provide confidence that shared nuclear technologies and expertise, as well as nuclear materials themselves, are not being diverted to nuclear-weapon programs.

[6] For a wide-ranging debate over the fate of the international nuclear order and the NPT regime, see the essays in *International Affairs 83*, (3, May 2007): 427–574.

In addition to strengthening safeguards and other traditional regime elements such as export controls, initiatives to address new and emerging threats, and unanticipated developments—from the end of the cold war to the rise of terrorism—have been especially prominent in the last two decades. Among these are critical initiatives involving threat reduction, detection and interdiction, such as programs for Cooperative Threat Reduction, Material Protection, Control and Accounting and Second Line of Defense, including the Megaports Initiative; the Proliferation Security Initiative and the Global Threat Reduction Initiative; the Global Initiative for Proliferation Prevention and the Global Initiative to Combat Nuclear Terrorism; and UNSC Resolution 1540, the Convention on the Suppression of Nuclear Terrorism and the amendments to the Convention on the Physical Protection of Nuclear Material.

Beyond these and other programs and initiatives, there are several ideas that are being considered, or rather reconsidered today to address proliferation and terrorism risks. Offering an assured supply of fresh nuclear fuel and spent-fuel take back are old ideas that are receiving new attention. They have become central to thinking about addressing emerging challenges, especially those involving the spread of enrichment and reprocessing capabilities. Multinational or multilateral ownership has been proposed by then IAEA Director General Mohammed ElBaradei as a means to address this issue.[7]

The difficulties of realizing these or any of the other proposals that have been put forward to minimize proliferation risks through reliable supply are significant and have bedeviled past efforts along these lines. Although such approaches have failed before, there are key differences in the situation today from that of the earlier considerations of various proposals, including a more widespread sense of insecurity; the rise of new, illegitimate sources of supply, including black marketeers; evidence of NPT noncompliance and the misuse of Article IV of the NPT to obtain the nuclear fuel cycle; and the prospect of nuclear terrorism. In any event, the viability of current proposals depends ultimately on common interests (commercial, political, industrial, etc.). They cannot be imposed from the top down, nor should they interfere with market mechanisms.

Finally, new attention to another old idea—proliferation resistance—has grown and can be expected to grow in the years ahead. Proliferation resistance was first raised in the Acheson-Lilienthal report's discussion of "denaturing," but the hope of finding a way to make the peaceful uses of nuclear energy resistant to proliferation appears and reappears in the history of nuclear power.

Proliferation resistance has often been oversold, and has had limited results to date; however, there are benefits that may yet be realized from reactors and other facilities designed to minimize risks coupled with effective safeguards and other

[7] See ElBaradei (2003). See also the report of experts that followed up the original ElBaradei proposal, *Multilateral approaches to the Fuel Cycle*, Expert Group Report submitted to the Director General of the International Atomic Energy Agency, issued as INFCIRC/640 at www.iaea.org Accessed Feb. 2014.

nonproliferation measures. The idea of proliferation-resistant small reactors with long-lived cores is among the new ideas for addressing underlying proliferation concerns, while expanding nuclear power to the developing world and increasing the attractiveness and acceptability of nonproliferation efforts. There is also a promising path ahead with reactors and other facilities designed to minimize proliferation risks coupled with effective safeguards and other non-proliferation measures that could have utility for states as well as for non-state actors. The novel approach involves new technologies and facility designs that embody cost-effective means to safeguard material or, where possible, refitting old facilities and technologies to make them more "safeguardable." The objective of safeguardability or safeguards-by-design, is to improve the application of safeguards by, for example, reducing or eliminating diversion or misuse pathways or increasing the prospects for detection along pathways by facilitating verification through, for example, technologies and approaches involving greater physical access as well as increased process and operating information.

Safeguards-by-design should be complemented by a specific "security-by-design" effort, and there have been discussions about such an integrated approach. It appears that fuel cycle design, facility layouts and protective measures can reduce or eliminate the feasibility of certain theft scenarios and make their success extremely unlikely. Such design measures could include steps to track and locate nuclear material accurately and cheaply throughout the cycle; to eliminate scenarios that are costly to address with external measures (i.e., additional guns, gates and guards); to make a facility inaccessible to outsiders (and to insiders for unauthorized actions); and to increase the role of "passive" security features (see Pilat and Budlong-Sylvester 2010).

However attractive, the feasibility, costs, operational impacts and other possible effects of such approaches have not yet been demonstrated and will need to be analyzed further.

There is a need to demonstrate a "business case" for safeguards- and security-by-design.

All of these responses to current and emerging threats are important, as efforts to reinforce and reform the global nonproliferation regime to address proliferation and terrorism risks. But not all are agreed or fully developed and implemented, and they may not be sufficient to address future threats.

Beyond the responses of today, we may expect the race to continue for the foreseeable future. What do we need to ensure we can respond to tomorrow's threats?

Our record of anticipating threats has not been great. Technological advances, their diffusion and their ultimate impacts are notoriously difficult to predict with any certainty. There is a need for improved capabilities in this regard. Even with the best threat assessment capabilities, however, we will likely still be surprised. The threat is dynamic and we must have the tools critical to respond effectively and rapidly to any new threats.

Beyond today's threats, the conditions under which one could envision widespread proliferation in the longer term depend on such factors as technology diffusion, globalization, regional and international security environments (particularly changes in Russia and China) and the like. Of these factors, which are all difficult to predict with any certainty, there are proliferation challenges we can anticipate today based on technology spread and on the development of certain technologies for civil use. The threats stemming from the spread of technology include:

- Technology diffusion via the Internet as well as through loose nukes, materials leakage and brain drain in the former Soviet Union, Pakistan and other states and through non-state actors like the A. Q. Khan network; and
- The security of nuclear and related technologies, materials and expertise in Russia and the other Soviet successor states, as well as in such states as South Africa.

Challenges based on technologies that are coming into use, or are expected to do so, include the following:

- Large, increasingly complex new facilities, with high material throughputs where improvements in current technology alone cannot meet detection goals;
- Difficult-to-measure materials and harsh environments with high dose rates, temperatures, etc.;
- The need to measure new isotopes and combinations of isotopes with, for example, separations outputs of pyroprocessing and electrorefining;
- The need to measure both continuous flows of nuclear materials and of non-nuclear process parameters (temperature, density, flow rate, etc.); and
- Possible diversions without physical change to plant through process controls, chemistry, etc.

History shows there will be many proliferation challenges we do not anticipate, including clandestine facilities, breakouts after abrogating treaty obligations, the emergence of new technologies and the like.

To address all of these challenges—both anticipated and unanticipated—a key (but certainly not the only) requirement is the further development of a defense-in-depth safeguards and security approach that can be more responsive to emerging threats. The ST&E investments done to maintain the US deterrent is absolutely necessary for this purpose.

The Stockpile Stewardship Program of the past 20 years—a science-based program of experimentation, improved diagnostics and increased computational capabilities, along with advances in modeling, the science of plutonium and other actinides and high explosives (HE) behavior have provided many of the tools necessary to assess issues within the stockpile through the resolution of outstanding science issues—has been a difficult and ongoing challenge. Its successes have been essential for ensuring a safe, secure and effective deterrent without further nuclear testing or the development of new nuclear warheads.

Not only has this capability allowed the United States to credibly extend deterrence to, and assure, Japan, South Korea and NATO, which promotes nonproliferation, but the ST&E has contributed to nonproliferation directly and indirectly, as it has in the past. The US weapon program developed many of the diagnostic and other technologies that have been instrumental to the development of safeguards and arms control verification. It is also critical to understanding current and future threats, developing effective responses and avoiding "technological surprise." Today, in the United States at least, the weapon program has been and remains essential to ensuring the capability to respond to pressing security concerns, including nonproliferation, counter-proliferation and counter-terrorism. Counter-terrorism, for example, benefits from weapon-program work on advanced sensing technology for enhanced detection, nuclear forensics for attribution and new information technology and capabilities to predict and track the flow of nuclear materials and knowledge.

7 Conclusions

At the dawn of the nuclear age, a small arsenal, even a single weapon, could change the strategic equation, it was believed. The threat did not depend only on delivery capabilities. It was intrinsic to the weapon. Only by eliminating nuclear weapons or by countering them by deterrence could the danger be reduced. The danger involving peaceful uses as well, which could be diverted to military applications, placed a premium on creating effective international controls.

There has been, in the US and Western perspective at least, a connection between nuclear technology development, including the use of nuclear energy for peaceful purposes, and measures to control the proliferation, or spread of nuclear weapons. Today, as at the dawn of the nuclear era, S&T involving weapons and nonproliferation and arms control efforts remain linked.

This perspective, at one level, is based on the laws of physics. As noted, the history of nuclear energy, the prospects of nuclear proliferation and the development of the nonproliferation regime, can be seen as elements of a race between nuclear technology diffusion on the one hand and international efforts to combat proliferation on the other. The perceptions of, and debates over, who was winning the race have, as suggested, changed over time.

Before Fukushima, the prospect of global nuclear expansion—in the view of some, a nuclear renaissance—appeared to be defining the future nuclear landscape as a result of increasing energy demand, growing concerns over energy security and climate change and other emerging issues. This prospect revived long-standing concerns that the growth of nuclear power will inevitably result in increased proliferation and terrorism risks.

Whatever the long-term impact of Fukushima on global interest in nuclear power, the pre-accident rationale for nuclear power remains, and it still appears at present that nuclear power will expand, perhaps at a slower rate and less widely, than would have been the case before the accident. We may expect that most of the growth will occur in states that already possess nuclear power programs and sensitive nuclear facilities, including China, Russia, India and France. Moreover, the Fukushima accident will not only limit the geographic spread of nuclear power but also likely slow its growth, which should limit any risks. Finally, the accident may also affect future fuel cycle choices by making multilateral, cradle-to-grave solutions more attractive, which could also have nonproliferation benefits. On the other hand, a global slowdown could in theory lead suppliers to sell to unstable or irresponsible states.

The risks associated with the development of nuclear power have often been exaggerated in the past—notably the charge that every reactor is a "bomb factory" (Markey 1982)—and the chosen path for proliferators to date has largely been dedicated weapon programs rather than diversions from or other misuse of peaceful programs. There is nonetheless an undoubted connection between the civil and military atom and the risks of the nuclear enterprise are real. Nuclear latency created by civilian programs poses a technological challenge that could affect future nuclear proliferation and the prospects for disarmament.

In this context, responding to emerging technological challenges with effective and efficient nonproliferation efforts does involve a race against time at one level. This level is critical and derives from the dual nature of the atom. However, it is not the only level on which the challenges which are critical to international security and stability need to be addressed. Addressing the threats of proliferation and terrorism of nuclear or other weapons of mass destruction will be critical, along with such issues as predicting and mitigating changes in the global climate and developing adequate, cost effective and sustainable sources of energy. These are all interrelated and among the most complex system-level challenges the world has faced.

Through the support to science and engineering that sustains nuclear deterrence at lower levels of weapons, the United States already has a base from which to address the diverse challenges that will shape the next century. On the basis of arguments presented in the Acheson-Lilienthal report, and on practice since then, the weapon program's foundation has been and remains critical to the effectiveness of threat reduction activities. The future will depend to a great extent on these capabilities and their continuing evolution in response to changing technological and geopolitical realities. The need to address incentives for proliferation is part of the evolution in thinking and is also critical, and can be approached in many ways. In addition to the supply-side focus on the existing regime, one way forward is the recognition that demand side considerations need greater attention. Intentions are difficult to divine and may change, but they must be addressed in some fashion in addition to the efforts to deal with emerging technical capabilities.

References

Atomic Archive. (1940, March). *The Frisch-Perls memorandum.* http://www.atomicarchive.com/Docs/Begin/FrischPeierls.shtml. Accessed February, 2014.

Baruch, B. (1946, June 14). Statement of United States policy, presented to the United Nations Atomic Energy Commission.

Bundy, M. G. (1990). *Danger and survival: Choices about the bomb in the first fifty years.* New York: Vintage Books.

Cohen, A. (1998). *Israel and the bomb.* New York: Columbia University Press.

Einstein, A. (1939, August 2). *Letter to president Franklin D. Roosevelt.* http://research.archives.gov/description/593374

Eisenhower, D. D. (1953, December 8). *Address by Mr. Dwight D. Eisenhower, President of the United States of America, to the 470th Plenary Meeting of the United Nations General Assembly.* http://www.iaea.org/About/atomsforpeace_speech.html. Accessed February, 2014.

ElBaradei, M. (2003, October 18). Toward a safer world. *The Economist.*

Fermi, E. (1934). Radioattività indotta da bombardamento di neutron. *La Ricerca scientifica, 1*(5), 283.

Fermi, E. et al. (1934). Artificial radioactivity produced by neutron bombardment. *Proceedings of the Royal Society A: Mathematical, Physical and Engineering Sciences 146*(857), 48–500

Gowing, M., & Arnold, L. (1974). *Independence and deterrence: Britain and atomic energy, 1945–1952* (Vol. 2). London: The Macmillan Press.

Hewlett, R. G., & Anderson, O. E. (1962). *The new world, 1939-1946.* University Park, PA: Pennsylvania State University Press.

Hewlett, R. G., & Duncan, F. (1969). *Atomic shield, 1947-1952.* University Park, PA: Pennsylvania State University Press.

Hewlett, R. G., & Holl, J. M. (1989). *Atoms for peace and war, 1953-1961: Eisenhower and the atomic energy commission.* Berkeley, CA: University of California Press.

Holloway, D. (1994). *Stalin and the bomb: the soviet union and atomic energy 1939-1956.* New Haven: Yale University Press.

International Atomic Energy Agency. (n.d). *Multilateral approaches to the Fuel Cycle*, Expert Group Report submitted to the Director General of the International Atomic Energy Agency, issued as INFCIRC/640 at http://www.iaea.org

International Control of Atomic Energy. (1946, March 16). *Report prepared for the secretary of state's committee on atomic energy.* Washington, DC: U.S. Government Printing Office.

Lewis, J., & Xue, L. (1988). *China builds the bomb.* Stanford, CA: Stanford University Press.

Lovelock, J. (2004, May 24). Nuclear power is the only green solution. *The Independent.* http://www.ecolo.org/media/articles/articles.in.english/love-indep-24-05-04.htm. Accessed February, 2014.

Markey, E. J. (1982). *Nuclear peril: The politics of proliferation.* Cambridge, MA: Ballinger.

Moore, P. (2008, April 12). Nuclear power: Massachusetts is facing up to carbon choices. *Patriot Ledger.* http://www.patriotledger.com/opinions/x1403477302. Accessed February, 2014.

Nader, R. (2007, September 11). *Nuclear power is not the answer.* http://www.commondreams.org/archive/2007/09/11/3761/. Accessed February, 2014.

Newhouse, J. (1989). *War and peace in the nuclear age.* New York: Knopf.

Perkovich, G. (2001). *India's nuclear bomb: The impact on global proliferation, updated with a new afterword.* Berkeley, CA: University of California Press.

Pilat, J. (Ed.). (2007). *Atoms for peace: A future after fifty years?* Baltimore, MD: Johns Hopkins University Press.

Pilat, J. F., & Budlong-Sylvester, K. W. (2010, July 11–15): *The role of proliferation resistance in addressing non-state actors.* Paper presented at the 51st Annual Meeting. Baltimore, MD: INMM.

Report of the Nuclear Energy Policy Subgroup. (1977). *Nuclear power: issues and choices.* Sponsored by the Ford Foundation, Administered by the MITRE Corporation. Cambridge, MA: Ballinger Publishing Company.

Rhodes, R. (1986). *The making of the atomic bomb.* New York: Simon & Schuster.

Scheinman, L. (1965). *Atomic energy policy in france under the fourth republic.* Princeton, NJ: Princeton University Press.

Schroeer, D. (1984). *Science, technology and the arms race.* New York: John Wiley & Sons.

Szilard, L. (1978). Letter from Leo Szilard to Sir Hugo Hirst, March 17, 1934. In: S.R. Weart, & G. W. Szilard. *Leo Szilard: His version of the facts selected recollections and correspondence* (pp. 38–40). Cambridge, MA: MIT Press.

Wells, H. G. (2014). *The world set free*. San Bernadino, CA: n. p.

Wohlstetter A, et al. (1976). *Moving toward life in a nuclear armed crowd?*. Report to the US Arms Control and Disarmament Agency. Los Angeles, CA: Pan Heuristics.

International Competition and Nanotechnology Policies: Discourse, Hegemony, and International Political Economy

Joscha Wullweber

Abstract The paper argues that the potential of post-positivist theories of International Relations should be more fully explored in the analysis of 'material'-economic processes of technology development. Taking the politics of nanotechnology as an example, the paper proposes an approach to innovation politics and technology development based on theories of hegemony, discourse and international political economy. The guiding hypothesis is that nanotechnology should be conceptualized not as a technology in its own right but as a broad and encompassing techno-socio-economic innovation strategy and, hence, a specific hegemonic project. The importance of such a project stems less from ongoing technological developments, but from its relation to the overall global socio-economic structure. As an innovation project, it creates a link between nanotechnology and the future of the industrial world. Above all, it aims to transform industrialized states into 'competition states' according to competitive criteria. Nanotechnology as a political project acts as a kind of driving force for economic expansion. In this capacity it serves to propel the pervasive reconstruction of the field of research and development in a competitive direction, especially in the more advanced industrialized countries. It is the term nanotechnology that brings different R&D politics together under a single cover, giving them a common identity. In a nutshell, the paper will develop a specific theoretical perspective on politics of technology and innovation within global political economy based on post-positivist theories of IR.

Keywords Global political economy • Science and Technology Studies • R&D policies • innovation policies • International Relations Theory • Nanotechnology

J. Wullweber (✉)
University of Kassel, Kassel, Germany
e-mail: joscha.wullweber@uni-kassel.de

M. Mayer et al. (eds.), *The Global Politics of Science and Technology - Vol. 1*, Global Power Shift, DOI 10.1007/978-3-642-55007-2_4,

1 I Spy with My Little Eye ...

This contribution aims at developing an analytical framework for technology and innovation policies based on theories of discourse and hegemony and international political economy. This framework of analysis is then applied to the politics of nanotechnology, focusing mainly on technology policies from the 1990s to the mid 2000s. Although there is a burgeoning body of literature and analyses on nanotechnology in the humanities and social sciences and although the realm of nanotechnology has been widely explored, the political and economic dimension remains cloudy. While many studies highlight the fantastic, visionary, and rhetorical character of nanotechnology and discuss the expectations, hopes, and fears of utopian scenarios (e.g. Coenen et al. 2004; Kaiser and Kurath 2007; Selin 2007), the global politics of nanotechnology is largely under-scrutinized. This lack of politico-economic analysis is surprising, given that 'nanotechnology has become a symbolic subject of international competition, much like the Cold War space program' (Schummer 2004: 67).[1]

Nanotechnology has been approached from diverse angles by many scholars across the humanities and social sciences. But rather than coming closer to elucidating the relevance of nanotechnology in global politics, they all struggle with its definition—explicitly or implicitly. There is no generally agreed definition and the concept of nanotechnology is still highly debated. It has even proven impossible thus far to agree on the appropriateness of the term nanotechnology. Most definitions of nanotechnology make reference to the nanoscale, which (usually) varies between 1 and100 nm (10^{-9} to 10^{-7} m). But even the range of scale is under debate. The most restricted definitions are those which refer to molecular manufacturing, the ability to understand, control, and manipulate matter at the level of individual atoms and molecules and to construct greater structures—atom by atom.

The question therefore arises: why is nanotechnology so difficult to define? The case is made in this article that nanotechnology is neither a specific technology or method, nor an array of applications or a research field, but must rather be understood as an encompassing political project or, more precisely, as diverse political and technological projects bound together only by the term nanotechnology. It will be argued that nanotechnology acts as a kind of 'carrier force'—as a techno-socio-political innovation strategy—for techno-economic expansion in general. In addition it serves for the reconstruction of the industrialised states along competitive criteria, especially in the advanced industrialised countries.

I shall begin the chapter by introducing the analytical framework based on theories of discourse and hegemony. Furthermore, I will shortly introduce different approaches to international political economy. Drawing on this framework, I will then scrutinize the politics of nanotechnology by locating these politics analytically within the notion of contemporary industrialized states as competition states.

[1] Parts of this contribution draw on Wullweber (2008). For a detailed analysis see Wullweber (2010).

Finally, I will connect innovation politics with different discourses, namely the narrative of the high-technology market, the discourse of a knowledge-based economy, and the pressure of international competition.

2 Discourse, Hegemony and International Political Economy

The theoretical approach is committed to an ontology based on theories of discourse, hegemony, and International Political Economy (IPE).

2.1 Discourse

Post-structural authors have emphasised the role of discourse as constitutive for politics. There are basically three different strands or generations of discourse theory in social science and IPE (Torfing 2005). The first has a strong linguistic bias, and includes areas such as socio-linguistics, theories of argumentation and conversation, and content analysis (based, for example on Austin 1962; Holsti 1969; Pêcheux 1982; Downes 1984; Atkinson et al. 1985). For the most part, methods are based on textual and linguistic analysis. Therefore, many approaches are confined to the structural analysis of text material, and largely ignore, or only casually deal with, the conflict-ridden and contradictory social process inscribed in the material as well as in the socio-political context. At worst, this leads to positivist empiricism—a quantitative counting of words and phrases. Such a discourse analysis 'tends to reduce discourse to a linguistic mediation of the events that are produced by the causal powers and mechanisms embedded in the independently existing structure of society' (Torfing 2005: 7).

The second generation broadens the concept of discourse including social practices such as those of Fairclough (1992; 1995) with his *critical discourse analysis*, Foucault (1973) with his studies on archaeology, or Jessop und Sum (2001, 2006a; Jessop 2004; 2005) with their cultural political economy approach.[2] With the third generation they share the notion that society is a contingent product of articulatory practices and that stabilized social forms are the result of hegemonic struggles (Jessop 1990: 297–298). The relation between discursive and non-discursive or extra-discursive practices, however, remains unclear in this strand. Especially critical discourse analysis 'tends to reduce discourse to a linguistic mediation of the events that are produced by the causal powers and mechanisms embedded in the independently existing structure of society' (Torfing 2005: 7).

[2] However, recent publications by Jessop and Sum are very close to the third strand (Sum and Jessop forthcoming).

The theory of hegemony and discourse elaborated by Laclau and Mouffe—as part of the third strand—offers an important analytical extension to such approaches. Here, a discourse reflects the sum of all verbal and non-verbal articulations on a particular issue, while shaping the perceptions, thought processes, and practices of individuals at the same time. According to this concept of discourse, action and meaning are closely related. Hence, *discourses* are constitutive of political, economic and social reality. Nonetheless, meaning is never fixed.

Following this third strand of discourse theories, language, social practices and meaning are closely connected. Articulation is understood as a 'practice establishing relations among elements such that their identity is modified as a result of the articulatory practice' (Laclau and Mouffe 1985: 105). Thus, a discourse is a social structure that has significance in social, economic, or political contexts. It can be seen as a relational ensemble of signifying sequences, which together constitute a more or less coherent framework of what can be said or done. Furthermore, a discourse fails to invoke a complete closure, since there is always something escaping the infinite processes of signification—an irreducible 'surplus of meaning' (Laclau and Mouffe 1985: 111).

Post-structural approaches highlight the importance of discourse for the construction of social reality. They stress the constructed nature of actors in politics and society, and the phenomenon of competing, conflicting, and often contradictory structures of meaning and expression. The pre-discursive meaning of social entities such as institutions, subjects of policymaking, and political identities is denied. Accordingly, the 'truth' of an event is the result and the contingent outcome of struggles among competing discourses and narratives, transforming 'what is out there' into a socially and politically relevant concept. The outcome of these struggles is contingent, insofar as no actor can anticipate the exact results of her action. However, the 'scope of possibilities' to determine a discourse differs much among the actors. Strategic articulations are an important part of discourses and can be understood as an attempt to establish a chain of equivalence between different discursive elements. They are modes to organise political, scientific, and economic reality. However, it is vital to stress the fact that 'structures rarely have a simple, unequivocal relation to a single strategy' (Jessop and Sum 2006b: 66). By introducing the concept of 'discourse-organisation', the argumentation intends to characterise a relatively stabilised spatio-temporal, socio-political and strategic-selective structure of a specific society, including general concepts and values of social order.

2.2 Hegemony

Following this reasoning, for political analysis the question arises whether certain articulations are able to prevail. With regard to nanotechnology this would imply that nanotechnology becomes widely accepted as a technology producing wealth for the society. For Antonio Gramsci hegemony means the ability of certain political groups to pursue their interests in such ways that the other political groups and people in society regard these interests as common or general interests. Hegemony is perceived as an active consent of the ruled (Gramsci 1971: 180–182). Neo-Gramscian approaches to international political economy introduce a concept of power that primarily rests on the ability to universalise the particular interests of a group as a socio-economic and political structure. A certain group is hegemonic, and not only dominant, if it succeeds to win approval of its authority among members of other social groups. The 'ruling groups' have to be responsive, at least to a certain degree, to the respective interests of other groups. This can be achieved by taking into account the interests of other groups in the formative processes of political institutionalization. These interests have to be merged, so that they become equated with the very institutions (Cox and Sinclair 1996: 99–100).

From a post-structural point of view, the concepts of discourse and hegemony are inextricably linked and mutually conditioned. Hegemonic practice shapes discourse, which in turn provides the conditions of possibility for hegemonic articulation (Laclau and Mouffe 1985). Framed like this, hegemony is a type of social relation. It can be described as the widening of a particular discourse—in the form of a socio-political and economic project—towards a horizon of social orientation and action, i.e. a discourse-organisation, through the articulation of unfixed elements into partially fixed moments. The ambit and the horizon of a particular discourse-organisation are constituted by the exclusion of competing discursive elements. The exclusion of alternative articulations and practices from the prevailing discourse is a substantial element of hegemonic practices.

To be successful, i.e. to become hegemonic, a socio-political project has to be articulated in relation to the common good. Since the common good only exists as an imaginary common good, and hence as an empty place, there are only particular interests, which try to occupy this empty space through strategic articulations (Jessop 2007: 11). Thus, a hegemonic project has to be articulated in a specific way: In these processes of articulation a multiplicity of subjects, actors and relevant social forces do act on the assumption that the implementation of the project is a prerequisite to achieve the common good. Within theories of IR and IPE, neo-Gramscian approaches are at the fore in the use of the concept of hegemony (e.g. Bieler and Morton 2007; Morton 2007; Ayers 2008; McNally and Schwarzmantel 2009). In general, two main differences can be highlighted between these approaches and a post-structural reading following Laclau and Mouffe: First, Laclau and Mouffe (1985, 1990) show that while Gramsci had critically revised some forms of essentialism within Marxian approaches (economism and etatism) one branch of essentialism is still present and this is classism. Accordingly, neo-Gramscian approaches are mostly interested in global class relations and transnational elite practices. While strong, rich and very detailed in empirical analysis, these approaches tend to give much leeway to political actors, especially political elites. Second, neo-Gramscian approaches tend to have a rather positivist epistemology, disregarding the possibility that social reality itself is being shaped by social struggles. However, there are far more similarities between neo-Gramscian and post-Gramscian approaches, especially concerning the focus on social struggles and the critical attitude towards global power relations (Gill 1993; Scherrer 2001; Jessop and Sum 2006a; Sum and Jessop forthcoming; and even in Bieler and Morton 2008).

2.3 International Political Economy

Theories and approaches to International Political Economy (IPE) are generally divided into three main paradigms: liberalism/neoliberal institutionalism, (neo-mercantilist) realism and Marxism/historical materialism (O'Brien and Williams 2010; Ravenhill 2011; Wullweber et al. 2013). Currently, these approaches are sometimes even more condensed into two research traditions: American IPE and British IPE (Cohen 2008). The former is said to be based on a positivist approach, relying heavily on quantitative data, mathematical models and methodological individualism (Keohane and Nye 1971; Gilpin 1987; Katzenstein 1978; Krasner 1983). The latter, in contrast, is described as being more post-positivist, interpretative and holistic (Strange 1986, 1988; Cox 1981, 1987).

In terms of political analysis the division between actor-oriented, structure-oriented and post-structural approaches is pivotal. Both structuralist and post-structuralist approaches argue strongly against notions of the actors, which by their very nature are perfectly informed, rational and self-interested individuals. For rational choice approaches, there is no society, but an 'invisible hand' and the natural workings of the market which reconciles the subject's aspirations into a system of common well-being. The same argument has been brought forward by neo-realists, which replace the individual actor with nation states, firms or other units. They all invoke an agent that acts rationally and brings order to the world. Hence, actor-centred theories conceptualize the actor with a free will, not constrained by any structure.

In contrast, structuralist as well as post-structuralist approaches emphasize that actors are always already bound within a social structure. They highlight that agency is always imbued by its social environment. Hence, there is no autonomous agency detached from its environment, acting only on behalf of pre-existing characteristics. But from a post-structuralist point of view, structuralist theories overshoot the mark, by turning the relation between agent and structure upside down. Instead of having no structure, for structuralists, structure is all there: "It [the social system] has the characteristics of an organism, in that it has a life-span over which its characteristics change in some respects and remain stable in others" (Wallerstein 1974: 347). Now action is (depending on the respective approach to different degrees) preordained by the function and operation of the structure: "The social fabric appears first and foremost as a problem of collective belonging, in the form of a system or systems in which individual interests are validated by the results they produce" (Aglietta 1998: 46). Functionalism here means that actors act to maintain the system as a whole without necessarily being aware of it. To have a function implies a line of thinking in which something only exists for the compliance of a certain (transcendental and ahistorical) task. That is why structuralist theories frequently argue in *functionalist* terms. Poulantzas for example, who theorized the state as the material condensation of a relationship of forces, argues that the state functions as a 'factor of cohesion between the levels of a social formation' (Poulantzas 1973: 44, Poulantzas 1978).

For post-structural approaches, social structures operate like a corridor of action. Social structures guide the actions of the actors to a certain degree, but they do not completely determine the actor's possibilities, because they are constantly dislocated (Laclau and Mouffe 1985). Hence, the contingency of structures as well as the struggles within the process of constructing structures is emphasized. Furthermore, structures only exist, because they are permanently produced, reproduced and altered by the agents. They are not operating "behind the backs" of the actors, detached from the actions of the agents; rather, they are constituted actively (though not necessarily consciously). To account for this mutual conditioning of structure and agency, the notion of discourse has been brought forward (Wullweber and Scherrer 2010).

Many studies have explored the transformation of the nation state in the era of globalisation. Almost all analyses share the notion that the form and structure of the state changed since the 1970s (Lipietz 1987; Jessop 2007). In this context, the notion of the internationalisation of the state refers to different and often contradictory political answers within the state administration to handle this new situation. In the following, I will argue that the hegemonic discourse to reconstruct the nation state in the advanced industrialised countries emphasises a state that has to become streamlined along competitive criteria. Technological innovation in general and nanotechnology in particular plays an important role in supporting this discourse.

Jessop (1990) emphasises that the nation state is social relation whose unity has to be constructed actively. The state is characterised by a selectivity, insofar as 'the state is not equally accessible to all social forces, cannot be controlled or resisted to the same extent by all strategies, and is not equally available for all purposes' (Jessop 1990: 317). The state can be seen as a terrain for struggles over political hegemony in terms of competing definitions of the common interest. Within this process of permanent reconstruction, technological policies, trade policies, and social policies are all mutually reinforcing discursive practices. The 'competition state' expresses the congealed form of these paradigms of competitiveness (Jessop 1990: 119f.).

A competition state aims at securing economic growth within its borders, while ensuring competitive advantages for investors on its territory. This can be achieved by promoting the economic and extra-economic conditions that are perceived vital for success. It emphasises strategies to create, restructure or reinforce the competitive advantages of its territory, social capital, social institutions and economic agents. This discourse-organisation highlights certain 'Schumpeterian' characteristics, 'because of its concern with technological change, innovation and enterprise and its attempt to develop techniques of government and governance to these ends'

(Jessop 1990: 96). For Joseph Schumpeter, entrepreneurial innovation can proceed in different ways (Lim 1990): via the introduction of a new good or a new quality of a good, via the introduction of a new method of production, via the opening of a new market, via the conquest of a new source of supply of raw materials or half-manufactured goods, and via the implementation of the new organisation of any industry. This approach highlights a prevailing thought of how society should be transformed in the light of a paradigm of innovation and competition. As it will be argued below, this discourse creates a link between nanotechnology and the industrial future of the advanced industrialised states. The notion of the competition state also has to be applied to competitive regions like the European Union.

3 Politics of High-Tech Innovation

The following analysis of nanotechnology politics integrates different theoretical approaches. While the concept of discourse-organization denotes forms of spatio-temporal socio-economic *structuration*, the concept of hegemony points to the always conflict-ridden process of generating these socio-economic structures. Furthermore, the analysis draws on the concept of the empty signifier, and the concept of competition state. Finally the study is based on a view of global politics as being interlinked with economic policies and power politics.

3.1 Nanotech as Empty Signifier

The 'nano-hype' plays an important role for the political contextualisation of nanotechnology. Both scientists and politicians promise revolutionary breakthroughs generated by nanotechnology: new ways of detection and treatment of diseases, in drug development, in the monitoring and protection of the environment (e.g. water decontamination), in the production and storage of energy, or in enhanced information and communication technologies. In their view, nanotechnology will enable to build complex structures as small as an electronic circuit or as large as an aeroplane and produce stronger and lighter material (Royal Society/ Royal Academy of Engineering 2004; Department of Trade and Industry 2002). Nanotechnology is perceived as an instrument to make powerful information technology available everywhere, to maximise productivity in agriculture, to increase health and longevity of human life, to provide abundant clean water globally, and to meet global energy needs with clean solutions.

To understand the nanotechnology hype, it is also necessary to take into account the most utopian expectations for future applications of nanotechnology. According to these visions, the most promising applications will stem from processes called 'self-assembly' or 'molecular manufacturing'. Self-assembly refers to the tendency that some materials are spontaneously 'able' to arrange themselves into ordered structures (Antón et al. 2001). The goal and aspiration is to build desired structures from atomic scratch. The idea is not only to manufacture individual particles with useful properties, but to manufacture complex and useful structures made from multiple molecules. Hence, the desired outcome of nanotechnology is the manipulation and assembly of nanoscale particles into supramolecular constructions and even larger structures. Some scientists believe that one day molecular manufacturing will be possible, i.e. to control atomic positioning so precisely that any object whose atomic composition is known could be assembled from its basic units (Drexler 1986).

Drawing on post-structural approaches, nanotechnology can be conceptualized as an empty signifier. An empty signifier is a signifier that has become detached from its previous particular content. Through (strategic) articulations this signifier tends to lose its particularity in order to become the name of fullness—a universality. Thus, an empty signifier is a hybrid of a particularity and a universality (Laclau 1996). The signifier nanotechnology denotes a universal technology that is able to solve the world's most pressing problems: the provision of clean water worldwide, the satisfaction of global energy needs (with 'clean' solutions), the maximisation of agricultural productivity, the creation of new jobs etc. Hence, nanotechnology can be seen as an innovation strategy—a strategy that offers a technological solution for socio-political problems. While Norio Taniguchi or, more probably, Eric Drexler coined the term, referring to the capability to precisely engineer materials on the nanometre scale, the term became more and more detached from that meaning—it became *tendentially empty*. By emptying the signifier from its 'original' meaning it was possible to refill the term with different contents and associate it with other positive connotations, such as the 'next industrial revolution', (economic) wealth, sustainable development and knowledge-based society. Therefore, an empty signifier emerges in the hegemonic process of signification. At the same time, it retroactively acts upon the system it denotes, giving identity to a previously non-existent field. Societal forces struggle to launch such signifiers and to fill their content (Laclau 1996).

The empty signifier nanotechnology is intimately connected with the emerging narrative of a nanotechnology industry and the fantastic expectations surrounding the nanotechnology market. They all construct the narrative of a technology that will bring wealth to the people and could serve as a competitive advantage in the global struggle for market shares. It follows that nanotechnology is neither a definite technology or method, nor an array of applications or a research field. It rather has to be understood as an encompassing political project or, more precisely, as different political projects that are kept together only by the empty signifier itself. Nanotechnology, as an ensemble of different technologies *and* as a political project, potentially changes the material livelihoods of many people (Wullweber 2006: 106–112). However, there is no *one coherent strategy* of a certain group guiding the nanotech project. Quite contrary, there are different interests and, to some extent, conflicting strategies competing.

3.2 *The High-Technology Market*

The nanotechnology discourse is supported by different strategies, such as the narrative of the nanotechnology market, the narrative of the knowledge-based economy and the narrative of the nanotechnology race. Economic interest in nanotechnology is not automatically given. Rather the interest itself is socially constructed and serves certain strategies. After years of basic research it is still uncertain whether nanotechnology will produce substantial goods for the market. Nevertheless, there are countless studies that assess the possible impact of nanotechnology for future markets. Immense expectations created an image of future markets, which is only achievable, if the industry branches move up the technological ladder and align their R&D policies with the nanoscale. For this reason all advanced industrialised countries, almost all Fortune 500 companies and two-thirds of the companies in the Dow Jones Industrial Average, convey nanotech research, development and investment in some way Roco 2007: 28).

The technological competence in nanotechnology is allegedly a compulsory condition to compete successfully with better procedures and products on future markets. At the same time, a view became hegemonic that does not permit any alternative to the development of nanotechnology, since those nations which fall behind will miss the junction to the future markets. Nanotechnology becomes a synonym for innovation within the competition states. As nano-materials and -processes apply to many manufactured goods, in almost all industry sectors, control and ownership of nanotechnology is decisive for virtually all governments and for the competitiveness of industry: in terms of attracting initial investment, and to ensure future revenue. With certain patents it could be possible to control complete chains of production. Apparently, nanotechnology is the first research field in which the basic ideas and applications are patented from the outset: The most basic ideas and fundamental building blocks in nanotechnology 'are either already patented or may well end up being patented' (ETC 2005: 10). Hence, intellectual property rights are a key element in both, the knowledge-based economy and the global competitive struggle for global market shares, since 'companies that hold pioneering patents could potentially put up tolls on entire industries' (Regalado 2004: 1). The 'race' for the nanotechnology patent 'gold rush' (ETC 2003: 24) has started among TNCs, leading academic labs, start-ups and universities.

3.3 *The Knowledge-Based Economy*

Within the competition state, the factor of 'knowledge' has becoming more important for international competitiveness: 'To create wealth and new employment in a globalised market and within a knowledge-based economy, the competitive production of new knowledge is essential' (European Commission 2004a: 9). Jessop

(2002: 96) describes this situation as a *knowledge-based economy*. The knowledge-based economy constitutes an important discourse within the competition state supporting and articulating today's process of primitive accumulation. The knowledge-based economy is the widely taken-for-granted focal point of economic strategies, flag ship projects and hegemonic visions. It is a discursive nodal point that partially fixes meaning within signifying chains (Laclau and Mouffe 1985: 112).

The nanotechnology discourse concurs with the issue of knowledge generation and the narrative of the knowledge-based economy in policy speeches, documents, and programmes. Philippe Busquin, European Commissioner for Research, states that 'nanotechnology provides a golden opportunity for the creation of new knowledge-based enterprises and has a 'revolutionary' potential that can open up new production routes' (European Commission 2004a: 1). Likewise, the EU Commission declares on its research homepage: 'Nanosciences and nanotechnologies are crucial to the establishment of a knowledge-based EU society and economy' (European Commission 2004b). According to this logic, 'Europe must (...) transform its world-class R&D in N&N [Nanosciences and Nanotechnologies] into useful wealth-generating products in line with the actions for growth and jobs' (European Commission 2005: 2). Together these narratives frame nanotechnology as a competitive advantage for the industrialised countries.

3.4 International Competition

Nanotechnology is framed as a technology which is *the* pre-eminent factor for achieving a nation's innovation: It does not only introduce new goods but also offers a new quality of goods, it ushers new methods of production, and it opens up new markets. And finally, it offers a new source of supply of raw materials. Nanotechnology is preordained as the magic tool leading to the production of ever smaller, faster and more efficient products with acceptable price-to-performance ratio. This has become an increasingly important success factor for many industrial branches in international competition.

The empty signifier nanotechnology serves as an innovation strategy: 'Because nanotechnology is of such critical importance to U.S. competitiveness, both economically and technologically, even at this early stage of development, it is a top priority within the Administration's R&D agenda' (NSET/NSTC 2003: 3). In a similar way, the European Commission argues: 'Advances across a wide range of sectors are being enabled through R&D and innovation in N&N [Nanotechnology & Nanoscience]. These advances can address the needs of citizens and contribute to the Union's competitiveness and sustainable development objectives and many of its policies including public health, employment and occupational safety and health, information society, energy, transport, security and space' (European Commission 2005: 2).

The ascription to nanotechnology of being an innovation and a competitive advantage gave rise to an enormous global nanotechnology race among the industrialised nations to win monopoly control of the expected huge nanotechnology market and a share of the two million nanotechnology workers that are said to be required by the nanotechnology industry (Roco 2003). States play a crucial role in promoting innovative capacities, technical competence and technology transfer. They hope that as many corporations and economic sectors as possible may benefit from the assumed new technological opportunities created by nanotechnology R&D activities. The nanotechnology discourse, combined with the narratives of the nanotechnology race and of a knowledge-based economy, supports the (re-) construction of the competition state. The competition state has to 'focus upon (. . .) knowledge-based industrial innovation ('nanomanufacturing'), integration at the macro-micro-nano interface and interdisciplinary ('converging') R&D. Appropriate synergy with the European Strategy on Life Sciences and Biotechnology may also be beneficial' (European Commission 2004a: 8).

However, a nanotechnology race is not simply existent. Rather, the creation of the discourse of a nanotechnology race is the outcome of political practices and struggles. The actors within the nanotechnology race are not acting independently from discourses, which in many ways have an influence on how these actors view the world, define their goals, and structure their actions. The perception of 'a highly competitive global economy' and the prevailing analysis that only those nations that thrive 'can compete on high technology and intellectual strength' (Her Majesty's Government 2005: 1) supports the discourse of the nanotechnology race.

4 Conclusion

The aim of this article is to outline an analytical framework based on theories of hegemony and discourse and international political economy. This framework of analysis has been applied to global high-tech nanotechnology politics. The theoretical approach provides a different perspective on the development of nanotechnology. While most of the current analyses treat nanotechnology as a definite technology, my thesis is that the term nanotechnology denotes an encompassing innovation project that is kept together only by the empty signifier itself.

A perspective is inaugurated that facilitates to delineate political interests and strategies within the process of nanotechnology development. Furthermore, it is possible to expound different discourses and policy strategies that have been associated with the nanotechnology discourse. While the discourse of international competition is fostered through the trope of the ongoing nanotechnology race, nanotechnology itself is presented as one of the most important strategies of innovation to win the battle for global market shares. To become hegemonic, a certain convergence of discursive elements is necessary. The success of the nanotechnology project derives from an alignment with the discourse-organisation of the competition state in general and the discourse of the knowledge-based economy, the nanotechnology race and the immense future markets for nanotechnology. These discourses are mutually reinforcing and strengthening.

Besides its embeddedness within global political economy structures, the success of a hegemonic project depends heavily on its public acceptance: "public trust and acceptance of nanotechnology will be crucial for its long-term development" (European Commission 2004a: 19). In the 2000s, the majority of people in the advanced industrialised states did not have much knowledge about nanotechnology (and still knowledge about nanotechnology is limited). This poses a problem for protagonists of nanotechnology, because "[w]ithout a serious communication effort, nanotechnology innovations could face an unjust negative public reception" (European Commission 2004a: 19). Accordingly, in many countries that run nanotechnology programmes advocates of nanotechnology started public debates, because "an open public dialogue with citizens and consumers is absolutely necessary as a basis for an objective judgement on nanotechnology and to avoid baseless fears" (Luther 2004: 94). Apparently, the overall goal of all these programmes is not to discuss possible problems but to achieve acceptance for nanotechnological development. Not the risk of nanotechnology is at stake but its "smooth" development (see Wullweber 2010: 233–298).

Nowadays, high-technology policies are predominantly characterised by an accelerating commercialisation. From this perspective, nanotechnology is *the* pre-eminent factor for achieving innovation and competitive advantages: It introduces new goods and offers a new quality of goods, it ushers new methods of production, it opens new markets, and, it offers a new source of supply of raw materials. However, the enforcement of new technologies is no automatic, self-evident process. Rather, it is embedded in social relations and has to be backed by political measures. Nanotechnology has to be embedded in modified governance structures, which has materialised in the political form of the competition state. For this new mode of socio-economic regulation the consent of the people is required. Therefore, many governments try to invoke different discursive elements in order to strengthen the perception that nanotechnology is indispensable for an economically viable society.

A post-structural theorising of nanotechnology puts emphasis on power struggles and interests and also takes into account contingent-accidental events as constitutive for technological development. It criticises the notion that technological development is an inevitable and automatic progress of science as well as the assumed progressive character of scientific development. In post-structural rendering neither the 'truth' of nanotechnology as the technology of the twenty-first century nor the policy problem 'nanotechnological risks' or the 'high-technology gap' are simply existent. Rather, the question is what constitutes a 'high-technology', whose interest does this kind of framing serve and what social forces try to articulate such tropes. Thus, science and political power are conceptualised as two strongly interconnected phenomena, and the a priori existence of stable boundaries between economy, politics and science is questioned.

References

Aglietta, M. (1998). Capitalism at the turn of the century: Regulation theory and the challenge of social change. *New Left Review, 1*(32), 41–90.

Antón, P. S., Silberglitt, R., & Schneider, J. (2001). *The global technology revolution: Bio/nano/materials trends and their synergies with information technology by 2015*. Santa Monica, CA: RAND.

Atkinson, M., Heritage, J., & Oatley, K. (1985). *Structures of social action: Studies in conversation analysis*. Cambridge: Cambridge University Press.

Austin, J. L. (1962). *How to do things with words*. Oxford: Clarendon.

Ayers, A. J. (Ed.). (2008). *Gramsci, political economy, and international relations theory: Modern princes and naked emperors*. Hampshire: Palgrave Macmillan.

Bieler, A., & Morton, A. D. (Eds.). (2007). *Images of Gramsci. Connections and contentions in political theory and international relations*. London: Routledge.

Bieler, A., & Morton, A. D. (2008). The deficits of discourse in IPE: Turning base metal into gold? *International Studies Quarterly, 52*, 103–128.

Coenen, C., Rader, M., & Fleischer, T. (2004). Of visions, dreams and nightmares: The debate on converging technologies. *Technologiefolgenabschätzung Theorie und Praxis, 13*(3), 118–125.

Cohen, B. J. (2008). *International political economy. An intellectual history*. Princeton/Oxford: Princeton University Press.

Cox, R. W. (1981). Social forces, states and world orders: Beyond international relations theory. *Millennium, 10*(2), 126–155.

Cox, R. W. (1987). *Power, production, and world order*. New York: Colombia University Press.

Cox, R. W., & Sinclair, T. (Eds.). (1996). *Approaches to world order*. Cambridge: Cambridge University Press.

Department of Trade and Industry. (2002). *New dimensions for manufacturing. A UK strategy for nanotechnology*. http://www.dti.gov.uk/innovation/nanotechnologyreport.pdf

Downes, W. (1984). *Language and society*. London: Fontana.

Drexler, E. K. (1986). *Engines of creation*. New York: Anchor Books.

ETC. (2003). *The big down: Atomtech—technologies converging at the nanoscale*. http://www.etcgroup.org/documents/TheBigDown.pdf. Accessed January 12, 2014.

ETC. (2005). *Nanotech's "second nature" patents: Implications for the global south, ETC Group Report*. http://www.etcgroup.org/upload/publication/pdf_file/54. Accessed January 12, 2014.

European Commission. (2004a). *Towards a European strategy for nanotechnology*. Luxembourg: Office for Official Publications of the European Communities.

European Commission. (2004b). *R&D is key to mastery of nanotechnologies*. http://ec.europa.eu/research/industrial_technologies/articles/article_800_en.html. Accessed January 12, 2014.

European Commission. (2005). *Nanosciences and nanotechnologies: An action plan for Europe 2005-2009*. COM (2005) 243 final. ftp://ftp.cordis.europa.eu/pub/nanotechnology/docs/nano_action_plan2005_en.pdf. Accessed February 10, 2014.

Fairclough, N. (1992). *Discourse and social change*. Oxford: Blackwell.

Fairclough, N. (1995). *Critical discourse analysis: The critical study of language*. London: Longman.

Foucault, M. (1973). *Archäologie des wissens*. Frankfurt am Main: Suhrkamp.

Gill, S. (1993). Epistemology, ontology and the 'Italien School'. In S. Gill (Ed.), *Gramsci, historical materialism and international relations* (pp. 21–48). Cambridge: Cambridge University Press.

Gilpin, R. (1987). *The political economy of international relations*. Princeton: Princeton University Press.

Gramsci, A. (1971). *Selections from the prison notebooks*. London: Lawrance & Wishart.

Her Majesty's Government. (2005). *Science & innovation investment framework 2004-2014*. London: The Stationery Office.

Holsti, O. R. (1969). *Content analysis for the social sciences and the humanities*. Reading, MA: Addisson-Wesley.
Jessop, B. (1990). *State theory. Putting the state in its place*. Cambridge: Polity Press.
Jessop, B. (2002). *The future of the capitalist state*. Cambridge: Cambridge University Press.
Jessop, B. (2004). Critical semiotic analysis and cultural political economy. *Critical Discourse Studies, 1*(2), 159–174.
Jessop, B. (2005). Critical realism and the strategic-relational approach. *New Formations, 56* (Autumn), 40–53.
Jessop, B. (2007). *State power. A strategic relational approach*. Cambridge: Polity Press.
Jessop, B., & Sum, N.-L. (2001). Pre-disciplinary and post-disciplinary perspectives. *New Political Economy, 6*(1), 89–101.
Jessop, B., & Sum, N.-L. (2006a). Towards a cultural international political economy: Poststructuralism and the Italian school. In M. de Goede (Ed.), *International political economy and poststructural politics* (pp. 157–176). Hampshire: Palgrave Macmillan.
Jessop, B., & Sum, N.-L. (2006b). *Beyond the regulation approach. Putting capitalist economies in their place*. Cheltenham: Elgar.
Kaiser, M., & Kurath, M. (2007). Nano visions: Implications for technology assessment, communication and regulation. *Technologiefolgeabschätzung. Theorie und Praxis, 16*(1), 118–122.
Katzenstein, P. J. (1978). *Between power and plenty: Foreign economic policies of advanced industrial states*. Madison, WI: University of Wisconsin Press.
Keohane, R. O., & Nye, J. (1971). Transnational relations and world politics. *International Organization, 25*(3), 329–349.
Krasner, S. (Ed.). (1983). *International regimes*. Ithaca, NY: Cornell University Press.
Laclau, E. (1996). Why do empty signifiers matter to politics? In E. Laclau (Ed.), *Emancipation(s)* (pp. 34–46). London: Verso.
Laclau, E., & Mouffe, C. (1985). *Hegemony and socialist strategy: Towards a radical democratic politics*. London/New York: Verso.
Laclau, E., & Mouffe, C. (1990). Post-marxism without apologies. In E. Laclau (Ed.), *New reflections on the revolutions of our time* (pp. 97–134). London/New York: Verso.
Lim, C. Y. (1990). The Schumpeterian road to affluence and communism. *Malaysian Journal of Economic Studies, 27*(1–2), 213–223.
Lipietz, A. (1987). *Mirages and miracles: The crisis of global fordism*. London/New York: Verso.
Luther, W. (Ed.). (2004). *Industrial applications of nanomaterials: Chances and risks—technological analysis*. Düsseldorf: Futures Technologies Division of VDI Technologiezentrum GmbH.
McNally, M., & Schwarzmantel, J. (Eds.). (2009). *Gramsci and global politics. Hegemony and resistance*. New York: Routledge.
Morton, A. D. (2007). *Unraveling Gramsci: Hegemony and passive revolution in the global political economy*. London: Pluto Press.
NSET/NSTC. (2003). *The national nanotechnology initiative. Research and development supporting the next industrial revolution. Supplement to the President's FY 2004 budget*. Arlington: National Nanotechnology Coordination Office.
O'Brien, R., & Williams, M. (2010). *Global political economy. Evolution and dynamics* (3 Auflage). Basingstoke: Palgrave Macmillan.
Pêcheux, M. (1982). *Language, semantics and ideology*. London: Macmillan.
Poulantzas, N. (1973). *Political power and social classes*. London: New Left Books.
Poulantzas, N. (1978). *State, power, socialism*. London: New Left Books.
Ravenhill, J. (Ed.) (2011). *Global political economy* (3 Auflage). Oxford: Oxford University Press.
Regalado, A. (2004). Nanotechnology patents surge as companies vie to take claim. *Wall Street Journal, 18*(July) 1.
Roco, M. C. (2003). Broader societal issues of nanotechnology. *Journal of Nanoparticle Research, 5*, 181–189.

Roco, M. C. (2007). National nanotechnology initiative—Past, present, future. In W. A. Goddard, D. Brenner, S. E. Lyshevski, & G. J. Iafrate (Eds.), *Handbook of nanoscience, engineering, and technology* (2nd ed., pp. 19–44). Boca Raton, FL: CRC Press.

Royal Society/Royal Academy of Engineering. (2004). *Nanoscience and nanotechnologies: Opportunities and uncertainties*. Plymouth: Latimer Trend.

Scherrer, C. (2001). 'Double Hegemony'? State and class in american foreign economic policymaking. *Amerikastudien, 46*(4), 573–591.

Schummer, J. (2004). "Societal and ethical implications of nanotechnology": Meanings, interest groups, and social dynamics. *Techné, 8*(2), 56–87.

Selin, C. (2007). Expectations and the emergence of nanotechnology. *Science Technology Human Values, 32*(2), 196–220.

Strange, S. (1986). *Casino capitalism*. Oxford: Basil Blackwell.

Strange, S. (1988). *States and markets*. London/New York: Pinter.

Sum, N-L., & Jessop, B., (forthcoming). *Towards a cultural political economy. Putting culture in its place in political economy*. Cheltenham: Edward Elgar.

Torfing, J. (2005). Discourse theory: Achievements, arguments, and challenges. In D. Howarth & J. Torfing (Eds.), *Discourse theory in European politics: identity, policy and governanc* (pp. 1–32). Hampshire: Palgrave Macmillan.

Wallerstein, I. (1974). *The modern world-system. Capitalist agriculture and the origins of the european world-economy in the sixteenth century*. New York: Academic.

Wullweber, J. (2006). Der Mythos Nanotechnologie. Die Entstehung und Durchsetzung einer neuen Inwertsetzungstechnologie. *Peripherie, 101*(102), 99–118.

Wullweber, J. (2008). Nanotechnology: An empty signifier à venir. A delineation of a techno-socio-economical innovation strategy. *Science, Technology and Innovation Studies, 4*(1), 27–45.

Wullweber, J. (2010). *Hegemonie Diskurs und Politische Ökonomie. Das Nanotechnologie-Projekt*. Nomos: Baden-Baden.

Wullweber, J., Graf, A., & Behrens, M. (Eds.). (2013). *Theorien der Internationalen Politischen Ökonomie*. Wiesbaden: Springer VS.

Wullweber, J., & Scherrer, C. (2010). Postmodern and poststructural international political economy. In Denemark R. A. (Ed.), *The international studies encyclopedia*. Oxford: Blackwell, Blackwell Reference Online.

Rise of a Cybered Westphalian Age: The Coming Decades

Chris C. Demchak and Peter J. Dombrowski

Abstract No FRONTIER lasts forever, and no global commons extends endlessly. Today we are witnessing the fence building process of the cyberspace. The days of limited cyber spying through software backdoors or betrayals by trusted insiders, vandalism, or even theft had evolved into the ability to deliver devastating blows from afar. Thus, states are establishing sovereign control in the virtual world in the name of security and economic sustainability. This paper analyses the emergence of what we call cybered territorial sovereignty as a consequence of states' efforts to secure their borders against cyber attacks. The case of Stuxnet is introduced as an instance of new threat faced in the cyberspace. We argue Stuxnet marks the beginning of a new cyber Westphalian world of virtual borders and national cyber commands. We have seen this phenomena before as states consolidated their position following the treaties of Onasbrucke and Munster. Existing theories about the emergence of international rules and institutions can help us understand how the Westphalian model will adapt to the demands of securing cyberspace.

Keywords Cyberspace • Cyber security • Cyber attack • Stuxnet • Sovereignty • Cyber Westphalia • Cybered conflict

No FRONTIER lasts forever, and no global commons extends endlessly. Sooner or later, good fences are erected to make good neighbors, and so it must be with cyberspace. Today we are witnessing the fence building process. From the Chinese intent to create their own controlled internal internet, to increasingly controlled access to the internet in less-democratic states, to the rise of internet filters in

---

This chapter is an updated and revised version of Chris C. Demchak and Peter Dombrowski. 2011. "Rise of a Cybered Westphalian Age". *Strategic Studies Quarterly*. 5(1), 31–62.

C.C. Demchak (✉) • P.J. Dombrowski
US Naval War College, Newport, RI, USA
e-mail: chris.demchak@usnwc.edu; dombrowp@usnwc.edu

M. Mayer et al. (eds.), *The Global Politics of Science and Technology - Vol. 1*, Global Power Shift, DOI 10.1007/978-3-642-55007-2_5,

Western democracies, states are establishing sovereign control in the virtual world in the name of security and economic sustainability. The topology of the internet is about to be changed forever—rationally, conflictually, or collaterally—by the decisions of states to raise virtual borders.

The emergence of Stuxnet, a sophisticated, precise, and expensive worm targeted against industrial control systems, ushered in an age of cybered conflict.[1] The worm's target was the program controlling centrifuges in Iranian nuclear reprocessing plants (Falliere et al. 2010). As a malicious piece of software, it came as a surprise despite having floated around a year doing nothing but stealthily copying itself. Spread by infected USB thumb drives and the software in printer spoolers, it bypassed the internet security controls and only behaved maliciously upon finding the precise computer DNA of Iranian nuclear reactors. When the worm infiltrated a heavily protected facility, Stuxnet authors demonstrated a new level of threat to cyber security. Despite early denials, the Iranian nuclear community admitted its plants had been infected and its centrifuges were unstable.

Stuxnet capped a period in which complexity of national security challenges posed by cyberspace created a new level of insecurity (Brenner 2004). From 2008 onward, a string of unsettling discoveries of massive theft of national data appeared via backdoors into otherwise secure national-level systems (e.g., GhostNet). Widespread stealthy infection of national systems occurred through sophisticated programs waiting to be connected to hidden remote servers, such as the Confiker worm and the wholesale copying of critical industrial technological advances by China. The age of vandals and burglars in cyberspace moved to the next level, resembling organized cyber mercenaries and pirates.

Until Stuxnet was discovered, few believed cyber intrusions could lead to strategic harm. Stuxnet's success thus changed notions of national vulnerability. If such malicious software can disrupt a heavily protected industrial facility, states have little choice but to protect their own agencies, corporations, and citizens (Porche 2010). The days of limited cyber spying through software backdoors or betrayals by trusted insiders, vandalism, or even theft had evolved into the ability to deliver devastating blows from afar. The ability to force nuclear centrifuges to oscillate out of control suggests that future innovations might be able to disrupt other critical infrastructures. The Stuxnet worm offered the possibility of distant enemies inserting such applications into oil pipelines, dam turbines, power plants or any other facility controlled by computers. As the designers of Stuxnet demonstrated, being disconnected from the internet will never again be a guarantee of security (Sanger 2010). If any part of a facility, transportation platform or system is

---

[1] *Cybered conflict* differs from *cyber war* or *cyber battle*. The latter is fully technological and could, in principle, be conducted entirely within a network. It is normally a component of the former. A cybered conflict is any conflict of national significance in which key events determining the path to the generally accepted outcome of the conflict could not have proceeded unless cyber means were non-substitutable and critically involved. The terms are distinctively and deliberately used in this article.

internally connected, if any electronic devices can connect to the system from the outside (even if the device must be hand-carried) the entire system is vulnerable.

Stuxnet is an exquisite example of the three major advantages afforded attackers in cyberspace. Attackers freely choose the scale of their own organization, the proximity of their targets, and the precision necessary to achieve their desired effects. They may use the internet itself to collect data on intended targets. As shown by the denial of government and banking service in Estonia in 2007, wholesale assaults can be deployed from one state to another by "patriotic hackers," while the originating state proclaims its inability to stop the assault (Michael et al. 2010). By 2008 alone, the daily attacks on simply the US ".gov" or ".mil" websites numbered in the millions.[2] In 2009, 75 % of global companies across 27 countries were the victims of cyber attack, with the average reported loss of $2 million (Kendall 2010). The losses have become exponentially worse since those early reports, cumulating to more than three trillion dollars annually according to the McKinsey 2014 annual cyber security survey of over 200 major global corporations [http://www.mckinsey.com/insights/business_technology/risk_and_responsibility_in_a_hyperconnected_world_implications_for_enterprises].

Protective measures have thus far been insufficient to repel cyber-attacks. Stuxnet's success ensured the rising perception of an all-source 24/7/360-degree national-level threat (Benson 2010). A "son of Stuxnet" might float for some time, seemingly harmless and unnoticed until triggered by a particular date, end-use, internet signal, or an encounter with a specific kind of computer or program. Critical systems might fail by sending destructive commands to others, or perhaps worse, replace true data with false. Even China recognizes the internal threat posed by its own support for cybered hacking talent. While promoting "patriotic hacking" against other states, Chinese authorities must contend with internal cybered threats to their own systems, despite severe sanctions against hacking (Wong 2010).

All states attempt to control what comes across their border even through cyberspace. The process of regulating cyberspace is already happening. A new "cybered Westphalian age" is emerging as states organize to protect their citizens and economies and thus initiate the path to borders in cyberspace. Not only are the major powers of China and the United States already demonstrating key elements of emerging cybered territorial sovereignty, other nations are showing similar trends. From India to Sweden, nations are demanding control over what happens electronically in their territory, even if it is to or from the computers of their citizens.

This process may be slow, but it is inevitable given the history of state formation and the subsequent emergence of the interstate system. Cyberspace is profoundly man-made; with sufficient motivation, states can construct virtual national borders. They are possible technologically, comfortable psychologically, and desirable politically. Small steps in securing against threats will lead to further steps over time.

---

[2] Threats are considered so serious that cyber-security officials are now expected to have training in known hacker methods (Gertz 2010).

In the process of border development, the singular marker of a new age of sovereignty is the creation of national cyber commands or their equivalents. To secure cyberspace, large, vulnerable states like the United States and China must anticipate and disrupt attacks far forward, outside their physical borders. Otherwise, attacks may spread too fast for effective defense. Just as conventional militaries still exist in the age of mass weapons, cyber warriors wielding cyber instruments will guard national borders in cyberspace. In so doing, they deepen national borders. We argue Stuxnet marks the beginning of a new cyber Westphalian world of virtual borders and national cyber commands. We have seen this phenomena before as states consolidated their position following the treaties of Onasbrucke and Munster. Existing theories about the emergence of international rules and institutions can help us understand how the Westphalian model will adapt to the demands of securing cyberspace.

1 The "Westphalian" Process

The Stuxnet worm marks a turning point into a new cybered conflict age in which states need to define territorial spaces of safety to reassure their citizens' safety and economic well-being. When it is widely accepted that critical systems are deeply vulnerable when they are open to the web, political leaders will demand ways to eliminate the threats from entering their territory. The cybered conflict age has begun, and it is natural for those hostile to any particular group to include cyber at key points in their plans, including debilitating entire systems. Equally expected, leaders of the threatened group will have to consider what responses keep critical functions secure.

Frontiers are places of conflict between groups, historically lightly and poorly governed, less populated, and risky—places where value is extracted for little cost. When a frontier starts to become a commons, productivity for all is imperiled. Those dependent on the frontier tend to form organizations to assert their claims. Modern democracies are in essence complex aggregates of large-scale organizations. Their leaders routinely reach out to absorb uncertainties to control them, if possible, or push them away. The rising perception of a national-level from threat cyberspace means that all states, in one way or another, will reach out to control what they fear from the internet—its frontier nature and the lack of sovereign control over what comes into their area of responsibility.

No commons extends endlessly nor lasts forever. It is normal for political leaders seeking relief from the interaction edges with other cultures or possible threats to look at reinforcing or installing borders. Establishing control is one hallmark of a functioning state. This need is true whether the border is enforced by passports for people, customs inspections for goods, or two-way filters for meta-tagged electronic bits. When states cannot protect their economic engines of growth and sustain-ability, the capacity of the state falls into question by those who control the resources under threat (O'Connell 1989; Tilly 1992).

The humans search for security has led to the formation of "fortress and badland" distinctions that marked territory for resource ownership for centuries, but until the 1,648 Treaties of Munster and Osnabruck (understood together as the Peace of Westphalia), borders did not stabilize over many generations. In this particular case, however, the Peace of Westphalia not only ended the 30 Years' War in Europe but also heralded the emergence of the modern interstate system. After the Westphalian peace, the nation-state became the dominant form of social organization. As a result, leading states of the period helped codify and set about more or less enforcing a collectively agreed upon set of rules, institutions, and norms by which they interacted with each other in international society (Krasner 2004).

Particularly useful for international stability was the effect of the treaties in creating conditions supporting the gradual hardening of borders between and among states, more or less, over the next 400 plus years. This process of settling on boundaries due to the mutual adjustments among states produced a concept of national territoriality that states could legitimately claim, and they could defend that territory against outside aggressors in just wars. With the rise of a general presumption of territoriality recognized by other external political leaders, modern states were able to stabilize internally and grow economically within established, increasingly fixed borders.

Westphalia provided a demonstration or a proof of concept. Over time, the more established a state became and the fewer ungoverned internal areas or frontiers it allowed to continue, the stronger and less existentially vulnerable the nascent state became (Tilly 1989). The significance of the Westphalian process for our argument is that the efforts of the modern state to cope with the emergence of the cyber sphere is in many respects similar to the processes by which states became the dominant form of social organization within the international system. The ability of the state to provide stability and security within the increasingly unchallenged borders was necessary to internal development of social and economic progress. Without a form of Westphalian borders, conflicts previously at the boundaries easily spill over in both directions from opportunistic resource appropriations by actors within and without. The wide variety of authorities, powers, and capabilities accruing to the modern state become difficult to employ, redirect, or even limit. Just as the ability of modern bureaucratic states to corral resources productively drove other less successful organization forms from the scene internationally, their ability to provide internal certainty in their domestic territory gradually came to define as civil society (Tandon and Mohanty 2000).

The uncertainties, predatory opportunism, resource conflicts, and changes to economic and social expectations reach directly into the domestic structures of the modern state. States are beginning to grapple with the difficulties inherent in incorporating new technologies into their societies. In particular, the cyber sphere is best seen as a 'substrate' which has extendi under and penetrated all modern societal functions, and now has challenged the security of individuals across the ordinary systems considered essential to the everyday functions of society. Citizens are often on the frontlines of fights for social stability in the face of predatory activities in cyberspace.

The international community has been relatively slow to come to grips with threats in and form cyberspace (Nye 2010). Initially surprised by the reach of the predatory behaviors made possible by cyberspace's unfettered global reach, states have been slow to reinforce their monopoly of violence over external threats entering their nations and harming citizens. Laws emerged over the early 2000s focused on the internal symptoms rather than the external sources of the uncertainties, many focused on the individual citizen or commercial internet service providers (ISP). In the United States, for example, financial liability to the individual defrauded online in credit card usage limited the amount the citizen would lose (Douglass 2009). In contrast, German law makes individual citizens responsible if they do not stop their personal computers from being used in spam or denial-of-service attacks (Anderson et al. 2009). Australia, however, enforces rules on the ISPs themselves to reduce malware to a minimum (Lindsay 2010).

Despite these efforts, social organizations have found their online presence vulnerable to attempts to extract information, prevent access, and even to disrupt operations. One US report stated the monthly number of attacks on the US Congress and government agencies had already reached 1.6 billion by 2010, largely from outside US borders (Evans and Whittell 2010). States are increasingly aware of the potential losses to firms, sub governments, and individuals. "It appears we can no longer see the internet as a friendly shared resource and restrict boundaries will have to be put in place," said Bert Hubert, founder of Dutch-based software provider PowerDNS.com (Mills 2010). The United and similarly threatened states recognize the need to respond. Modern states intend to put in place a buffers and borders in cyberspace to relieve the pressure of the barrage of assaults from outside actors.

2 Borders Decrease the Ease of Cybered Offense

Creating cyber borders make it more difficult for bad actors in cyberspace to cause harm cheaply, easily, and frequently. Borders make it necessary for attacker to get around them often physically. Forcing attackers to move people rather than bytes means higher operational barriers to entry: more costs, more coordination efforts, and many more opportunities for intelligence, law enforcement and military organizations to detect potential intrusions.[3] Borders may also slow the pace at which bad actors regroup from failures, capitalize on new information or, coordinate operations across jurisdictions.

[3] This effect, according to John Mallery (2010), is a national cyber security means of increasing the "work factors"—conceptualized as computational complexity, cost, cognitive difficulty, risk and uncertainty, cultural factors, and information differentials—for bad actors.

2.1 *Virtual Borders: Feasible, Comfortable, and Manageable*

Efforts to develop a Westphalian-style system overlayed on cyberspace will eventually succeed. There are few impediments to a neo-Westphalian process. A cybered national border is technologically possible, psychologically comfortable, and politically manageable. Technologists are arguing for separation of critical systems to protect them from internet predators. Even if national policymakers prefer an open internet, they will face technical arguments against maintaining that position. Borders are psychologically normal for citizens focused on continuing their safe access to internet services. Users already expect government sanctions against those who harm individuals via the Internet (Gibler 2007). Finally, a cyber border fits easily with the institutional compromises and allocations of responsibilities already existing in modern governance structures.

First, the technology of cyberspace is man-made. It is not, as described by the early "cyber prophets" of the 1990s, an entirely new environment which operates outside human control, like tides or gravity (Shiode 1997). Rather, as its base, the grid is a vast complex system of machines, software code and services, cables, accepted protocols for compatibility, graphical pictures for human eyes, input/output connections, and electrical supports. It operates precisely across narrow electronic bands but as a socio-technical system (Mayntz 2008) with such an amalgamation of redundancies, substitutions, workarounds, and quick go-to fixes that disruptions can be handled relatively well as long as everyone wants the system to work as planned.

States are learning that everything about today's grid can be regulated using the very technologies that allowed cyberspace to be created in the first place. There are many opportunities for national government to control what ends up being received on internet desktops, laptops, mobile devices, or even independent appliances. While connectivity is global—now increasingly found everywhere like land, air, sea, and even space—what is known as cyberspace *is* and *will* remain always man-made. And, unlike the sea, land, air, or space, it can be unmade. Furthermore, land expanses, seas, air, and space quadrants do not exist only if information is flowing. Seeing a mountain does not automatically connect one individual to the next or even offer one useful clues about it, yet being on one node does connect individuals to others in this cybered underlayment, even if only with some hacking. Air masses are air masses, but strings of cyber bytes already have information in the way they connect from node to node in protocols. It would be as if a car could not continue on the freeway without broadcasting its VIN number, license, weight, and other data each time it approached an exit. If not approved to continue by the owner of that freeway node, the car would be forced off" onto another road.

Someone, firm or government agency built, bought or runs every single connection on the internet. Even peer-to-peer (P2P) networks require people to connect and maintain them. Firms develop the software and applications that allow connections and the use of data and information. The filters between networks are largely under the control of private or semi-private institutional intermediaries.

Internet Service Providers (ISPs) are providing filtering services determined by large, state-level regulatory authorities. ISPs operate almost as regulated utilities and governments use them to control what enters or leaves the nation from overseas cables. It is even possible for governments to require source tagging of bytes at some point to assure the passage of legally acceptable streams of data.

Second, physical borders are known, accepted, and desired by citizens in modern civil societies, and that psychological comfort will be no different with the creation of borders in cyberspace. The relevant emphasis is on "borders," not on universal control of all cybered transactions occurring entirely within the boundaries of a state. Historically, citizens accepted borders as a security-enhancing necessity against external uncertainties undermining internally accepted rules of interaction. Without such limits, the collective sense of belonging is more easily undermined, as are the rules of civil behavior. Even a willingness to abide by norms of trust and nonthreatening behavior is tied to security. To live in ungoverned societies is not only insecure; it is also a psychologically palpable existential threat.

Third, borders fit institutionally into the existing architecture of national and interstate systems. Most nations make a distinction between the forces defending the borders from attack (militaries) and those protecting the individual citizens inside the nation from attack (police). This distinction is a direct outcome of the rise of the modern state. But as we have discussed, the distinction between military and law enforcement functions is severely challenged by the topology of cyberspace. Militaries, police, and intelligence organizations are challenged both by cyber attacks and by the lack of jurisdictional clarity. Both state and nonstate competitors have used the inter-connectivity inherent to the web to attack and disrupt operations and gather intelligence about capabilities and intentions across borders. This is especially true for nations like the United States that are highly dependent on telecommunications for logistics, command and control; intelligence, surveillance, and reconnaissance. Moreover, many military and intelligence organizations have grasped the offensive possibilities of the cybersphere to reach past the borders of other states directly, in concept at least, into the homes of an opposing state's citizens. Across the military communities of the more modern states, information operations and strategic communications programs have been developed to influence adversaries and allies. Physical or "kinetic" attacks are now routinely facilitated by exploiting enemy cyber vulnerabilities (Libicki 2009).

Without the legitimating and bureaucratic clarity of a virtual border, for example, jurisdictional disputes in nations observing centuries of criminal versus national security civil society laws are hamstrung. As its designers intended, Stuxnet easily crossed borders. If bad actor promulgating malware or a virus is a nonstate actor, then the action is criminal, invoking the powers of police forces. If it is state, then militaries should be involved. Unfortunately it is seldom clear which groups are involved, in large part because the difficulty of attribution, a problem made worse because with regard to cyber attacks a state has no obligation to reveal or sanction bad behavior emanating from its territory.

Once the virtual limits of sovereign power can be demarcated in cybersphere, states ignoring or supporting massive denial-of-service attacks from their territories will be held internationally responsible. Domestic legal systems that do not currently have internal laws criminalizing predatory cyber behavior affecting other states will have to initiate the kinds of internal controls required for law enforcement. If they do not impose internal controls or if they knowingly allow external attacks from their territory, they will have to accept the right of the attacked states to defend them. Distinguishing criminal activities from national security activities becomes more manageable when the jurisdictional lines are drawn and recognized.

Managing the bordered virtual sphere will also enable us to identify a unique cyberspace as well—ungoverned badlands equivalent to the territory of failed or failing states. As civil society extends into cyberspace with rules of accepted behavior and reinforced by modern state institutions, it becomes easier to invoke international law and organizations to curb, if not cure, the disruptive cyber activities. Institutions will adapt while replicating in cyberspace the functions of existing rules of behavior. What is happening today is the slow civilizing of cyberspace, a process that depends on how governments perceive the threat and what institutions they develop respond to the threats (Demchak 2011). For all, the beginnings of a need to control the sovereign, albeit digital, national territory is already present. None are controlling the harm, transmission, laws, or sanctions emerging on the sovereign territory of another state; rather, each is operating under the modern notion of monopoly of power on the territory already demarcated and looking to its own laws and control of actions on its territory.

2.2 Emergent Virtual Borders

Indications of emergent borders within the cybersphere are appearing at many levels. So far, some are quite singular. China leads the authoritarian states in a more ubiquitous cyberspace regulation model aimed at controlling information from outside and circulating inside its borders. In this "all points" model, the border boils down to gateways largely filtering information with the ability, in principle, to curtail the internet connections, either between internal regions or between China and the rest of the world. It is a technological (limited gateways), institutional (regulated telecoms), and psychological (cyber self-censors and vigilantes) model operating on many levels at once.

In this model, China is expressing a long-standing concern for the stability and security of its territory and citizenry. In the 1990s, the Chinese Communist Party recognized the power of unfettered access for Chinese citizens and declared the internet to be a fifth area of territoriality to be secured. They built the "Golden Shield" that employs an estimated 40,000 internet police who in 2009 shut down about 7,000 websites, deleted 1.25 million pieces of information, and arrested 3,500 people, including 70 dissidents and bloggers now in jail. About 30,000 netizens are employed part-time to intervene in online forum discussions and redirect conversations away from sensitive topics. The Chinese leadership routinely characterizes Westernized social media as subversive tools and sees the hand of the United States in diplomatic subversion in open internet. "Whether we can cope with the internet is a matter that affects the development of socialist culture, the security of information, and the stability of the state," President Hu said in 2007 (Wines et al. 2010). China implemented the strong, technologically sophisticated, heavily intelligence collection-driven second phase of the Golden Shield in 2010 (Cheong 2010).

China has also been working on constructing its own internet. In what is called China's Next Generation internet (CNGI), the current limited number of internet addresses add enough digits (IPv6)[4] to provide every single machine connecting to the internet its own unique web address. This addressing protocol means every single web transaction can be tracked from the original machine to any other, allowing a massive societal control advantage when linked to other rapidly emerging advances in the raw computing speed and storage of computer systems. Not only will three-dimensional online worlds move faster and more realistically, but also every interaction in those worlds can be recorded or individually tracked in real time to the specific machine (Worthen 2006).

A new, more surveillance-friendly addressing system is useful to all governments desiring to control their own borders without using proxies or agents. The so-called Great Firewall that Google declined to support in 2010 was in reality the imposition of liability onto ISPs if one of their users accessed forbidden sites or topics (MacKinnon 2010). As Google demonstrated, this "intermediary liability" approach to control has its limitations (Masnick 2010).

The justification of these measures as essential for citizen safety against social disharmony, false information, fraud, piracy, and social ills such as pornography is a common theme in the oft-times bumpy path to creating a sovereign cyber border. For example, in 2005 the Chinese announced an upgrade to the national text messaging filtering system with automatic police alerts when false information, reactionary remarks, or harmful activities such as fraud and scams are found in cell phone texts. In December 2005 the vice-minister of the Ministry of Public Safety announced that the upgraded system's 2,800 surveillance centers had tracked about 107,000 illegal cell phone text messages in November 2005. With about 33 % of the texts associated with criminal fraud activities, 9,700 cell phone accounts were shut down over the month (New York Times 2005). At the time (2004), Chinese citizens annually already sent 218 billion text messages, against which an objectionable number of 107,000 is not even a drop in the bucket. By 2010, however, the addition of supercomputers which can move terabytes per second (or more) provided advanced capabilities to filter cell phone text messages centrally. The police, using undisclosed criteria, create lists that cell phone companies must use to scan all customer text messages. Companies must automatically suspend the accounts and report the incidents to police if banned terms are found.

---

[4] Three of the world's largest sites are banding together with two of the largest content distribution networks, Akamai and Limelight, coordinated by the internet Society, to declare 8 June 2011 World IPv6 (internet Protocol version 6) Day.

The new technologies have enabled not only massive increases in the intrusive and comprehensive search mechanisms but also more punitive measures against those found to violate the restrictions. To help gain control over all communications media, the Chinese authorities have closed websites, especially those able to share files, and increased the difficulty for citizens to have their own sites (La Franiere 2010). Already the Chinese government has channeled the physical access of all web traffic in or out of China through three major gateways in Beijing, Shanghai, and Guangzhou (Wines et al. 2010). Whether or not the international community approves, China's government is using internal controls on content as a consistent part of a state asserting sovereignty over key aspects of its internal social territory (Hille 2009).

Several democratic nations have charted a "key firm" model of regulating the large telecoms, albeit loosely, with the goal of curbing malicious or thieving activity, not information flows. These include Australia and, to some extent, Germany. Many westernized, largely European, democracies are enacting internet control measures to prevent theft or abuse of their citizens' personal information and the economic assets of their countries. Others, such as the United Kingdom, turned initially to pan-agency coordinating economic or social, but not security, institutions to encourage, monitor, and guide internal internet transactions. The goal is to curb foreign and local theft of national economic assets and private personal information (Cornish et al. 2009). More recently, however, even European nations have tended to see a role for national security controls. In 2008, Sweden passed legislation allowing its national police force's intelligence section to monitor all internet traffic in and out of the country, whether by Swedish citizens or others. It was challenged by privacy advocates, but the law withstood challenges as a central piece of anti-terror legislation and was implemented in late 2009 (Constantin 2009). The model is still firm based but is increasingly focused more directly on security.

The path to a national border in cyberspace may not prove as difficult for EU nations as it would for other sectors because cyberspace policies are currently left largely to member states. The level of security varies greatly across nations; it is unlikely the United Kingdom or France, for example, will wait for an EU-wide solution to cyber threats (Evans and Whittell 2010). The United Kingdom, in particular, is laying the foundation for a national border, in part for political reasons having little to do with cyberspace, such as administering a national identity card to curb illegal immigration. The rise of serious intrusions into sensitive government networks—at least 300 over the course of 2009—has pushed the United Kingdom to construct two agencies with the specific missions of coordinating the tools, tactics, and targets of cyber security across all governmental agencies (Lloyd 2010). Trends suggest the United Kingdom will follow the American lead in developing the elements of a national cyber border because it is a close security partner of the United States, more of a target for external hackers, and more informed about its vulnerabilities than other EU nations.

The singular marker of an emerging border, however, is the creation of a military organization—a cyber command. When a nation establishes and declares the existence of such a unit, it that state is explicitly saying it has territory to defend from potentially serious, even existential threats. Such a unit marks the acknowledgement of a nationally owned space that the nation values and will protect using available and appropriate resources, including regulatory, law enforcement, and military capabilities. That the borders have not yet been recognized by other nations—an eventual outcome of the long Westphalian process—does not diminish the significance of this institutional declaration of sovereignty to be defended, by definition, in cyberspace itself. While not as advanced as either China in controlling its domestic internet access or policing its key telecommunications firms, the United States in 2010 established its US Cyber Command that lays the cornerstone necessary for a national cyber border. The United States has made clear its intention to use military resources to repel or prevent whatever threats come across its cyber border. The declaratory aspect of this unit is a symbol of a new cyber-Westphalian international system. China has government organizations with what Western observers presume are the same missions as Western cyber commands, but they are not publicly named as military defenders of the nation. The “cyber command” model rests on the use of national security institutions for cyber defense at and beyond a border.

3 Cyber Command: The US Model

In the fall of 2010, the US Cyber Command became operational after an exceptionally rapid year of institutional and legal preparation (Whitney 2010). This institutional response to the rise of the cybered conflict age emerged to anchor a future cybered border for the whole nation. Its initial mission was to protect only military organizations from cyber attack, but as soon as a military unit existed to create a cyber safety wrapper around US critical military assets, political statements emerged about creating the same protection for the whole nation (Lynn 2010).

From the RMA to net-centric warfare, the United States has long provided innovative models for national security that diffuse internationally (Golfman and Eliason 2003). For the United States to announce a new national cyber command automatically provokes a new debate in the international military and legal communities (Shackelford 2009). Whether or not other nations need, want, or can afford to have a singular military unit focused on cybered conflict, their leaders, doctrine writers, and strategic thinkers will contemplate the potential benefits of the model.

If patterns of military emulation hold true, many nations will develop organizations that look like a national cyber command. Already we have seen nations closely associated with the United States either creating their own cyber commands or declaring an interest in approximating the functions of US Cyber Command. South Korea, for example, now has a military cyber command after enduring a massive cyber assault in early July 2009 (Channel News Asia 2009). In its own internal strategy discussions, the United Kingdom, while focused on the cyber protection of the entire society, also began discussing closer integration of its military and intelligence cyber resources (Aldrich 2010).

The US model of a national cyber command has several distinctive elements. First, the unit chosen by national leaders as their initial foray into strategic national security in cyberspace was a military, not a civilianized, internal security agency built for law enforcement or disaster mitigation. With most US resources dedicate initially to a strategy of purely defensive mitigation, officials decided that waiting for the attacks to hit and then mitigating the effects is devastatingly insufficient. Militaries operate at the national borders or deployed forward to prevent attacks; US Cyber Command and similar organizations in other countries will adopt the same approach. Using the military to defend the nation against cyber attacks also reinforces the seriousness of the threat because the military is historically the last resort for national survival. Creating US Cyber Command has redirected much of the global conversation about cyber security from merely blunting attacks to repelling or disrupting the attacks before they cause great harm. If cyber security involves military-like actions in repelling attackers, then borders will guide when and where these actions occur (Nakashima 2010a, b).

Second, the structure of the new cyber command is intended to blend offensive and defensive cybered operations with the collection of global intelligence. In cybered conflict, the offensive advantages of the attacker lie in preemptive surprise. The result is that a good defense requires the ability to preempt the preemptive attack or to, at least, anticipate it with sufficient time to prepare and mitigate its effects. The traditional peace versus war distinction has very little meaning operationally in cyberspace. The US Cyber Command model directly acknowledges the loss of this widely accepted distinction by dual-hatting its commander as the head of the premier electronic intelligence agency, the National Security Agency (NSA), and the military commander of the new cyber command.[5] The blend of intelligence and a decision to act results from the internal deliberations of one man subject to national laws but able to act quickly and knowledgeably if necessary.[6]

That the cyber command has the ability to attack, defend, and collect information globally is an important innovation for the United States but also for the wider international community. The model demonstrates a conclusion—that offense, defense, and extensive knowledge collection are needed to be secure—and a hypothesis that the best way forward is to build on military organizational structures. For the vast majority of European democracies which have a great deal of difficulty in publicly and politically endorsing offensive measures in cyberspace, cyber security institutional adaptations have been mired in lengthy debates on civil liberties. The rapid implementation of the cyber command model by the United States helped break the allies' collective cognitive logjam. Now, whether or not

[5] US laws enable "authorities" to draw legislative lines between offense (a military "Title 10" authority), defense (of military, "Title 18"; or of the wider government, a DHS mission), and the collection of national intelligence (a "Title 50" mission given the National Security Agency as primus inter pares electronic collector among other intelligence agencies).

[6] This structural compromise was unusual for the United States, and it was hotly debated in the Congress before the first commander, General Keith B. Alexander, was confirmed as head both of agencies (Nakashima 2010a, b).

senior leaders agree in principal with the solution, they are discussing new organizations and responses for repelling cyber threats. Stuxnet, the Confiker worm, and growing cyber crime sparked a strong interest in the US model.

In late 2008 France published a defense white paper that elevated cyber security to one of four key national threats. The mission was to create an institution capable of guiding the other agencies in protecting the entire nation's national cyberspace. In the process a small, formerly secretive organization has become its central and publicly discussed Agency for National Information Security (ANSSI). In the first year of existence, 2009–2010, ANSSI has helped research and justify legislation to allow further central control of defensive and, if necessary, offensive national cyber means.[7] Other nations, especially those with limited cyber resources such as the Baltic States, have pushed for NATO to be designated as the guarantor of their national cyber security (Computerworld 2010).

Third, by making US Cyber Command across rather than separate from the four military services, the new organization carries within it the seeds of its future elevation in importance for the nation. As concepts for repelling attacks aimed beyond military forces at the heart of the United States have begun to coalesce politically, critical practical decisions will be made about where the tripwires are to be virtually drawn and maintained. The model does not make a small unit that simply supports other government actors in the military. Rather, its size, prominence, and position atop subordinate service-only cyber commands reinforce the importance of the task to the whole nation beyond the .mil community. All the services are required to contribute to a coordinated national response to a major event. Only a few threats—such as nuclear war and terrorism—have forced such rapid, unequivocally collective and ubiquitous responses beyond traditional physical domains of land, air, sea, and space.

Fourth, the offensive operations mission of any cyber command working for a democracy underscores the need for other democracies to establish their own borders in cyberspace to demand noninterference in practice as well as de jure. The US Cyber Command model leaves unanswered the question of bad actors operating from within one democracy operating outward to harm other democracies. Precise rules of engagement and reach were left unresolved in part because a debate on that legal authority could have stalled the creation of the cyber command. Leaving the debate open to discussion with allies allows for parsing out the actions of allies, especially in NATO.

While the United States does not view its cyber command as threatening to allies, this benign assessment is not universally shared. Many on the European left are concerned, with good historical reasons, about the concentration of power in government hands. For example, until legislatively interrupted by the Greens Party in the Bundestag in 2011, Germany was creating a centralized cyber-crime facility that could support de facto if not de jure all-source cyber-crime service. Its potential

[7] See the website http://www.ssi.gouv.fr/site_rubrique97.html, hosted by ANSSI, which rather openly discusses its successes in strengthening cyber defenses.

for unified analysis is unlikely to occur because the Green Party feared concentrated data on citizen actions in the hands of the federal government. As a result, the facility was designed to be more of a repository that individual agencies may consult as needed. The deliberate dispersal of organizational interaction defeats the concept intrinsic to an organization such as US Cyber Command or, for that matter, a centralized cyber security operations center (CSOC) as it was originally set up in the UK.[8] This fear reinforces European skepticism about the military cyber command model. In particular, they are likely to be interested in creating their own cyber borders to constrain US government actions in cyberspace.[9]

The creation US Cyber Command is only one mark of the global transformation toward demarcating virtual borders in cyberspace, but it further accelerates the interest of other states in acquiring greater control over cyberspace. This transformation is not only natural, it may be desirable for a future civil global society.

4 International Relations Theory and History

With the establishment of borders in cyberspace, much of what we know about wars, conflict, and security will make sense again as practical and historical guides to state actions. With a border in and enforced by technological means, also essential will be the means to monitor who is electronically crossing the line in the virtual sand and whether that passage of bytes is permitted by national law, either criminal, civil, or national security. These means will have to be maintained and adapted to emerging new threats. These mechanisms will be a combination of encryption, unique machine/user identifiers centrally controlled, and local hardware-human "bio"-metrics. No more would the near-Herculean task of tracking bad cyber actors on a massive scale hinder a normal civil society's desire for a functioning mechanism to deter that source of harm. A border in cyberspace presumes verifiable and current originating data for everything trying to pass into the nation, from bytes to malware to phishing or mass assaults. The nature of connectivity means bad data will necessarily come from the territory of a state with overarching responsibility for allowing such transmissions to continue. A state will no longer be able to claim it is not harboring those attacking .mil addresses while encouraging their internal development of "patriotic" hacking skills (Markoff et al. 2010).

In the digitally bordered future, actual hot war will also be forced into expressions that can be recognized. Cross-border attacks will be regarded as such, even if

[8] Private conversation with senior civilian cyber-security police official in Germany, October 2010.

[9] Even the Chinese government has felt the need to have a cyber command equivalent and publicly announced its creation of a cyber warfare unit as a defensive measure in response to the provocative actions of the US government in creating a cyber command (Branigan 2010).

largely cybered in their characteristics. If the sponsoring state refuses to stop the attacks or to allow the defending state to reach inside its territory to stop them, then the sponsoring state can be presumed to support them. Conditions much like the onset of war can then be said to exist. Cyber wars will include familiar characteristics collateral damage, revenge myths, and arms races. We will deal with war as well as its phases in warmth, cooling, and even termination en route to civil or at least calm relations as well as we were likely to do without the internet.

International norms regarding cybered conflict will be negotiated state by state, region by region, coalition by coalition, and international regime by international regime. The vaguely commons-like characteristics of the man-made environment of cyberspace will be negotiated just like food production and safety, trade subsidies, banking reserves, and even whaling. Life on, around, and through the virtual borders will be as turbulent, semi-stable, and prone to smugglers, free riders, would-be upstarts, and annoyances as the physical borders are now in harbors, airports, land crossings, and maritime lines of control. According to former British prime minister Gordon Brown, "just as in the nineteenth century we had to secure the seas for our national safety and prosperity, and in the 20th century we had to secure the air, in the twenty-first century we also have to secure our position in cyberspace in order to give people and businesses the confidence they need to operate safely there." (Espiner 2009)

Many unique national concerns will continue. Germany, for example, closes its ports to neo-Nazis and chases smuggled peer-to-peer internet sites that encourage attacks on brown-skinned people, just as Saudi Arabia will close off pictures of women in positions of power, P2P porn sites and dissidents. China will not agree to international constraints on its right to execute online game fanatics who commit crimes or jail those who smuggled pictures of the Dalai Lama. Tunisia former autocratic leaders would simply not discuss their internal controls.

The United Nations will provide mechanisms for nations to quietly and practically cooperate even if they publicly are at odds. When cyberspace becomes a more normalized international system for modern states, one might see cyber ambassadors at UN agencies or cyber attaches at embassies to physically and rapidly calm crises or to coordinate responses if cyber systems are under assault (Vijayan 2010). Rules of conflict resolution and acceptable cybered civil society engagement are collectively, not individually, developed and enforced. When states establish sovereign boundaries in cyberspace and can defend themselves against cyber attacks, a less predatory and chaotic era has a better chance of emerging.

5 Coming Decades

In the not-too-distant future, about a generation or so, states will delineate defensible borders in the formerly ungoverned, even chaotic cybersphere. They will legislate the domestic laws giving them the necessary authority, establish the competent military and law enforcement organizations, negotiate international treaties, and perhaps even modify existing international organization to assert control. The creation of cyber commands is only a first, albeit significant, step.

States will thus meet the challenge of cyber threats just as they expanded from the limited competencies implied by Munster and Osnabruck to the full set of functions enjoyed by the modern nation-state. International relations scholars and historians have demonstrated how state gradually overwhelmed their institutional competitors city-states, trading federations, and even empires. They have also shown how states gradually assumed more control over international commerce (for example, slavery, finance, and trade) and even how interstate war is regulated (for example, the laws of war and arms control regimes). Slowly but surely leading states will take steps, internally and externally, to defend hard fought prerogatives. They will do so because it is in their economic and military self-interest. Despite the rear-guard actions of cyber-utopians, most, but not all, citizens will willingly trade freedom for greater safety and security in cyberspace, especially insofar as cyber activities have material consequences. Commercial self-regulation, domestically and internationally, will be tried and fail, as usual, because the incentives for firms will lead to underinvestment (in prudential technologies), cheating, and pressures to sink to the lowest common denominator.

The process of establishing borders will be nonlinear. As technologies evolve 'wicked' actors will seek advantages in each new software and hardware developments. Technical, organizational/institutional, and cultural weaknesses will be exploited by individuals, firms, non-state actors like terrorists and criminal syndicates as well as states not fully committed to the emerging international rule set. Over time however, progress will be made.

The transition—from now to the point where legitimate authorities gain the upper hand—is the period of maximum vulnerability and thus danger. As has been seen in Europe and North America already, individual firms and entire industries are vulnerable to hacking on a wide and often overwhelming scale. Disruptions to day-to-day operations and economic espionage resulting in the theft of valuable intellectual property rights have hit the financial services and defense industries among others. States have taken increasing notice, not simply because their own information systems are compromised but because government officials recognize their dependence on the private sector which has become the weaker link in national security for a deeply cybered nation. This is particularly true for sectors like public utilities and privately provided critical infrastructure where commercial firms hold enormous sway over the economic future of society. It is also true with regard to reports to weapons designs and the technical specifications of key military systems may have been compromised. Until proper domestic and legal frameworks have been devised and more resilient socio-technical systems have emerged by design or evolution, states and their citizenry will be vulnerable to attacks by both state and nonstate actors.

However, although "cyber war" has entered into common lexicon, we are not among those who worry that cyber 'war' is an immediate threat between states or that it will never be possible. Instead, we believe that today and the immediate future will be subject to a transition era of 'cybered conflict'. In this 20 year period, cyber activities by military forces will occur in consort with intelligence agencies, law enforcement organizations, and other associated departments, matching out of necessity the connectedness of the nonstate and state actors. For the foreseeable future, the tactical, operational, and strategic use of cyber means will blend and be present in all types of conflict, during all phases of military operations, and at all levels of competition from peace to war. From our perspective, cybered conflict characterizes the whole spectrum of old and new forms of conflict borne of, enabled through, or dramatically altered by, cyberspace.

The transition period to a cyber Westphalian accommodation will be fraught with complications as all parties who can will "test" the limits of what effects can, and cannot, be accomplished using cyber operations without escalating to overt, declared, kinetic exchanges typical of the older industrial era 'war'.[10] How long the transition era will last until a new international system of cybered jurisdictions becomes the accepted and legitimate topology is unclear. Signs of progress are mixed. On one hand, China and the United States, two of the most politically, economically, and technologically, powerful states seem increasingly willing to engage with each the other dialogue. On the other hand, little progress has been made in installing major power collective control of nonstate actors or even in mutually acceptable cessation of a mass scale of state-sponsored cyber penetrations pursued by one state in particular. Furthermore, there is little evidence that criminals and perhaps even terrorists have been deterred by state action so far, or that defensive/protective technologies have caught up with offensive tools.

The pre-cybered era since World War II in which the United States set the international agenda, especially for the system outside the Soviet bloc, is reaching an end and turbulence lies ahead. The relative economic and military decline of the United States, or perhaps more accurately, the economic and military rise of China as a regional and increasingly global power poses special challenges for the United States and the current global system. Many western analysts believe that Chinese firms, perhaps with official backing, have been the most aggressive sources of cyber penetrations and theft for military and economic gain. However, regardless of official culpability for such intrusions, China has relatively little experience helping govern the international system by agenda setting and collaboration (Shambaugh 2013). The 'everyman empowerment' aspect of cyberspace up to now continues to make it difficult for all parties, but in particular hard for the leading powers aside from the US, to readily or smoothly establish the new rules of the game for the emergent cyber Westphalia state interstate system. This fact alone will complicate and perhaps even delay the emergence of a set of new international arrangement governing cybered conflict and reasserting the primacy of states in defining and defending emerging borders.

[10] For more explanation of cybered conflicts see unpublished manuscript, currently under editorial review, Peter Dombrowski and Chris Demchak, "Cyber war, Cybered Conflict and the Maritime Domain."

In the meantime the community of democratic civil societies can expect to experience confusion and multi spectrum high and low intensity conflict. Some negative effects will be accidentals as the actions of as states, firms, and other non-state actors, not to mention citizen-users, collide when each resorts to a variety of self-help mechanisms. Firms may seek to implement more active defensive measures to proactively and preemptively thwart intrusions, and while protecting themselves, disrupt others or large communities. Private vigilantism could readily pose threats of inadvertent or even opportunistic escalations into state-on-state engagement.

This transition will be unsettled and unsettling for modernized societies used to relatively open interactions, unfettered economic globalization, and rule of law. Developing the original Westphalian system needed about 400 years to be recognized and more or less observed. In the process, there were many missteps, conflicts, and squandered or misspent resources and lives. The newer version of an international system will take a good deal less time, but may not be any less decisive for the system that emerges as the transition ends. Global economics of technology and knowledge diffusion will be increasingly disjointed as states seek to defend their national wellbeing from the onslaughts of knowledge extraction occurring today. Several states including westernized nations such as Germany are already asserting a state interest in 'technological sovereignty' that helps state defense of 'national' cyberspace by funding or encouraging in-state R&D and sourcing of national network technologies from design, to manufacture, to implementation, and operation. Over this unprecedented period, some states will try belatedly or weakly to reassert a measure of control and governance over their increasingly cybered resources and societies, and make only partial success. Others will resolutely seek more explicit controls of everything that crosses their territorial borders, including all electronic packets or cables or satellite transmissions (Tuutti 2010). Still others will implement fragmented policies, perhaps only relying on disruption of perceived attacks and limited to no resilience, or only relying on centrally orchestrated resilience and no forward disruption of attacks, or neither well.

As with the older Westphalian system, the new one will not guarantee universal peace or economic security, or a low risk environment in general. States are already attempting to fend for their own cyber security as best they can, and occasionally cooperating when it seems advantageous to do so. Some states will finally be dealt their own cyber jurisdiction simply by dint of the collective decisions of other states, and yet find themselves ill prepared to defend what they have been given. Cyber-confident but unethical states may still ignore the rules of good cyber citizens and continue engage in multispectrum attacks on less capable states and their economies despite international condemnation. Events may then move along the full spectrum of cybered conflict to erupt into intentional or accidental cybered and kinetic exchanges. Depending on how often, overt, and destructive cybered exchanges occur or are seen to be pending, the new system will evolve more or less rapidly in response.

The more turbulent this transition era becomes, the more interested states may be in collective defense with like-minded others. More advanced states may reach out to less advanced states out of self-interest to make larger, more dispersed targets across many states, or to orchestrate more cyber substrate sensors against adversaries. States may even deeply cooperate to increase the human or technological capital resources on which the cooperating nations may draw for better technological advances in economics as well as security. The emergent topology could be populated with a wide variety of socio-technical-economic alliances, cyber mutual defense leagues, or even regional technological sovereignty coalitions reminiscent of the collective defense and development arrangements of the 1700s, late 1800s, or even the post World War II period.

Over the long run, It is also possible over time that there develops a convergence of policies, tools, ideas, and common practice norms across like-minded democratic civil societies who then establish, at least for themselves, a newer cybered version of the shared expectations—and collective ability to enforce rules—of the older, pre-cyber liberal international system. This outcome, however optimistic, is for the far future given today's lack of knowledge, capacities, consensus, and trust across the cyberspace substrate today.

At the end of the day, frontiers do not last forever, and we have begun the cybered conflict transition era. What states do during this period to ensure their own robust resilience and disruption capacities and thereby their national cyber power will matter greatly for how prepared they are to defend their own cyber jurisdiction as the new global order takes concrete shape. It is not clear the older system's major powers will be the newer system's major powers if these strategies, institutions, or actions are inadequate. Much lies before us.

References

Aldrich, R. J. (2010). *GCHQ: The uncensored story of Britain's most secret intelligence agency*. London: HarperCollins.

Anderson, R., Böhme, R., Clayton, R., & Moore, T. (2009). Security economics and European policy. In E. Johnson (Ed.), *Managing information risk and the economics of security* (pp. 55–80). New York: Springer.

Baker, J. (2010, November 23). EU and US join NATO cyber security pact. *Computerworld UK online*. http://www.computerworlduk.com/news/security/3249914/eu-and-us-join-nato-cyber-security-pact/. Accessed June 25, 2013.

Benson, P. (2010, November 18). Computer virus Stuxnet a 'game changer,' DHS official tells Senate. *CNN online*. http://edition.cnn.com/2010/TECH/web/11/17/stuxnet.virus/index.html. Accessed June 25, 2013.

Branigan, T. (2010, July 22). Chinese army to target cyber war threat. *Guardian* online. http://www.guardian.co.uk/world/2010/jul/22/chinese-army-cyber-war-department. Accessed June 25, 2013.

Brenner, S. W. (2004). Distributed security: Moving away from reactive law enforcement. *International Journal of Communications Law & Policy, 9*(2), 1–43.

Channel New Asia. (2009, July 9). South Korea to set up cyber command against North Korea—two years earlier than planned. *Channel New Asia online*.

Cheong, C. (2010, April 5). Fighting the Digital War with the great firewall (op-ed). *Straits Times*. http://wanderingchina.org/2010/04/05/fighting-the-digital-war-with-the-great-firewall-straits-times/. Accessed June 25, 2013.

Constantin, L. (2009, November 6). Attack hits Swedish signals intelligence agency's website. *Softpedia News*. http://news.softpedia.com/news/Attack-Hits-Swedish-Signals-Intelligence-Agency-s-Website-126289.shtml. Accessed June 25, 2013.

Cornish, P., Hughes, R., & Livingstone, D. (2009). *Cyberspace and the National Security of the United Kingdom*. London: Chatham House.

Demchak, C. (2011). *Wars of disruption and resilience: Cybered conflict, power, and national security*. Athens: University of Georgia Press.

Douglass, D. B. (2009). An examination of the fraud liability shift in consumer card-based payment systems. *Economic Perspectives, 33*(1), 43–49.

Espiner, T. (2009, June 25). UK launches dedicated cyber security agency. *ZDNet UK* online. http://www.zdnet.com/uk-launches-dedicated-cybersecurity-agency-3039667231/. Accessed June 25, 2013.

Evans, M., & Whittell G. (2010, March 8). Cyber war declared as China hunts for the West's intelligence secrets. *The Times*. http://www.thetimes.co.uk/tto/technology/article2463349.ece. Accessed June 25, 2013.

Falliere, N., Murchu, L. O., & Chien, E. (2010). *W32.Stuxnet Dossier. Symantec security response*. http://www.symantec.com/content/en/us/enterprise/media/security_response/whitepapers/w32_stuxnet_dossier.pdf. Accessed June 25, 2013.

Gertz, B. (2010, March 4). Inside the ring: Hacker training. *The Washington Times*. http://www.washingtontimes.com/news/2010/mar/04/inside-the-ring-94616149/?page=all. Accessed June 25, 2013.

Gibler, D. M. (2007). Bordering on peace: Democracy, territorial issues, and conflict. *International Studies Quarterly, 51*(3), 509–532.

Golfman, E., & Eliason, L. (Eds.). (2003). *The diffusion of military technology and ideas Stanford*. Stanford, CA: Stanford University Press.

Hille, K. (2009, July 17). How China polices the internet. *Financial Times*. http://www.ft.com/intl/cms/s/0/e716cfc6-71a1-11de-a821-00144feabdc0.html#axzz2XDiG0U7M. Accessed June 25, 2013.

Kendall, N. (2010, February 22). Global cyber attacks on the rise. Report. *The Times*. http://www.thetimes.co.uk/tto/technology/article1859912.ece. Accessed June 25, 2013.

Krasner, S. (2004). Shared sovereignty: New institutions for collapsed and failing status. *International Security, 29*(2), 85–120.

La Franiere, S. (2010, January 19). China to scan text messages to spot 'unhealthy content'. *New York Times*. http://www.nytimes.com/2010/01/20/world/20text.html. Accessed June 25, 2013.

Libicki, M. C. (2009). *Cyber deterrence and Cyber war*. Washington: RAND. www.rand.org/pubs/monographs/2009/RAND_MG877.pdf. Accessed June 25, 2013.

Lindsay, D. (2010). Liability of ISPs for end-user copyright infringements: The first instance decision in Roadshow Films Pty Ltd v iiNet Ltd (No 3). *Telecommunication Journal of Australia, 60*(2), 1–21.

Lloyd, A (2010, March 8). Britain applies military thinking to the growing spectre of cyber war. *The Times* (London). http://www.thetimes.co.uk/tto/technology/article2463358.ece. Accessed June 25, 2013.

Lynn, W. J. (2010). Defending a new domain: The Pentagon's cyber strategy. *Foreign Affairs, 89* (5), 97–108.

MacKinnon, R. (2010, January 18). Commentary: Are China's demands for internet 'self-discipline' spreading to the West? *McClatchy Report: Washington Bureau*. http://www.mcclatchydc.com/2010/01/18/82469/commentary-are-chinas-demands.html. Accessed 25 June 2013.

Mallery, J. C. (2010, September 17). Towards a strategy for Cyber Defense. *Paper presented at the US Naval War College*, Newport, RI.

Markoff, J., Sanger, D. E. & Shanker T. (2010, January 26). Cyber war: In digital combat, U.S. finds no easy deterrent. *New York Times*. http://www.nytimes.com/2010/01/26/world/26cyber.html?pagewanted=all. Accessed June 25, 2013.

Masnick, M. (2010, January 20). The similarity between ACTA and Chinese internet censorship. *Tech-Dirt*. http://www.techdirt.com/articles/20100120/0216537828.shtml. Accessed June 25, 2013.

Mayntz, R. (2008, May 30–31). *The changing governance of large technical infrastructure systems (LTS)*. Paper presented at the Complexity and Large Technical Systems Conference, Michigan State University and University of Konstanz, Meersburg.

Michael, J. B., Sarkasain, J. F., Dementis, G., & Dementis, G. (2010). Integrating legal and policy factors in cyber preparedness. *Computer, 43*(4), 90–92.

Mills, E. (2010, March 25). Web traffic redirected to China in mystery mix-up. *CNET*. http://news.cnet.com/8301-27080_3-20001227-245.html. Accessed June 25, 2013.

Nakashima, E. (2010a, May 11). Gen. Keith Alexander confirmed to head cyber-command. *Washington Post*. http://www.washingtonpost.com/wp-dyn/content/article/2010/05/10/AR2010051005251.html. Accessed June 25, 2013.

Nakashima, E. (2010b, November 6). Pentagon's Cyber Command seeks authority to expand its battlefield. *Washington Post*. http://www.washingtonpost.com/wp-dyn/content/article/2010/11/05/AR2010110507304.html. Accessed June 25, 2013.

Nye, J. S., Jr. (2010). *Cyber power*. Paper, Belfer Center for Science and International Affairs, Harvard Kennedy School. http://belfercenter.ksg.harvard.edu/publication/20162/cyber_power.html. Accessed June 25, 2013.

NYT Technology Section. (2005, December 6). China keeping closer eye on phone text messages. *New York Times*. http://www.nytimes.com/2005/12/06/technology/06iht-sms.html. Accessed June 25, 2013.

O'Connell, R. L. (1989). *Of arms and Men: A history of war, weapons, and aggression*. London: Oxford University Press.

Porche, I. (2010, December 9). Stuxnet is the world's problem. *Bulletin of the Atomic Scientists*. http://www.thebulletin.org/stuxnet-worlds-problem. Accessed June 25, 2013.

Sanger, D. E. (2010, September 25). Iran fights strong virus attacking computers. *New York Times*. http://www.nytimes.com/2010/09/26/world/middleeast/26iran.html?_r=0. Accessed June 25, 2013.

Shackelford, S. J. (2009). From nuclear war to net war: analogizing cyber attacks in international law. *Berkeley Journal of International Law, 25*(3), 191–250.

Shambaugh, D. (2013). *China goes global: The partial power*. Oxford: Oxford University Press.

Shiode, N. (1997). Toward the construction of cyber cities with the application of unique characteristics of cyberspace. *Online Planning Journal*. http://www.casa.ucl.ac.uk/planning/articles21/urban.htm. Accessed June 25, 2013.

Tandon, R., & Mohanty, R. (2000). *Civil society and governance: A research study in India. Part of a global comparative research study on civil society and governance*. Sussex: Society for Participatory Research in Asia.

Tilly, C. (1989). Cities and states in Europe, 1000-1800. *Theory and Society, 18*(5), 563–584.

Tilly, C. (1992). *Coercion, capital, and European states, AD 990-1992*. Malden, MA: Blackwell.

Tuutti, C. (2010, October 1). European Union considers stronger cyber security, stricter penalties for hackers. *New internet (TNNI) online*. http://www.thenewnewinternet.com/2010/10/01/european-union-considers-stronger-cybersecurity-stricter-penalties-for-hackers/. Accessed 25 June 2013.

Vijayan, J. (2010, April 7). After Google-China dust-up, cyber war emerges as a threat: The episode highlighted cyber threats facing the U.S., but it's not a war—yet. *Computerworld*. http://www.computerworld.com/s/article/9174558/After_Google_China_dust_up_cyberwar_emerges_as_a_threat. Accessed June 25, 2013.

Whitney, L. (2010, March 23). U.S. Cyber command prepped to launch. *CNET News-Security*. http://news.cnet.com/8301-1009_3-10470186-83.html. Accessed June 25, 2013.

Wines, M., LaFraniere, S., & Ansfield, J. (2010, April 7). China's censors tackle and trip over the internet. *New York Times*. http://www.nytimes.com/2010/04/08/world/asia/08censor.html?pagewanted=all. Accessed June 25, 2013.

Wong, G. (2010, February 8). Chinese police shut down hacker training business. *The Guardian*. http://www.guardian.co.uk/world/feedarticle/8935515. Accessed June 25, 2013.

Worthen, B. (2006, July 15). Internet strategy: China's next generation internet. *CIO.com*. http://www.cio.com/article/22985/internet_Strategy_China_s_Next_Generation_internet. Accessed June 25, 2013.

Conceptualizing the Ambivalent Role of Technology in International Relations: Between Systemic Change and Continuity

Stefan Fritsch

Abstract Departing from Kranzberg's thesis that technology is neither good, nor bad, nor neutral, this chapter argues that technology as a multi-facetted—and ultimately deeply political—phenomenon impacts the global system in a variety of ways, thereby representing a formidable conceptual challenge to the disciplines of International Relations and International Political Economy. In order to theorize questions of systemic evolution between transformation and continuity, the goals and capabilities of actors, system conditions as well as technology-inherent aspects have to be analyzed in an integrated—and often interdisciplinary—way. The chapter argues that a combination of major perspectives of Science and Technology Studies and different theories of International Relations/International Political Economy enable us to highlight the ambivalence of technology and technology-related aspects of global politics and economics, such as power, interdependence, systemic interaction capacity and the socially constructed meaning/interpretation of technology. While focusing on the case of information and communication technologies, the chapter will demonstrate that by defining technology as an ambivalent endogenous core component of the global system, instead of relegating it to the role of a value-neutral external and residual variable, technology itself increasingly becomes the contested terrain on which security, economic and identity struggles play out.

Keywords Technology • Science and Technology Studies • IR/IPE Theory • System • Transformation • Interdisciplinarity

S. Fritsch (✉)
Bowling Green State University, Bowling Green, OH, USA
e-mail: sfritsc@bgsu.edu

M. Mayer et al. (eds.), *The Global Politics of Science and Technology - Vol. 1*, Global Power Shift, DOI 10.1007/978-3-642-55007-2_6,

1 Introduction

The contemporary narrative in International Relations (IR) and International Political Economy (IPE) has been that rapidly diffusing new information and communication technologies (ICTs), from the printing press to international mail systems to the radio, TV, VCR, satellite networks, the Internet and quite recently social media paired with mobile communication and satellite networks have contributed to a remarkable transformation of global affairs in the last 150 years. Global security, economics and culture have been radically transformed by the results of technological evolution. The emergence of complex interdependence between countries and their societies as well as time-space compression of social interaction on the local, national and global level are the two defining characteristics of the global system that have been deeply impacted by technology.

New forms of socio-economic and political interaction, largely based on new technologies, have also caused some scholars to reassess the role of the individual and non-state actors in global affairs. According to some, especially the last two decades have witnessed a "skill revolution", understood as the increasing ability of individuals to use ICTs for the collection and dissemination of information as well as for the transnational and global mobilization of like-minded individuals for various socio-political purposes. The assumption is, that ICTs and other technologies have empowered individuals and other non-state actors to play a more pro-active role in global affairs and to influence its outcomes in interesting new ways.

Many IR and IPE scholars have identified technology as one of the main driving forces for change and transformation in the structure of the global system and the interactions among states and a plethora of non-state actors. However, the paper argues that—compared to other social science disciplines such as history, philosophy and sociology—IR/IPE is a latecomer with regard to its theoretical incorporation of technology, which is primarily a result of the discipline's neglect of technology as explanans.

Typically, it has been conceptualized as an external factor, which represents a residual variable for the explanation of structural and process change in global affairs. This paper argues that technology should finally be considered a vital core component of the global system. It is shaped by and shapes the system and its units, embedding them in dense global socio-technical systems. Those systems become the material backbone for processes such as globalization, transnationalization, but also fragmentation as new polities develop through and within these new networks and challenge and/or overlap with older existing polities that also react to the new opportunities and challenges posed by those same technologies.

The paper further argues that due to predominantly narrow conceptions of technology as apolitical "tool" (Pacey 1983) our discipline underestimates its ambivalent character in global affairs. As Kranzberg (1986) pointed out, "Technology is neither good, nor bad, nor is it neutral." As sociologists and historians of technology have remind us for a long time, a certain technology can often be used in various ways, depending on the user's values, preferences and interests, and often quite differently from the inventor's original intent. Starting from the observation that global society's reproduction increasingly relies on large socio-technical systems, the chapter argues that those seamless socio-technical systems evolve over time as result of political, economic and cultural interest conflicts, which play out at the local, national and global level. Therefore, technological evolution must be understood as deeply political phenomenon as technology becomes woven into the global social fabric and is adapted to context-specific needs. To develop the argument further, the chapter will first explore the status of efforts to explore technology as political phenomenon, primarily within Science and Technology Studies (S&TS), and its incorporation into research on global affairs. Therefore, the paper will briefly discuss how different IR/IPE theories have addressed technology and how their respective insights help us to conceptualize technology and its impact on actors, structures and interaction processes on the global level. Drawing from many different examples, the following section will then briefly discuss the interrelationship between technology (mostly in the area of ICTs), system and actors in the policy areas of security and economics, to illustrate the potential pitfalls of overgeneralization and deterministic explanations for technology-driven systemic evolution.

2 International Relations and Technology: The Need for Interdisciplinarity

IR scholarship has only recently begun to seriously explore the impact of technological evolution on global affairs in their broadest sense. Early works, dating back to the 1940s did indeed pay attention to the international implications of new technologies such as the telegraph and telephone as well as shipping and aviation and—of course—the atomic bomb (Brodie 1946, 1947; Ogburn 1949). However, they did rarely engage in a more thorough systematic exploration of the interdependency between technological evolution and system change. Despite a significant number of publications devoted to the investigation of political economic issues associated with particular technologies (Codding 1972; Cowhey 1990; Krasner 1991), a rather small number of studies has tried to address the political nature of technological evolution, particularly with regard to its global dimensions (Castells 2000; Kraft and Vig 1988; McNeill 1982; Sanders 1983; Skolnikoff 1993; Rosenau 1990, 2003).

Recent efforts to incorporate insights from S&TS into IR and IPE theory have explored more nuanced conceptions of technology and ways to incorporate S&TS findings via interdisciplinary approaches (Fritsch 2011; Herrera 2006; Talalay et al. 1997). S&TS is a highly interdisciplinary research field that has experienced remarkable growth in recent decades. For Sismondo (2004: VI) "the field is a result of the intersection of work by sociologists, historians, philosophers, and anthropologists studying the processes and outcomes of science and technology. Because it examines science and technology, its findings and debates have repercussions for almost every understanding of the modern world," including scholars of IR/IPE who so far have not systematically investigated those issues on a broader scale, despite their often proclaimed awareness of them.

3 Main Perspectives in S&TS

Broadly speaking, one can distinguish two major perspectives on technology and society that have influenced S&TS for the last decades: technological determinism and social constructivism (Marx and Smith 1994; Kaplan 2004). Both perspectives highlight different aspects of the technology-society nexus and come to different conclusions regarding technology's impact on society and our ability to govern technological evolution.

3.1 Technological Determinism

Technological determinists emphasize the pervasive expansion of technology into every sphere of human life, which ultimately results in a thorough reorganization of society along technological imperatives (Mumford 1934, 1966, 1970; Ellul 1964). Technology is isolated as the central driving force behind societal change. Furthermore, determinists assume that technology is characterized by an automatic self-drive towards maximum efficiency, which at least partly inhibits humanity's ability to fully control its evolutionary direction. While a potentially complete loss of control was rejected in later works, a more nuanced concept that emerged from such perspectives was the concept of unintended side effects, which states that human engineering and risk assessment can never predict every possible consequence related to the selection of particular technological options/paths over others (Winner 1977, 1986). To compensate for the perceived totality and oppressive tendencies of the mega-machine, technological determinists vehemently argued for the democratization of technology-related decision-making to enable society to discuss its impact and future direction (Smith 1994).

3.2 The Social Construction of Technology

In reaction to technological determinism and its transcendental philosophy, the 1970s witnessed an empirical shift towards more practical and contextual interpretations of technology that explored their interrelations with politics, economics and society at large (Kaplan 2004). Labeled as social constructivism of technology

Table 1 Conceptions of technology in S&TS

	Technological Determinism	Social Constructivism	Middle Ground
Character of technology	Automatic drive towards the "one best solution"; realization of technology-inherent rationality of ever increasing efficiency	Socially constructed artifact to satisfy human needs and wants, interdependence between society and technology	Technological evolution is the result of divers technological options and societal interests/decisions; time-sensitivity
Relationship between technology and society	Technological evolution determines society; society loses control over the mega-machine, defined as system of systems	Human agency alone determines technological evolution	Mutual dependence; societal impact largest in the early stages of new technological developments; in later stages of matured technology, path-dependencies and unintended side-effects impact available future options and choices (time-sensitivity)
Consequences/ impact of technology	Path dependencies and potential loss of control over technological evolution; danger of unintended side-effects	Result of political decisions made by societal actors with specific interests that materialize in the form of physical artifacts	From open systems with various technological options to closure and path dependencies
Representatives	Jacques Ellul Martin Heidegger Robert Heilbroner Herbert Marcuse Lewis Mumford David Noble Langdon Winner	Wiebe Bijker Michel Callon Donna Haraway Trevor Pinch Günther Ropohl Peter Weingart	Thomas Hughes David Landes Nathan Rosenberg

(SCOT), constructivists reject determinism's pessimistic assessment of a singular technological rationality and understand the relationship between societal and technological evolution as interdependent (Lutz 1986). It is ultimately social actors who decide how technology evolves and is applied in various socio-economic contexts, opening technology for socio-political control and steering measures (Rip et al. 1995; Teusch 2000) (Table 1).

Technology cannot be adequately understood without the context of its genesis and concrete application, which represents specific social, political or economic interests (Ropohl 1979, 1985). More radical constructivists reject any (analytical) separation between technology and society, or "subjects and objects" (Suchman 2008) and define technology as manifestation of cultural, gender, political, societal and economic power structures. Together, these spheres—as well as the actors dominating them—constitute a seamless web that impacts technological evolution (Pinch and Bijker 1987; Bijker and Law 1994).

3.3 The Middle Ground: Time Sensitivity and Path-Dependence

By synthesizing findings from both sides, a newly emerging "middle ground" (Hughes 1986, 1994) has provided more balanced explanations that offer a pragmatic avenue for incorporating technology into IR/IPE explanations for change in global affairs. For IR and IPE, the middle ground seems well suited to be integrated into a global framework for several reasons: (1) The middle ground allows for technological evolution to be conceptualized at the intersection of human agency and technological path-dependencies (Hughes 1969). (2) It emphasizes a historical perspective in order to retrace the impact of human decision-making for technological evolution as well as the impact of certain technologies on system structures and system units and vice versa (Hughes 1983). (3) Recent work has begun to apply certain concepts form S&TS to IR/IPE, namely aspects of the invention/innovation process, the system-wide diffusion of technologies, and finally the issue of time-space compression with consequences for the interaction capacity in the global system (Buzan and Little 2000; Castells 2000; Fritsch 2011; Herrera 2006), a concept that was originally developed by Durkheim (1933/1984) to conceptualize the structural and functional changes taking place in modernizing societies.

The emphasis on globally relevant socio-technical systems such as energy, information and communication, transportation, and certain weapons systems derives from their network character. Particularly ICTs' global expansion can be explained with network effects. Networked digital ICTs are characterized by high fixed costs, such as investments in infrastructure or other technical components such as microchips, or communication satellites and low variable costs, e.g. addition of new network participants or copying of music CDs from an original digital recording. I.e., the exploitation by a number of parties does not degrade a digital artifact's quality (Mowery 1996). This explains, why the networks ideally should be standardized and grow as much as possible in order to amortize the high initial investments while lowering the costs for each participant (Katz and Shapiro 1985, 1994). These cost savings have enabled many less developed regions to catch up to the developed world with regard to digital interconnectivity. It has also empowered a wide range of state and non-state actors to acquire new interaction capacities thereby becoming more effective and influential players on the global stage.

Moreover, the increasing digitization of global ICTs has driven the convergence process of once functionally separate sectors such as speech, text, images, data and music or video into one seamless global communication environment (Singh 2002). Global networks also represent an excellent case to illustrate time-sensitivity, i.e. the higher impact of human agency during the early development stages, while path-dependencies—such as technological lock-in effects—become more prominent during stages of maturity. Based on these insights, it is now possible to provide a brief discussion of major IR/IPE theories, their technology conceptions and contributions to a better understanding of technology's role in global affairs.

4 Theories of IR/IPE and Technology

Theoretical pluralism in IR/IPE complicates any discussion of technology's fundamental role in global affairs. Varying understandings of what technology entails and how it impacts system structures, actors and interaction processes prevent a clear and simple answer. However, this could also be regarded as an advantage, since theoretical pluralism also allows for a nuanced appreciation of each perspective's potential contribution to technology-related investigations. Limited space prevents an in-depth analysis of this interesting intra- as well as inter-disciplinary cross-fertilization, therefore this section shall concentrate on the very essentials (for extensive analysis see Fritsch 2006, 2011; Herrera 2006).

4.1 Realism and Neo-Realism

For Realism and Neo-Realism technology simply represents one component of their definition of power as "the military, economic and technological capabilities of states" (Gilpin 1981: 13). The anarchical self-help character of the international system defines the behavior of states as the only relevant actors in that it forces them to develop survival strategies (Waltz 1979; Ashley 1984). Power distribution between states defines the system structure and it can only change due to a redistribution of power to which states are very sensitive. Particularly for Neo-realists a' la Waltz, the basic character of the international system, the interaction processes among its units and their interests don't change due to technological evolution (Waltz 1959). In his view this holds even true with regard to nuclear weapons of mass destruction (Waltz 1981).

Classical Realism has generally been more sensitive to technology-driven systemic transformations. Particularly scholars who addressed the relationship between politics, nuclear weapons and mutually assured destruction questioned Realist cornerstones such as state sovereignty, the rationality of nuclear war and the relationship between means and end in foreign policy (Aron 1966; Morgenthau 1961). Overall, though, the instrumental and exogenous conception of technology as a passive tool for the realization of state goals has not changed dramatically during the last two decades. New technologies such as ICTs and transportation are mostly considered as yet another tool in the power arsenal of states, which has to be mastered and used to guarantee military security and economic success in an increasingly global economy. The power concerns of states might have switched in relative terms from military to economic categories, the basic issues defining global affairs, though, remain largely unchanged (Brzezinski 1997; Gilpin 1987, 2001; Kennedy 1987, 1993; Rice 2000; Rosecrance 1986, 1999).

In the context of the globalization discussion, Realists and Neo-Realist clearly belong into the globalization sceptic camp. Globalization sceptics doubt that technology-driven globalization has transformed the structures and processes of the global system and reject the observation that states' sovereignty—i.e., their autonomous ability to control economic and processes—has eroded and is increasingly shared with various international organizations and non-state actors on multiple levels (Weiss 1998; Hirst and Thompson 1999; Rugman 2001; Gilpin 2002).

4.2 Liberal Interdependent Globalism

Technology takes a central role in liberal discussions about the current state of the global system, since it has generally been identified as the main driver of globalization, defined as growing complex military, economic, cultural and ultimately political interdependence between societies (Morse 1976; Keohane and Nye 1998; 2001; van Creveld 1999). Held et al., who among other issues investigate the relationship between technological evolution and military globalization observe that

> the most significant forces driving this process, aside from geopolitical and domestic factors, is military technological innovation. Military technological innovation is central to the arms dynamic and has important global consequences. [...] The operation of the arms dynamic therefore exhibits an *inherent globalizing* logic in that it directly encourages the transnational proliferation of military capabilities and technology (Held et al. 1999: 103; emphasis by the author).

And Castells (2000: 1) observes with regard to the transformational impact of ICTs that

> [t]oward the end of the second millenium of the Christian era, several events of historical significance transformed the social landscape of human life. A technological revolution, centered around information technologies, began to reshape, at accelerated pace, the material basis of society. Economies throughout the world have become globally interdependent, introducing a new form of relationship between economy, state, and society, in a system of variable geometry.

With regard to the question of system transformation and technology's role therein, one can broadly distinguish two broad camps. Hyper-globalists emphasize the dramatic transformation processes in the global system driven by technological progress, mainly new ICTs, which undermine system-building and system-stabilizing principles of political sovereignty. Territoriality and autonomy have been significantly eroded, according to some even rendered obsolete, thereby making it necessary to search for new modes of political organization (Ohmae

1995; Reich 1992). Hyperglobalism displays clear elements of technological determinism. New technologies change the system and establish new realities to which everyone has to adapt.

According to the second or transformationalist perspective, globalization furthers the establishment of new, as well as re-strengthening of older, spheres of authority or polities, which cause increasingly complex structures on the global level (Held et al. 1999; Rosenau 2003; Ferguson and Mansbach 2004; Held and McGrew 2007). For transformationalists, the state-centric world, represented by the "Westphalian" state, is enriched by a multi-centric world consisting of actors who are less bound by sovereignty or territory, but rather organize along the logic of global networks in which concepts of time and space are radically modified (Innis 1950/2007, Carnoy and Castells 2001). In some aspects the resulting patterns of global politics resemble earlier forms of political organization during the Middle Ages, with complex and overlapping patterns of politico-economic governance (Friedrichs 2001; Slaughter 2004). Rather than wither away, or become irrelevant, states play a variety of very influential roles in the global politics and economics of ICTs. While integration processes are certainly enabled by rapidly diffusing technologies, fragmentation of polities in reaction to those very globalization processes and/or to reduced output-legitimacy of political systems in the face of various policy-choices such as liberalization, retreat of welfare states, loss of cultural cohesion etc. can also be observed (Rosenau 1997). The results are new state and non-state polities that overlap, cut-across or are nested within larger/older polities (Ferguson and Mansbach 1996, 2012). Therefore, transformationalists are more nuanced in their account of technology's role for the explanation of systemic evolution.

Furthermore, Liberalism has also been concerned with the question of how the tensions between the increasing demand for global governance solutions (Savage 1989; Zacher and Sutton 1996) and democratic control of those should be solved (Risse-Kappen 1995; Karns and Mingst 2010). Another interesting research strand is concerned with strategic alliances, networked knowledge-oligopolies and completely new industry governance mechanisms that are more informal (Kim and Hart 2002; Mytelka 1991). This is a special aspect of the larger globalization debate and the impact of ICTs on globalization. Research suggests that ICTs have a double-edged character. They can lead to greater homogenization, but at the same time, provide the means to conserve surprising cultural and political diversity (Held et al. 1999; Hanson 2008). While some strands of Liberalism exhibit a clearer awareness of technology's ambivalent impact on the structures and actors of the global system, it mostly displays an instrumental understanding of technology as passive tool in the hands of diverse actors with differing goals and as an exogenous variable, thereby neglecting its quality as political actor sui generis and its endogenous role in the global system.

4.3 Social Constructivism

Social constructivism has—similar to Liberalism—experienced a dramatic diversification since its emergence in the late 1980s (Adler 1997). What unites most of its exponents, are two fundamental assumptions: Namely that knowledge and meaning are socially constructed, as is social "reality" (Guzzini 2000). Most versions also converge towards an ontology that understands the social world as intersubjectively and collectively meaningful structures and processes in which "material resources only require meaning for human action through the structure of shared knowledge in which they are embedded" (Wendt 1995: 73). Material facts, to which technology belongs, only acquire specific meaning through discursive negotiation processes between social actors.

Constructivists have criticized Neorealism as well as Neoliberal Institutionalism as materialist theories that almost completely neglect the role of ideas and identities for the analysis of global affairs. Theories acquire such materialist status when "it accounts for the effects of power, interests, or institutions by reference to brute material forces—things which exist and have certain causal powers, independent of ideas, like human nature, the physical environment, and, perhaps technological artefacts" (Wendt 1999: 94).

Constructivist scholarship has recently begun to investigate the meaning of various technologies for global military relations (Tannenwald 1999) and economic globalization (Shaw 2000; Palan et al. 1996). The unifying point is that technology is defined as a structural factor that "may increase the potential of events to happen in a certain way, human intervention may cause the events to happen quite differently" (Adler 1986: 676). Ultimately, technological evolution is not decided by physical or technological determinants, but by human choice (Ruggie 1975). However, some constructivists concede that human choices are not unlimited. Agency is quite often limited by material restrictions (Onuf 1998). The growing density of global interactions, driven by technology, is—of course—an important prerequisite for social interaction between individuals and societies. It therefore represents an important structural factor that can trigger changes in ideas, norms, identities and ultimately the character of global affairs.

Table 2 summarizes this brief discussion of major IR/IPE theories. But how do technologies now impact the main actors and policy areas of security and welfare? Can one find examples for its ambivalent impact?

5 The State in the Digital Age

Until the invention of the printing press with moveable type by Johann Gutenberg, the production, storage, and dissemination of information in Latin through a European network of seminars and monasteries was largely controlled by the Catholic Church (Dudley 1991). The cheap mass production of books in common

Table 2 Technology conceptions in major IR/IPE perspectives

	Realism and Neorealism	Interdependent Globalism	Social Constructivism
Technology	Material artifact; instrumental understanding; concrete application as power source; exogenous to the international system	Material artifact; instrumental understanding; driver of interdependence; exogenous to the international system; skill revolution of non-state actors; system transformation	Technology is the material representation of human values, ideas or interests; its meaning derives from its embeddedness in societal ideas/norms
Actors/system units	States	State and non-state actors	Individuals and collective actors
System characteristics	Anarchy Self-Help System structure defined by power distribution between states	Anarchy Self-Help Rational actors decide to cooperate to manage Interest conflict caused by interdependence	Values and ideas are the basis for interpretations of material reality; anarchy and globality are social constructions
System transformation	Redistribution of power between states; in the absence of world government, basic system features remain unchanged	Interdependence driven by technological evolution; potential erosion or transformation of state sovereignty; system transformation defined by fragmentation; change vs. continuity	Result of changing ideas and norms, which impact interpretations of material reality and can ultimately lead to new actor identities, behavior and structure (via institutional adaptation)

languages not only challenged the monopoly of the church, but also made the reduction of language pluralism as well as the formation of national identities much easier (Eisenstein 1980; Anderson 1991). Worldly rulers quickly realized the usefulness of this new medium to extend their control over citizens and strengthen state authority. The following centuries would further strengthen the state as a central organizing agent of information and communication.

While the discussion concerning the role of the Westphalian state in global affairs of the late twentieth and early twenty-first century is far from over (see the discussions above), one can certainly find plenty of anecdotal evidence to support the argument that oversimplifying generalizations concerning a dramatic erosion of sovereignty are as unwarranted as the counter-argument that nothing has changed at all. ICT-related issues illustrate the point. For example, states collect an ever-growing multitude of data on their own as well as foreign citizens, criminal and terrorist networks or other countries. In some instances, such as during the Egyptian Revolution of 2011, the government of Hosni Mubarak successfully shut down practically most of Egypt's online traffic for 2 days, before global and domestic pressure forced the government to reopen all connections (Wired, 1/28/2011).

The widespread application of ICTs in national military structures and offensive as well as defensive strategies has contributed to a so-called revolution in military affairs and the evolution of new conflict forms (Arquilla and Ronfeld 1997; Combs 2006). Several new conflict forms are directly related to the growing spread of global information and communication networks such as the Internet, mobile communication devices, satellite technology, fiber-optic cables, global positioning systems, social networks or robots etc. Over the last two decades, the armies of all major powers have included information and cyber war fighting and defense strategies into their conflict strategies (New York Times, 1/26/2010, A6; New York Times, 10/17/2011, A1). Practical examples for these strategies are to be found in cases such as the First and Second Iraq War, the humanitarian intervention in Kosovo 1999, the Russian-Georgian conflict of 2008, the Stuxnet computer virus attack on the Iranian nuclear program in 2010 and repeated cyber-incidents involving China and the US as well as Russia. It is sometimes hard, though, to assess the exact extend to which cyber warfare capabilities have impacted the actual outcomes of the above mentioned conflict examples (Carr 2011).

Many observers have also pointed out the growing relevance of information and communication-related issues for state survival in a broad sense. With regard to the relationship between power/influence/welfare and ICTs, Strange (1994) developed the concept of structural power, which emphasizes the importance for states to control knowledge generation, storage and distribution in order to stay at the cutting edge of technological innovation and competitiveness.

In this context, another form of power, which according to many scholars gains more importance in the knowledge society, is soft or cultural-communicative power, which is defined as a state's ability to disseminate its cultural, societal, ideological, and political core values into other countries so that their societies might accept and internalize them (Nye 2004). Instead of classical power categories such as military power, new power categories such as patents, technological standards, size and capacity of communication networks (broadband width, PCs per capita, number of cell phones or landlines, etc.), and levels of interconnectivity or access between states become vital for measuring state power in a globalized knowledge economy (Rifkin 2000).

Scholars from diverse theoretical backgrounds agree that the central role of states in the governance of global information and communication has changed dramatically over the last 150 years and generated a large body of scholarly literature (for a more detailed overview see Fritsch 2010). States have played a central role in the creation of all major ICTs—the telegraph, radio, telephone, satellite technology, internet and mobile communication (Deibert 1997; Abbate 1999; Hanson 2008)—as well as in their domestic and global regulation (Aronson and Cowhey 1988; Zacher 2002), thereby covering a wide range of issues from online commerce, intellectual property protection to the defense of critical national infrastructure as well as investments in education, research and development and infrastructure to remain on the cutting edge of technological evolution in the twenty-first century (Dosi et al. 1988; Nelson 1993; Lundvall 2010).

They were responsible for extending them globally, e.g. to improve control over colonial territories and to further their national-security and economic interests (Headrick 1988; Hugill 1999). Importantly, though, state bureaucracies often controlled ICTs directly in the form of public postal, telegraph, and telephone administrations (PTTs) such as in many European countries (Great Britain, Germany, Austria, Italy, France etc.), while in the US the government allowed AT&T to develop the national telephone system on the basis of a private monopoly.

Implemented by states around the world liberalization and deregulation during the 1980s and 1990s shifted the control of technological networks and consumer markets from public agencies to private firms (Schiller 1999). Due to ICTs' dual-use character, they can be dedicated to civilian as well as military purposes, ultimately required new forms of regulation, particularly with regard to records keeping by telecommunications providers after the terrorist attacks of 9/11, raising questions about the protection of citizens' privacy, net-neutrality and the freedom of the Internet. Other concerns relate to the increasing corporate control of the Internet with potentially negative consequences for culture, creativity and innovation (Lessig 2004). A core component of the global liberalization of ICTs were intergovernmental trade negotiations in the World Trade Organization (WTO) to liberalize trade in information and communication services and the breakup of former state monopolies in the mid-1990s as well as efforts to standardize technological standards within the International Telecommunication Union (ITU) (Drake and Noam 1997; Salter 1999).

With regard to questions of national macro-economic governance, innovation patterns and the institutional settings in which these activities take place, countries do not necessarily converge towards one ideal model of a political economy of market relations or national innovation system in reaction to common technological pressures, as postulated by neoclassical economics and its IR/IPE pendant of hyperglobalism (Fukuyama 1992; Reich 1992). Instead, a rapidly growing comparative political economy literature points to persistent institutional and policy differences among countries and regions (Hall and Soskice 2001; Hollingsworth et al. 1994). The persistence of these varieties of capitalism and innovation outcomes calls into question deterministic assumptions about techno-globalization's homogenizing powers. These differences have mainly been explained with institutional and historical path dependencies, country-specific labour relations, extend of government regulation and societal norms and values. Technology might exert similar adaptive pressures around the world, yet each society filters those pressures differently.

6 Individuals and Other Collective Non-State Actors and Technology

6.1 *Not-for-Profit Actors*

Of particular recent interest is the role of individuals in global politics. Better education and particularly the broad application of information and communication technologies as well as cheaper transportation enables people to interact with each other, share ideas and coordinate political activities more efficiently on the global level. This skill revolution (Rosenau and Fagan 1997) has empowered individuals on a scale not witnessed before in global affairs, with interesting ontological and epistemological consequences for the disciplines of IR and IPE. Instead of focusing on traditional actors such as states, IGOs or MNCs, the fair trade movement challenges us to think about ways to prominently bring the individual back into the analytical frameworks of IR/IPE scholarship and to realize that individual people, their ideas and actions do actually count (Rosenau 2008; Tetreault and Lipschutz 2009).

One of the earliest examples for this process of empowering of individuals and other groups through new ICTs was the elaborate online campaign of the Zapatistas in Oaxaca, Mexico. Negatively effected by the liberalization of corn imports through NAFTA, and faced with political marginalization within Mexico, the guerillas applied a two-track strategy including limited military conflict and intense online campaigning thereby generating a global support base of sympathizers (Arquila et al. 2000; Castells 1997). ICTs are also used extensively by other non-state actors, such as terrorist groups, to coordinate, plan, and execute attacks and for new member recruiting (Homer-Dixon 2002). The increasing asymmetry of conflicts between states and non-state actors has ironically also extended the lifetime of older technologies such as video or music tapes, which are harder to locate than mobile phones. Even older communication forms such as human messengers have seen a renaissance in various conflicts. The higher the level of asymmetry, the stronger the imperative of the technologically weaker conflict party to rely on low-tech tools (Headrick 2010).

Various political issues ranging from human rights or arms-control issues to environmental problems, and concerns about the uneven distribution of globalization benefits, e.g. the digital divide between information haves and have-nots, have gained tremendous salience, due to their constant presence within the digital cosmos. Countless initiatives, from the International Coalition to Ban Landmines, to the International Criminal Court have largely come into existence because concerned citizens increasingly network globally to create countless alliances with other likeminded individuals and NGOs (Bob 2005; Slaughter 2004). Other NGOs use ICTs to organize global epistemic communities and support networks of local activists and groups in areas such as land and water use, environmental destruction etc. (Warkentin 2001; Litfin 2002).

Recent political revolutions in North Africa and the Middle East are another example of social networks' impact on citizen mobilization, although as recent discussions have highlighted, it remains unclear to which extend those revolutions were driven or rather supported by ICTs and what this means for their role in the creation of political space and real world action (Hofheinz 2011). Nevertheless, these recent events serve as example for the argument that interconnectivity has reached such levels that almost no country can completely shut down communication between its citizens and the outside world, an interesting indicator for the whole discussion about the ambivalent role of technology in global affairs.

Another indicator of the growing relevance of ICT-related issues for the "average voter" are the rather surprising recent election victories of political parties that promote freedom of the Internet and the abolishing of restrictive intellectual property rights regimes in several European countries (New York Times 9/19/2011, A5) and even in the European Parliament (Wired.com 6/8/2009). Another prominent example would be the media coups of loosely organized watchdog groups such as Wikileaks or anarchist hacker-collectives such as Anonymous, which opens academic discussion on information monopolies of states or state agencies and how the ability of those actors to collect and disseminate information impacts public discussion or even legal proceedings.

6.2 *The Multinational Corporation: Driver of and Driven by Technology*

The MNC itself has been described as a result of industrial modernization processes, which required new organizational structures to effectively manage processes of mass production and distribution by eliminating problems such as imperfect market information and market insecurities (Chandler 1977; Beniger 1986). MNCs have been of central importance to the development of global information and communication industries as (1) producers of telecommunication equipment and information and communication networks (Chandler 2001), (2) producers and distributors of media content (Mosco 1996) and (3) large-scale consumers of various information and communication services (Junne 1997).

The microelectronic revolution of the 1970s and accelerated technological innovations have resulted in a dramatic fall of prices for international communication services, which enabled MNCs to reorganize their economic activities on a global scale and thereby improve their productivity. Those productivity increases were largely based upon a transformation of industrial production processes from vertically integrated Fordist modes of standardized mass production to transnational production networks based on "flexible specialization" (Gereffi et al. 2005; Piore and Sabel 1984). ICTs thereby enabled companies to respond rapidly to changing consumer demand and concentrate on their core business (Harrison 1994).

MNCs have also played a well-documented role in the communication, media, and high-technology industries, as main drivers of liberalization and privatization of domestic and global communication since the 1970s (McChesney 2008). Their role in these dramatic policy-changes has been investigated for the media industry (Smith 1991) as well as for the producers of telecommunication equipment and consumer electronics (Peterson and Sharp 1998).

Other studies have discussed the dominant role of corporate demand for the development of wireless signal transmission (radio telegraphy) and telephone services (Edelmann 1950; Smythe 1957). The increasingly private and global (domestic and international) corporate control of communication and media industries, however, also nurtured fears of reduced public access to information and communication networks and services and the ability to cross-subsidize between local and international services in order to lower the prices of local service access, a main goal of state policies during the era of state-owned or state-controlled communication systems, all of which "translates into less societal control and reduced democratic accountability" (Schiller 2007: 96). The merger-mania of the mid- to late 1990s in the telecommunication and media industries was a clear signal for market and capital concentration, which resulted in ever larger and powerful MNCs or "behemoths" (Korten 2001).

ICTs provide MNCs with unprecedented opportunities to restructure their activities globally. Outsourcing and off-shoring labor-intensive production processes allows them to focus on their core business. Hierarchical structures emphasizing vertical integration have given way to flat horizontal and globally integrated network structures consisting of many value-generating entities within larger profit-generating networks (Scholte 2000). ICTs thereby enable MNCs to make use of time arbitrage, defined as the exploitation of "time discrepancies between geographic labour markets to make a profit" (Nadeem 2009: 21), as well as lower labor costs or geographically concentrated technological competence in the form of innovation clusters (Saxenian 1994; Koski et al. 2002). However, as has been discussed earlier, MNCs still rely on locally or nationally rooted skill-pools, research potentials or institutional and regulatory environments, factors which cannot easily be replicated elsewhere, even in the era of global flows. Moreover, many of the ICT-related opportunities can also be enjoyed by small and medium enterprises, so that it becomes increasingly harder to define the typical MNC as well as make generalizations with regard to MNCs' impact on the global political economy (Cohen 2007).

7 Conclusion: The Ambivalent Impact of Technology on Global Affairs

Our discipline has just begun to address technology more seriously in its exploration of system evolution and the issue of change vs. continuity. IR/IPE scholarship, interested in questions pertaining to the mutual relationship between technology and the structures, actors, interaction patterns and policy issues of global affairs, must begin to systematically engage with S&TS to better understand the driving forces behind technological evolution and global social change. S&TS can offer useful concepts that should be incorporated into our own theoretical frameworks and help us to highlight the often ambivalent character of technology as demonstrated in the previous sections. Once technology becomes defined as a highly politicized core component of the global system one can begin to understand technology itself as the location of, oftentimes, intense political struggles over a wide range of technology-related issues (availability, access, definition of technical standards and architectures, regulations, security, market shares etc.).

The discussed contemporary IR/IPE theories each contribute isolated parts to a fuller understanding of the complex relationship between technology in its totality and global affairs. While Realism/Neorealism highlights the power aspects of technology, Liberalism focuses on the interdependence driving quality of many technologies as well as on the divers usage of technologies resulting in simultaneous integration and fragmentation processes. Social Constructivism illuminates the normative and value aspects of technological evolution. However, each of these theories still lacks a clearer definition of technology as endogenous core component as well as the acknowledgement of technology as a potential actor sui generis. Only such a politicized definition of technology would allow the analysis of its ambivalent impact on global affairs that defies social as well as technological determinism.

A global perspective, as potentially provided by a sensitized IR and IPE highlights the interrelationship between technological evolution and modified system parameters on a local, national and worldwide scale. In its global totality, large socio-technical systems especially in their advanced development stages, which are often defined by path-dependencies, take on quasi actor qualities in that they profoundly shape systemic interaction capacities and opportunities, time-space relations and the ability of various actors to act and react in this increasingly techno-mediated world. Within this general development path towards growing complexity, various actors (states, companies, NGOs, individuals) can use or try to shape technology in different ways, depending on their own interests, needs and wants within technology-based limitations, thereby furthering the ambivalence of technology in global affairs.

In this context and with regard to ICTs, two examples shall illuminate current and future policy issues that might also form focuses of scholarly research. One revolves around the growing tensions between the concepts of intellectual property protection vs. freedom of information/public access and the integration of selective or temporal access/ownership into the technological architecture of communication networks and consumer products (Lessig 2004). As pointed out earlier, these issues increasingly represent not only a politicization of new technology-related issues, but also a raising awareness of states, companies, voters and consumers for issues of privacy, online security, censorship, access, commodification etc.

Another problem, which will likely become more important, is the issue of "digital gaps" between information-rich and information-poor people, states and regions (UNDP 2001). How do information technologies or their lack affect socio-economic development? Who defines the digital gap and its relevance for development? How and by whom is the discourse surrounding ICT and development framed? Does it only consist of technical aspects or is this issue embedded in larger challenges such as social capital formation, definition of access as entitlement vs. public/private service etc. (Mansell 2002; Selwyn 2004; Thompson 2004). This is not only a specific aspect of the larger North-South problematic, but also is of growing domestic concern for industrialized countries as well as a larger issue within the globalization debate.

References

Abbate, J. (1999). *Inventing the Internet*. Cambridge, MA: MIT Press.

Adler, E. (1986). Ideological "Guerrillas" and the quest for technological autonomy: Brazil's domestic computer industry. *International Organization, 40*, 673–705.

Adler, E. (1997). Seizing the Middle Ground: Constructivism in world politics. *European Journal of International Relations, 3*, 319–363.

Anderson, B. (1991). *Imagined communities*. New York: Verso.

Aron, R. (1966). *Peace and war: A theory of international relations*. Garden City, NY: Doubleday.

Aronson, J. D., & Cowhey, P. F. (1988). *When countries talk: International trade in telecommunications services*. Cambridge, MA: Ballinger Publishing.

Arquilla, J., & Ronfeld, D. (Eds.). (1997). *In Athena's camp: Preparing for conflict in the information age*. Santa Monica, CA: Rand.

Arquila, J., Ronfeldt, D., & Zanini, M. (2000). Information-age terrorism. *Current History, 86*, 179–185.

Ashley, R. K. (1984). The poverty of Neorealism. *International Organization, 38*(2), 225–286.

Beniger, J. (1986). *The control revolution: Technological and economic origins of the information society*. Cambridge, MA: Harvard University Press.

Bijker, W. E., & Law, J. (1994). General introduction. In W. E. Bijker & J. Law (Eds.), *Shaping technology/building society: Studies in sociotechnical change*. Cambridge, MA: MIT Press.

Bob, C. (2005). *The marketing of rebellion: Insurgents, media and international activism*. Cambridge: Cambridge University Press.

Brodie, B. (1946). The atomic bomb as policy marker. *Foreign Affairs, 27*, 17–33.

Brodie, B. (1947). *The absolute weapon: Atomic power and world order*. New York: Harcourt Brace.

Brzezinski, Z. (1997). *The grand chessboard: American primacy and its geostrategic imperatives*. New York: Basic Books.

Buzan, B., & Little, R. (2000). *International systems in world history. Remaking the study of international relations*. Oxford: Oxford University Press.

Carnoy, M., & Castells, M. (2001). Globalization, the knowledge society, and the network state: Poulantzas at the millenium. *Global Networks, 1*, 1–18.
Carr, J. (2011). *Inside Cyberwar: Mapping the cyber underworld*. Sebastopol, CA: O'Reilly.
Castells, M. (1997). *The information age: Economy, society and culture. Part 2: The power of identity*. Cambridge, MA: Blackwell.
Castells, M. (2000). *The information age: Economy, society and culture* (Pt. 1: The rise of the network society). Oxford: Blackwell Publishers.
Chandler, A. (1977). *The visible hand: The managerial revolution in American business*. Cambridge, MA: Belknap Press.
Chandler, A. D. (2001). *Inventing the electronic century: The epic story of the consumer electronics and computer industries*. New York: Free Press.
Codding, G. A. (1972). *The international telecommunication union: An experiment in international cooperation*. Leiden: Brill.
Cohen, S. D. (2007). *Multinational corporations and foreign direct investment: Avoiding simplicity, embracing complexity*. Oxford: Oxford University Press.
Combs, C. C. (2006). *Terrorism in twenty-first century*. Upper Saddle River, NJ: Prentice-Hall.
Cowhey, P. F. (1990). The international telecommunications regime: The political roots of regimes for high technology. *International Organization, 44*, 169–199.
Deibert, R. J. (1997). *Parchment, printing and hypermedia: Communication in world order transformation*. New York: Columbia University Press.
Dosi, G., Freeman, C., Nelson, R., Silverberg, G., & Soete, L. (Eds.). (1988). *Technical change and economic theory*. London: Pinter.
Drake, W. J., & Noam, E. M. (1997). The WTO deal on basic telecommunications. Big bang or little whimper? *Telecommunication Policy, 21*, 799–818.
Dudley, L. (1991). *The word and the sword: How the technologies of information and violence have shaped our world*. Cambridge, MA: Basil Blackwell.
Durkheim, E. (1933/1984). *The division of labor in society*. New York: Free Press.
Edelmann, M. (1950). *The licensing of radio services in the United States, 1927-1947: A study in administrative formulation of policy*. Urbana, IL: University of Illinois Press.
Eisenstein, E. (1980). *The printing press as an agent of change*. Cambridge: Cambridge University Press.
Ellul, J. (1964). *The technological society*. New York: Vintage Books.
Ferguson, Y. H., & Mansbach, R. W. (1996). *Polities: Authority, identity and change*. Columbia, SC: University of South Carolina Press.
Ferguson, Y. H., & Mansbach, R. W. (2004). *Remapping global politics: History's revenge and future shock*. Cambridge, MA: Cambridge University Press.
Ferguson, Y. H., & Mansbach, R. W. (2012). *Globalization: The return of borders in a borderless world?* London: Routledge.
Friedrichs, J. (2001). The meaning of new medievalism. *European Journal of International Relations, 7*, 475–501.
Fritsch, S. (2006). *Technologie und Internationale Politische Ökonomie im Zeitalter der Informationsgesellschaft*. Baden-Baden: Nomos.
Fritsch, S. (2010). The economics of international communication. In R. A. Denemark (Ed.), *International studies encyclopedia* (pp. 1301–1319). Oxford: Wiley-Blackwell.
Fritsch, S. (2011). Technology and global affairs. *International Studies Perspectives, 12*, 27–45.
Fukuyama, F. (1992). *The end of history and the last man standing*. London: Penguin.
Gereffi, G., Humphrey, J., & Sturgeon, T. (2005). The governance of global value chains. *Review of International Political Economy, 12*, 78–104.
Gilpin, R. (1981). *War and change in world politics*. Cambridge: Cambridge University Press.
Gilpin, R. (1987). *The political economy of international relations*. Princeton, NJ: Princeton University Press.
Gilpin, R. (2001). *Global political economy: Understanding the international economic order*. Princeton, NJ: Princeton University Press.

Gilpin, R. (2002). *The challenge of global capitalism.* Princeton, NJ: Princeton University Press.
Guzzini, S. (2000). A reconstruction of constructivism in international relations. *European Journal of International Relations, 6*, 147–182.
Hall, P. A., & Soskice, D. (Eds.). (2001). *Varieties of capitalism: The institutional foundations of comparative advantage.* Oxford: Oxford University Press.
Hanson, E. C. (2008). *The information revolution and world politics.* Lanham, MD: Rowman and Littlefield.
Harrison, B. (1994). *Lean and mean: The changing landscape of corporate power in the age of flexibility.* New York: Basic Books.
Headrick, D. R. (1988). *The tentacles of progress: Technology transfer in the age of imperialism, 1850-1940.* New York: Oxford University Press.
Headrick, D. R. (2010). *Power over peoples: Technology, environments, and Western imperialism, 1400 to the present.* Princeton, NJ: Princeton University Press.
Held, D., & McGrew, A. (2007). *Globalization anti-globalization: Beyond the great divide.* Cambridge, MA: Polity.
Held, D., McGrew, A., Goldblatt, D., & Perraton, A. (1999). *Global transformations: Politics, economics, and culture.* Cambridge, MA: Polity.
Herrera, G. L. (2006). *Technology and international transformation: The railroad, the atom bomb, and the politics of technological change.* Albany, NY: State University of New York Press.
Hirst, P., & Thompson, G. (1999). *Globalization in question.* Cambridge, MA: Polity.
Hofheinz, A. (2011). Nextopia? Beyond revolution 2.0. *International Journal of Communication, 5*, 1417–1434.
Hollingsworth, J. R., Schmitter, P. C., & Streek, W. (Eds.). (1994). *Governing capitalist economies: Performance and control of economic sectors.* New York: Oxford University Press.
Hughes, T. P. (1969). Technological momentum in history: Hydrogenation in Germany 1898–1933. *Past and Present, 44*, 106–132.
Hughes, T. P. (1983). *Networks of power: Electrification of Western society 1880–1930.* Baltimore, MD: John Hopkins University Press.
Hughes, T. P. (1986). The seamless Web: Technology, science, etcetera, etcetera. *Social Studies of Science, 16*, 281–292.
Hughes, T. P. (1994). Technological momentum. In M. R. Smith & L. Marx (Eds.), *Does Technology drive history? The dilemma of technological determinism.* Cambridge, MA: The MIT Press.
Hugill, P. J. (1999). *Global communications since 1844: Geopolitics and technology.* Baltimore, MD: Johns Hopkins University.
Innis, H. (1950). *Empire and communication.* Oxford: Oxford University Press.
Junne, G. (1997). The end of Dinosaurs? Do new technologies lead to the decline of multinationals? In M. Talalay, R. Tooze, & C. Farrands (Eds.), *Technology, culture and competitiveness: Change and the world political economy* (pp. 57–72). London: Routledge.
Kaplan, D. M. (2004). *Readings in the philosophy of technology.* Lanham, MD: Rowman & Littlefield.
Karns, M. P., & Mingst, K. A. (2010). *International organizations: The politics and processes of global governance.* Boulder, CO: Lynne Rienner.
Katz, M. L., & Shapiro, C. (1985). Network externalities, competition, and compatibility. *The American Economic Review, 75*, 424–440.
Katz, M. L., & Shapiro, C. (1994). Systems competition and network effects. *Journal of Economic Perspectives, 8*, 93–115.
Kennedy, P. (1987). *The rise and fall of great nations.* New York: Random House.
Kennedy, P. (1993). *Preparing for the 21st century.* New York: Random House.
Keohane, R. O., & Nye, J. S. (1998). Power and interdependence in the information age. *Foreign Affairs, 77*, 81–94.
Keohane, R. O., & Nye, J. S. (2001). *Power and interdependence.* Upper Saddle River, NJ: Longman.

Kim, S., & Hart, J. A. (2002). The global political economy of Wintelism: A new mode of power and governance in the global computer industry. In J. N. Rosenau & J. P. Singh (Eds.), *Information technologies and global politics* (pp. 143–168). Albany, NY: State University Press of New York.
Korten, D. C. (2001). *When corporations rule the world*. Bloomfield, CT: Kumarian Press.
Koski, H., Petri Rouvinen, P., & Ylä-Anttila, P. (2002). ICT clusters in Europe: The Great Central banana and the small Nordic potato. *Information Economics and Policy, 14*, 145–165.
Kraft, M. E., & Vig, N. J. (Eds.). (1988). *Technology and politics*. Durham: Duke University Press.
Kranzberg, M. (1986). Technology and history: "Kranzberg's laws". *Technology and Culture, 27*, 544–560.
Krasner, S. (1991). Global communications and national power. Life on the Pareto Frontier. *World Politics, 43*, 336–366.
Lessig, L. (2004). *Free culture: How big media uses technology and the law to lock down culture and control creativity*. New York: Penguin.
Litfin, K. T. (2002). Public eyes: Satellite imagery, the globalization of transparency, and new networks of surveillance. In J. N. Rosenau & J. P. Singh (Eds.), *Information technologies and global politics. The changing scope of power and governance* (pp. 65–89). Albany, NY: State University of New York Press.
Lundvall, B.-A. (Ed.). (2010). *National systems of innovation: Toward a theory of innovation and interactive learning*. London: Anthem Press.
Lutz, B. (1986). Das Ende des Technikdeterminismus und die Folgen: Soziologische Technikforschung vor neuen Aufgaben und neuen Problemen. In *Technik und sozialer Wandel: Verhandlungen des 23*. Deutschen Soziologentages *in Hamburg 1986*. Frankfurt am Main: Campus Verlag.
Mansell, R. (2002). From digital divides to digital entitlements in knowledge societies. *Current Sociology, 50*, 407–426.
McChesney, R. W. (2008). *The political economy of media: Enduring issues, emerging dilemmas*. New York: Monthly Review Press.
McNeill, W. H. (1982). *The pursuit of power: Technology, armed force, and society since A.D. 1000*. Chicago: The University of Chicago Press.
Morgenthau, H. (1961). Western values and total war. *Commentary, 32*, 277–304.
Morse, E. L. (1976). *Modernization and the transformation of international relations*. New York: Free Press.
Mosco, V. (1996). *The political economy of communication: Rethinking and renewal*. London: Sage.
Mowery, D. C. (Ed.). (1996). *The international computer software industry: A comparative study of industry evolution and structure*. Oxford: Oxford University Press.
Mumford, L. (1934). *Technics and civilization*. New York: Harcourt Brace.
Mumford, L. (1966). *The myth of the machine: Technics and human development*. New York: Harcourt Brace.
Mumford, L. (1970). *The myth of the machine: The pentagon of power*. New York: Harcourt Brace.
Mytelka, L. K. (1991). *Strategic partnerships in the world economy*. London: Pinter.
Nadeem, S. (2009). The uses and abuses of time: Globalization and time arbitrage in India's outsourcing industries. *Global Networks, 9*, 20–40.
Nelson, R. R. (1993). *National innovation systems: A comparative analysis*. Oxford: Oxford University Press.
Nye, J. S. (2004). *Soft power: The means to succeed in world politics*. New York: Perseus.
Ogburn, W. (Ed.). (1949). *Technology and international relations*. Chicago: Chicago University Press.
Ohmae, K. (1995). The rise of the region state. *Foreign Affairs, 72*, 78–87.
Onuf, N. (1998). Constructivism: A user's manual. In V. Kubalkova, N. Onuf, & P. Kowert (Eds.), *International relations in a constructed world* (pp. 58–78). New York: M. E. Sharp.

Pacey, A. (1983). *The culture of technology*. Cambridge, MA: The MIT Press.

Palan, R., Abbott, J., & Deans, P. (1996). *State strategies in the global political economy*. London: Pinter.

Peterson, J., & Sharp, M. (1998). *Technology policy in the European Union*. Basingstoke: MacMillan.

Pinch, T. J., & Bijker, W. E. (1987). The social construction of facts and artifacts: Or how the sociology of science and the sociology of technology might benefit each other. In W. E. Bijker, T. P. Hughes, & T. Pinch (Eds.), *The social construction of technological systems. New directions in the sociology and history of technology* (pp. 17–48). Cambridge, MA: The MIT Press.

Piore, M. J., & Sabel, C. F. (1984). *The second industrial divide: Possibilities for prosperity*. New York: Basic Books.

Reich, R. (1992). *The work of nations: Preparing ourselves for 21st century capitalism*. New York: Knopf Doubleday.

Rice, C. (2000). Promoting the national interest. *Foreign Affairs, 79*, 45–62.

Rifkin, J. (2000). *The age of access: The new culture of hypercapitalism: Where all of life is a paid-for experience*. New York: Penguin.

Rip, A., Misa, T. J., & Schot, J. (Eds.). (1995). *Managing technology in society: The approach of constructive technology assessment*. London: Pinter.

Risse-Kappen, T. (Ed.). (1995). *Bringing transnational relations back in: Non-state actors, domestic structures and international institutions*. Cambridge, MA: Cambridge University Press.

Ropohl, G. (1979). *Eine Systemtheorie der Technik: Zur Grundlegung der Allgemeinen Technologie*. Munich: Hanser.

Ropohl, G. (1985). *Die unvollkommene Technik*. Suhrkamp: Frankfurt am Main.

Rosecrance, R. (1986). *The rise of the trading state: Commerce and conquest in the modern world*. New York: Basic Books.

Rosecrance, R. (1999). *The rise of the virtual state: Wealth and power in the coming century*. New York: Basic Books.

Rosenau, J. N. (1990). *Turbulence in world politics: A theory of change and continuity*. Princeton, NJ: Princeton University Press.

Rosenau, J. N. (1997). *Along the domestic-foreign frontier: Exploring governance in a turbulent world*. Cambridge, MA: Cambridge University Press.

Rosenau, J. N. (2003). *Distant proximities: Dynamics beyond globalisation*. Princeton, NJ: Princeton University Press.

Rosenau, J. N. (2008). *People count! Networked individuals in global politics*. Boulder, CO: Paradigm.

Rosenau, J. N., & Fagan, M. W. (1997). A new dynamism in world politics: Increasingly skillful citizens. *International Studies Quarterly, 41*, 655–686.

Ruggie, J. G. (1975). International responses to technology: Concepts and trends. *International Organization, 29*, 557–583.

Rugman, A. (2001). *The end of globalization*. New York: Random House.

Sanders, R. (1983). *International dynamics of technology*. Westport, CT: Greenwood Press.

Salter, L. (1999). Regime for communication and information technologies. In C. A. Cutler, V. Haufler, & T. Porter (Eds.), *Private authority and international relations* (pp. 97–127). Albany, NY: State University of New York Press.

Savage, J. G. (1989). *The politics of international telecommunications regulation*. Boulder, CO: Westview.

Saxenian, A. L. (1994). *Regional advantage: Culture and competition in silicon valley and route 128*. Cambridge, MA: Harvard University Press.

Schiller, D. (1999). *Digital capitalism: Networking the global market system*. Cambridge, MA: The MIT Press.

Schiller, D. (2007). *How to think about information*. Urbana, IL: University of Illinois Press.

Scholte, J. A. (2000). *Globalization: A critical introduction*. New York: Palgrave.
Selwyn, N. (2004). Reconsidering political and popular understandings of the digital divide. *New Media & Society, 6*, 341–362.
Shaw, M. (2000). *Theory of the global state: Globality as unfinished revolution*. Cambridge: Cambridge University Press.
Singh, J. P. (2002). Introduction: Information technologies and the changing scope of global power and governance. In J. N. Rosenau & J. P. Singh (Eds.), *Information Technologies and global power: the changing scope of power and governance* (pp. 1–38). Albany, NY: State University of New York Press.
Sismondo, S. (2004). *An introduction to science and technology studies*. Oxford: Blackwell.
Skolnikoff, E. B. (1993). *The elusive transformation: Science, technology, and the evolution of international politics*. Princeton, NJ: Princeton University Press.
Slaughter, A.-M. (2004). *A new world order*. Princeton, NJ: Princeton University Press.
Smith, A. (1991). *The age of Behemoths*. New York: Priority Press.
Smith, M. R. (1994). Technological determinism in American culture. In M. R. Smith & L. Marx (Eds.), *Does technology drive history? The dilemma of technological determinism* (pp. 1–36). Cambridge, MA: MIT Press.
Smythe, D. (1957). *The structure and policy of electronic communications*. Urbana, IL: University of Illinois Press.
Strange, S. (1994). *States and markets*. London: Pinter.
Suchman, L. (2008). Feminist STS and the sciences of the artificial. In E. J. Hackett, O. Amsterdamska, M. Lynch, & J. Wajcman (Eds.), *The handbook of science and technology studies* (pp. 139–163). Cambridge, MA: The MIT Press.
Talalay, M., Tooze, R., & Farrands, C. (Eds.). (1997). *Technology, culture and competitiveness: Change and the world political economy*. London: Routledge.
Tannenwald, N. (1999). The nuclear taboo: The United States and the normative basis of nuclear non-use. *International Organization, 53*, 433–468.
Tetreault, M. A., & Lipschutz, R. D. (2009). *Global politics as if people mattered*. Lanham: Rowman and Littlefield.
Teusch, U. (2000). Technik und Internationale Politik. In W. Woyke (Ed.), *Handwörterbuch Internationale Politik*. Leske und Budrich: Opladen.
Thompson, M. (2004). Discourse, 'development' & the 'digital divide': ICT and the World Bank. *Review of African Political Economy, 31*, 103–123.
UNDP. (2001). *Human Development Report 2001: Making new technologies work for human development*. New York: Author.
Van Creveld, M. (1999). *The rise and decline of the state*. Cambridge, MA: Cambridge University Press.
Waltz, K. (1959). *Man, the state and war: A theoretical analysis*. New York: Columbia University Press.
Waltz, K. N. (1979). *Theory of international politics*. New York: McGraw-Hill.
Waltz, K. N. (1981). The spread of nuclear weapons. More may be better. *Adelphi Papers 171*. London: International Institute for Strategic Studies.
Warkentin, C. (2001). *Reshaping world politics: NGOs, the internet, and global civil society*. Lanham, MD: Rowman & Littlefield.
Weiss, L. (1998). *The myth of the powerless state*. Ithaca, NY: Cornell University Press.
Wendt, A. (1995). Constructing international politics. *International Security, 20*, 71–81.
Wendt, A. (1999). *Social theory of international politics*. Cambridge, MA: Cambridge University Press.
Winner, L. (1977). *Autonomous technology: Technics-out-of-control as a theme in political thought*. Cambridge, MA: MIT Press.
Winner, L. (1986). *The whale and the reactor: A search for limits in an age of high technology*. Chicago, IL: The University of Chicago Press.

Zacher, M. W. (2002). Capitalism, technology, and liberalization: The international telecommunications regime 1865-1998. In J. N. Rosenau & J. P. Singh (Eds.), *Information technologies and global power: The changing scope of power and governance* (pp. 189–210). Albany, NY: State University of New York Press.

Zacher, M. W., & Sutton, B. A. (1996). *Governing global networks: International regimes for transportation and communication.* Cambridge, MA: Cambridge University Press.

Part II
Interventions from STS, History, Innovation Studies and Geography

Petroleum Supply, Marine Transportation Technology, and the Emerging International Order of the Post World War One Period

Peter J. Hugill

Abstract At Versailles in 1919 President Wilson described three arenas of economic struggle, mediated by technology and geopolitics, that he saw developing between America and Britain as a new international order emerged: struggles over international transportation, petroleum, and international communication. The case of international transportation, almost all of it maritime in the 1920s, is intimately connected with that of the problem many observers saw with an insecure and possibly faltering global supply of petroleum. America dominated the early period of global petroleum supply, especially after the strike at Spindletop in Texas in 1901, but Spindletop was faltering by 1920. Even before World War One broke out German and British War Aims thus began to focus on the likely oil reserves of the Middle East that America could not control, Germany via the mineral rights the company was granted along the line of the Berlin to Baghdad Railway concession of 1903 and Britain via the Sykes-Picot-Sazanoff agreement of 1916, the latter resulting in the eventual creation of Iraq. In the run-up to the war oil-firing of steam turbine engine warships was seen as a technology that would substantially increase the fighting capacity of naval vessels through, amongst other things, radically easier refueling and a substantial reduction in stokehold crew size. Operators of merchant vessels followed a similar line of reasoning for a new generation of steam turbine powered ships coming to dominate the Atlantic shuttle by virtue of their speed. The war, however, proved a huge technology forcer in the area of diesel engines, which even in their early form were ideally suited to powering merchant ships, offering substantial savings in fuel and/or labor costs over steamships. Commentators in the 1920s were very clear on the relationship between the petroleum and marine transportation struggles, arguing that whoever controlled the right mix of marine transportation technology, in particular that of the merchant marine, and petroleum supply would control the emerging post-war international order. Of the major powers neither Britain nor Germany did particularly well in shifting their merchant

P.J. Hugill (✉)
Texas A&M University, College Station, TX, USA
e-mail: pjhugill@tamu.edu

M. Mayer et al. (eds.), *The Global Politics of Science and Technology - Vol. 1*, Global Power Shift, DOI 10.1007/978-3-642-55007-2_7,

marines to diesel power, America did very poorly. America's strength was in continuing to control the global oil supply.

Keywords Hegemonic transition • Oil • Naval power • Diesel • Merchant marine

Hegemonic transitions are intimately tied up with geopolitics and, as initially put forward by Mackinder, geopolitical change was driven by technological change (1904). For Mackinder the technologies of the railroad and telegraph shifted the balance of global geopolitics away from the maritime states towards a new generation of land-based states. In such a model changes in transportation and telecommunications technologies were thus clearly significant. In addition, scholars who deal with hegemonic transitions tend to theorize such transitions as culminating in military struggle (for example, Goldstein 1988; Modelski 1978; Tilly 1990; Wallerstein 1974). Such theories pay little or no attention to state type or the polities within them that control state decisions and the impact of those decisions on state behavior, including military and economic struggles. Other crucial decisions concern the development of educational systems and the encouragement or discouragement by state policies of the new technologies that can drive geopolitical shifts. A more complete theory describes three types of transitions between the two main types of states, territorial (land-based) and trading (maritime) states, and includes analysis of the fact that some states contain both trading and territorial polities (Hugill 2005, 2009b).

By 1919 Mackinder had developed his model of geopolitics further and had incorporated into his analysis the problems posed by internal struggles between polities embedded within states: "Berlin committed a fundamental mistake [in 1914]; she fought on two fronts without fully making up her mind on which front she meant to win. Berlin had not decided between her political objectives: Hamburg and overseas dominion [the aim of its trading polity] or Bagdad [sic] and the Heartland [the aim of its territorial polity] and therefore her strategical aim was uncertain" (Mackinder 1919: 154).

In similar vein the American historian, Edward Whiting Fox, would later elaborate his theory of "two Frances," one a trading polity, the other a territorial one, to account for persistent French failure to defeat Britain, a highly focused trading state (Fox 1971).

1 Hegemonic Transitions and Technological Innovations

The last hegemonic transition, which was ultimately between a declining Britain run entirely by a trading polity and a rising America in which the interests of its trading polity predominated, involved multiple other trading and territorial polities and a long period of military struggle, conventionally World Wars One and Two. Overshadowed by that military struggle, which focused mostly on the rise of a

German state run predominately by its territorial polity, a state that created an effective system of higher education designed to force new technologies, was the central and entirely economic struggle between Britain and America. American victory in both the military and economic struggles came from both its (eventual) technological advantages and its ability to shape the global policy environment in a series of treaties and agreements that often emerged out of those technological advantages. America's scientific and technical education lagged badly behind that of Wilhelmine Germany and even behind that of Britain (Hugill and Bachmann 2005), but began to catch up in critical areas such as petroleum engineering and geology in the 1930s. At Versailles in 1919 President Wilson, in evaluating the likely future of international relations, described three technological arenas he believed would dominate the struggle between Britain and America: international transportation, petroleum, and international communication. Between 1944 and 1947 a series of treaties and agreements effectively ended that struggle and resulted in overall but not complete American victory (Hugill 2009a).

Innovations, whether in the form of software as ideas or, more conventionally understood, in the form of hardware, and the transfer or diffusion of those innovations away from their area of origin have geopolitical consequences (Hugill and Dickson 1988; Hugill 1993). "Between 1860 and 1913 . . . attaining and sustaining a position of leadership in the world economy depended largely on the domestication of new technology" (Tuma 1987: 408). Innovatory technologies drive Kondratiev cycles in the economy and two Kondratiev cycles combine to create a world leadership cycle (Modelski 1978). In later work, Modelski and Thompson (1988) tied such cycles to a strongly navalist and technological reading of the history of capitalism, arguing that world leadership cycles were periods in which the power that could afford the world's dominant navy enjoyed global hegemony. I followed this structure of world history in *World Trade since 1431*, arguing that world leadership cycles are better understood as hegemonic cycles and that hegemons have achieved hegemony through commitment to high levels of technological innovation (Hugill 1993, 2003).

Wilson's evaluation at Versailles was prophetic. The economic struggle between Britain and America that began with the Morrill Tariff of 1861 and that was seemingly damped down by the period of rapprochement leading up to World War One broke out with renewed vigor after Versailles, at which point only two powerful actors seemed left standing on the global stage: an up-and-coming America and a British Empire acquiring significant new territories under League of Nations' mandate (Hugill 2009a). Wilson saw maritime transportation as crucial to the struggle over international transportation, stating at the beginning of the war "we have grossly erred in the way we have stunted and hindered the development of our merchant marine. . .and now when we need ships, we have not got them" (quoted in Denny 1930: 350). Unfortunately Wilson failed in three areas. First, he lacked understanding of the remarkable technological changes then underway in the field of marine propulsion systems as triple expansion steam engines gave way to steam turbines and, more significantly, diesel engines. In today's world merchant shipping is almost exclusively powered by diesels (Smil 2010). Second, in 1920

Wilson signed into law the protectionist Jones Act: all goods transported by water between American ports had to be in American flagged, constructed, owned, and crewed ships. Such water routes were usually short haul, around the American coast, through the Panama Canal, and throughout the Caribbean, with long runs only to Hawaii and the Philippines, routes where re-fuelling was easy and the long range of diesels scarcely mattered. The Jones Act could not protect American shipping on long hauls in international waters where the diesel came into its own and by encouraging American ship owners to stay with steam technology worked against America's long-term interests. Third, Wilson failed to see the rapidly increasing importance of the aerial component of international transportation. This was the component of international transportation where, by 1947, America would convincingly out-perform Britain (Hugill 2009a).

Wilson further failed to grasp the intimate connection that had emerged between maritime transportation technology and petroleum supply in the first 20 years of the century, and in 1919 he believed America would continue to dominate the arena of petroleum supply just as it had since the late 1860s. In fact, by the late teens informed observers began to believe that the global supply of petroleum was insecure and theorized "oil exhaustion" was on the horizon, especially in an America that, according to Canadian banker E. Mackay Edgar in an influential comment, had "recklessly and in 60 years run through a legacy that, properly conserved, should have lasted her for at least a century and a half" (Edgar 1919). To observers at the time the link between the struggles over maritime transportation and petroleum supply was also clear: "Britain waged an exhausting war with the main object of ruining German shipping for ever. . .Yet now, from the very war which destroyed that competitor, a new one has arisen, twice as formidable as the old, for, in addition to a superiority in tonnage, it enjoys the practical monopoly of a fuel of which Britain has none" (Delaisi 1922: 21). In shipping "as a means of raising steam oil was about twice as economical [as coal], and was four times so if used in internal combustion engines" (Mejcher 2007: 8). By the beginning of World War Two about half the world's merchant ships were oil fueled: "of these ships, half were steam and half diesel powered" (Winkler 2003: 145). Despite the gloomy forecasts of oil exhaustion America continued, then deepened its practical oil monopoly in the 1920s and 1930s, but it signally failed to capitalize on the superiority in merchant ship tonnage it had developed during World War One.

2 Global Oil Supply Issues, 1901–1945

This section encompasses three main time periods: from 1901 to around 1919, from 1919 to around 1931, and from 1931 to 1945. In the first period American production predominated, Russia was an important secondary source, with the Dutch East Indies a distant third and Mexico and Rumania coming into play (Table 1). Most contemporaries believed that the Middle East promised a huge though unproven

Table 1 Global oil production, 1861–1941, in 42 gallon barrels (Fanning 1947: 256–259)

Year	Burma	Dutch East Indies	Persia/Iran	Mesopotamia/Iraq	Mexico	Rumania	Russia	Saudi Arabia	USA	Venezuela	World
1861						47			5,791		5,838
1871						247	452		14,260		15,699
1881						334	9,866		75,784		87,652
1891						1,337	94,721		148,746		249,589
1901		10,977			27	4,597	233,337		190,108		458,740
1911		33,351			34,392	30,433	181,326		603,969		943,455
1921		47,992	45,679		549,544	23,391	79,364		1,293,652	4,104	2,120,383
1931		93,913	121,578	2,000	90,335	136,900	446,685		2,331,729	325,696	3,764,998
1941	20,959	141,918	138,704	31,632	117,488	109,749	647,000	11,800	3,841,721	625,017	6,075,989

bonanza. The Bolshevik Revolution withdrew Russian oil from the world market after 1917.

In the second period, fears of American "oil exhaustion" arose, especially as mass adoption of the automobile accelerated domestic demand. Throughout the 1920s there was an intense policy struggle over access to the Middle East, which precluded much production but seemed to be resolved largely in Britain's favor both by a series of treaties settling the disposition of the former Ottoman possessions in the region and by the Red-Line and Achnacarry Agreements of 1928 between the major oil producers (Venn 1986; Yergin 1991). The main American company to sign these agreements was Standard Oil of New Jersey. In the third period the discovery and development of two major new fields in Texas allowed American domestic production to skyrocket, production finally began in the Middle East, and new technologies exploited by a company that was not a party to the Red-Line Agreement gave an American company access to Saudi Arabia's massive reserves. This amounted by 1945 to a massive American end-run around Britain's previous and largely policy driven dominance of Middle Eastern oil reserves.

2.1 1901–1919

Before 1901 the main use of petroleum in America was as a replacement for whale oil in domestic lighting and as a source of fuel for other domestic uses, such as cooking stoves. But in the late 1800s it became obvious that it had great potential use in transportation: being burned to develop power directly in the rapidly developing automobile and indirectly to generate steam for electric power plants, shipping, and railroad locomotives. Its direct use was pioneered most heavily in Germany, through the work of engineers such as Nicolaus Otto, who developed a working four-cycle, spark-ignition, gas engine, and in the work of Rudolf Diesel, who theorized a very much more efficient engine using heavy oil and compression ignition. Otto was an empiricist. Diesel was very much a product of the first class technological universities that Germany had begun developing in the late 1800s. In Wilhelmine Germany technological education was based on excellent education in mathematics and science, and that put it, at least temporarily, ahead of such states as Britain and America that lagged in this regard, maintaining a much more empirical educational system for their engineers.

By 1901 all the major industrial states had recognized that petroleum was the coming source of power, especially for the steam engines that powered their navies, but only America and Russia had proven and developed substantial domestic supplies, and only America was a maritime power. Through exploitation of its resources in Pennsylvania, California, Oklahoma, and Texas, and through its invocation of the Monroe doctrine to control the increasingly important resources of Mexico and, after World War One, Venezuela, America completely dominated global oil supply for most of the period before World War One broke out. Only the Russian fields around Baku seemed competitive, and only briefly, between 1898 and 1901 did they exceed American output. The Dutch East Indies was the third main supplier by 1911, but much smaller, and oil from the region was shipped by Shell, a British company that got its start shipping exotic seashells for European collectors. By 1911 Mexico and Rumania had also become as important as the Dutch East Indies. The other powers with maritime interests, in particular Britain, but also Germany, began to search for geostrategically secure sources of oil.

Well before World War One broke out Germany and Britain began to focus on the likely oil reserves of the Middle East that neither America nor Russia could control. The area was already geostrategically important to Britain since the main maritime route from Europe to Asia ran through it via the Suez Canal. Before oil was discovered in the region Germany had begun to contest its control with Britain. This contestation was sharpened by the increasing incompetence and corruption of a collapsing Ottoman Empire that, nominally at least, controlled all of the Middle East except Persia, itself an area of contestation between Britain and Russia as part of the "Great Game." Persia was effectively removed from that contestation when Russia and Britain agreed to the partition of the country into a Russian "sphere of influence," a British sphere, and a neutral zone, in 1907.

The collapsing Ottoman Empire was another matter. By 1900 it was pretty obvious that several Ottoman provinces held significant amounts of oil. British military officers reported on this to the Royal Geographical Society as early as 1886 (Ardagh 1886; Stewart 1886). By 1901 a German technical commission would describe Mesopotamia as a "lake of petroleum" which it would be advisable to develop the break the stranglehold on world supplies held by the Americans (Earle 1923: 15). In 1903 Germany negotiated what amounted to a promise of mineral rights in Ottoman territory along the line of the Berlin to Baghdad Railway (BBR) concession, though the BBR never fulfilled its obligations to the government. These rights focused on the three Ottoman provinces that made up Mesopotamia: Mosul, Baghdad, and Basra, modern Iraq. As the Ottoman Empire collapsed and was superseded by the Turkish state, the German interest in BBR evolved into a partnership with Britain in the Turkish Petroleum Company (TPC) of 1912, but the Turks had not approved any concession before war began in 1914 (Earle 1924).

Britain's interests were convoluted. Their focus on the Middle East began with the D'Arcy concession of 1901 from the Shah of Persia, which evolved after the 1907 rapprochement with Russia into the Anglo-Persian Oil Company (APOC) of 1908. In searching for a geostrategically secure source of oil within the Empire Britain first focused on Burma, but Burma's oil was waxy and became almost solid at North Sea winter temperatures. The Burma Oil Company was re-formed in 1908 as APOC, and Britain's interests then ran through the effective merger of APOC with TPC and BBR in 1912 into the secret agreement signed by Sykes, Picot, and Sazanoff of 1916 dividing the former Ottoman provinces among Britain, France, and Russia when the war ended. When Russia left the war this became Sykes-Picot, and it amounted to a French agreement to cede to Britain the Ottoman provinces of Baghdad and Basra and a significant part of Mosul, including all its oil resources, expanded just before Versailles to all of Mosul (Fitzgerald 1994). This resulted in the creation of Mesopotamia under League of Nations mandate after the war was over and, after Britain took the TPC as a war prize, the alignment of APOC with TPC (Fromkin 1989; MacMillan 2001). Burma, Persia, and Mesopotamia produced no oil before World War One and only Persia much before World War Two (Table 1).

The massive strike at Spindletop in Texas in 1901 secured America's position as the dominant global producer, with a threefold increase in production 1901–1911 and a doubling from 1911 to 1921. American oilfields before 1901 were, for the most part, developed and marketed by Standard Oil and its associated companies. But Standard Oil failed to see Spindletop's potential and, in fact, thought of it as a fluke. Petroleum engineering did not yet exist as a discipline and the company's geologists, all trained as hard-rock mining engineers, believed oil was only found trapped in the folds of porous sandstones or in fractured rocks such as limestones. Spindletop's oil was trapped in a salt-dome and it was not until the 1930s that geologists at Louisiana State University and companies such as Standard Oil of California (SOCAL) began to understand such structures. SOCAL had been broken away from Standard Oil proper by anti-trust legislation in 1911 and had gone very much its own way.

2.2 *1919–1931*

Following French complaints that they had been denied Mesopotamian oil, especially after the cession of Mosul, the British agreed at San Remo in 1920 to cut the French in for 25 % of the TPC. An important consequence of San Remo was that Britain and France between them effectively excluded American companies from the Middle East. This pushed Standard Oil of New Jersey into fomenting a storm of American diplomatic protest that eventually forced Britain into giving American signatories to the Red-Line Agreement of 1928 a further 25 % share in TPC. That still left Britain with rights to all of Persia's oil via APOC and half of Mesopotamia's via TPC, which changed its name to the Iraq Petroleum Company (IPC) in 1929. By 1931 it seemed that such policies would allow Britain to control much of the oil of the Middle East (Fromkin 1989). An important innovation in Britain by 1910 was the development by John Cadman at Birmingham University of Petroleum Engineering as an academic discipline, but Cadman was trained as a hard-rock mining engineer and thus had the same shortcomings as the geologists at America's Standard Oil.

One of the main problems in the early period of oil exploitation, basically through 1931, was that there was no proration policy to protect landowners from reservoir depletion by their neighbors. Massive numbers of wells could be drilled in close geographical proximity based simply on possession of title to the land. This first came to a head in America when the output from Spindletop dropped in the late teens from massive over-drilling. In 1917 the Texas Railroad Commission began controlling oil and gas production via its regulation of pipelines as common carriers and in 1919 it expanded well spacing to reduce fire hazards. In 1930 it adopted its crucial proration policy (Texas State Handbook 2013; Herald 1957). These were empirical responses uninformed by any science as yet. Most Texas wells were drilled by wildcatters with no academic training. Initially no major companies were involved. More generous well spacing combined with deeper and more careful drilling allowed Spindletop's output to recover somewhat by the mid-1920s but the growth of the American supply was far from certain until the huge East Texas field, the largest oilfield in American history, was discovered in October 1930. The other major Texas field, in the Permian Basin of West Texas, had been explored in the late 1920s, but where East Texas was easily connected into the pipeline hub that Spindletop had created in Houston the cost of connecting West Texas was initially regarded as prohibitive (Flis and Price 1990).

The history of American oil output from 1901 through 1931, especially the substantial increase in output from 1901 to 1921, is pre-eminently the history of Spindletop. But that period also saw considerable concern over the amount of oil available and the rate at which it was being used up as automobile related demand rose. There was no sense of how large the country's reserves might be, merely figures for production and consumption. Although the Oklahoma field helped compensate for lower output at Spindletop in the 1920s and could be easily connected to the Houston pipeline complex, the real recovery came after 1931. The combination of the massive output of the East Texas and Permian Basin fields, conservationist policies, especially that of proration, better survey technologies, better geological understanding developed in the universities of the region, and the development of petroleum engineering as an academic discipline all combined to end fears of American domestic shortages by the mid-1930s. New wells and new production were added faster than old wells were being closed.

Modern understanding of the size and depletion rate of oil reservoirs began with the innovation of well-head logging by the French company, Schlumberger, in 1926, and the development using the accumulated data from well-head logging of the concept of "peak oil" in 1956 by the Houston based geoscientist, M. King Hubbert (Deffeyes 2006). Neither Schlumberger's data nor Hubbert's model were available to inform those worried about the collapse of American output in the period 1919–1931, hence the peculiar potency in that period of the "oil exhaustion" thesis. Modern scholars with modern knowledge dismiss this history and analyze it as a period in which American oil companies were engaged in a "search for markets" (Bromley 1991; Stivers 1982). Yet at the time the Norwegian political geographer, Anton Mohr, could describe the likely coming global conflict over oil resources as *The Oil War* (1926) and noted the potential of the Middle East to rise to dominance over the global oil supply. Mohr's work is one of a substantial series of accounts from the 1920s and 1930s that follow the "oil exhaustion" thesis. These are exemplified by Ludwell Denny's *We Fight for Oil* (1928), which constructed the fight as between Britain and America. They also include a substantial number of French and German works, which took an explicitly anti-British stance (Delaisi 1922; Fischer 1926; Hanighen 1934; de la Tramerye 1924; Marcosson 1924; Zischka 1933, 1934, 1939). The first of these accounts was Delaisi's book, published initially in Paris in 1920 as *La Pétrole*. There are two "odd men out" in this list: Anton Zischka and Frank Hanighen published their books much later than the rest, after the Depression had bitten. Zischka was a Viennese born journalist who later took up with the Nazis and whose writing focused heavily on the role of energy, commodities, and technology in world power, usually from a strongly German perspective. His first book, *La Guerre Secrete pour le Pétrole* (1933) was republished in German in 1934 as *Der Kampf um die Weltmacht Öl* and a second book on oil followed in 1939. Zischka's work is intertwined with that of Hanighen, and hard to disentangle. Hanighen was an American muck-raking journalist and his *The Secret War* was reissued in England under the expanded title of *The Oil War, The Secret War* with Zischka listed as co-author.

2.3 1931–1945

After 1930 American domestic production surged and the struggle over access to Middle Eastern oil came down to a strong British rearguard action fought primarily on the grounds of policy and a remarkable American end-run around British interests on the basis of technology developed in American fields, though actual output in the Middle East was initially slight (Table 1). APOC started to produce oil in the 1930s, but the British were concerned that such oil had to travel through the Suez Canal, which was geostrategically vulnerable. At San Remo the British had ceded the right to the French to build a pipeline to get Mesopotamian oil to the Mediterranean, which meant the British could still be denied access to Mesopotamian oil. The eventual solution, which involved a long row between Britain and France that culminated in intervention by Standard Oil, was to bifurcate the pipeline at Haditha, with one branch to Tripoli in French controlled Syria and the other through British controlled Transjordan to Haifa in British controlled Palestine (Cadman 1934; Fitzgerald 1993).

SOCAL, later the Arab American Oil Company (Aramco), now Chevron, recognized that oil was present in Bahrain in salt-domes of the sort it was familiar with in California and which one of its partners, the Texas Company, later Texaco, had come to understand empirically at Spindletop. SOCAL cleverly circumvented the Red-Line and Achnacarry Agreements, recognized that Saudi Arabia was a massive potential site for salt-dome deposits, and was able to persuade King Ibn Saud, who was no friend of Britain, to lease Saudi Arabia's potential oil fields to them. Like Standard Oil of New Jersey APOC and IPC had no understanding of salt-dome geology, believing oil was exclusively found trapped in porous rock formations. Shell, one of the main constituent companies of IPC, refused to participate in any lease negotiations in Saudi Arabia since it "did not regard the prospects as favourable from either a geological or a political point of view" (Fitzgerald 1991: 458). SOCAL's final success was in policy. They were able to get President Franklin D. Roosevelt to visit Saudi Arabia on his return from the Yalta Conference in 1945, cementing a close relationship between the Saudi monarchy, America, and Aramco that guaranteed America access to the Middle East's most prolific oilfields (Yergin 1991).

3 Ideas and Reality: Oil Fuel, Naval Power, and the Merchant Marine

Tied up with the "oil exhaustion" argument of the 1920s was the realization that oil-fuelling of ships had become central to the development and retention of global power at both a military and an economic level. By the 1920s three main strands in the innovation of oil-fuelling were apparent. The first was in naval vessels, where the advantages of oil fuel were quickly seen as out-weighing the costs. The second was in oil-fuelling of merchant steamships. The third was in the adoption of a radically new form of powerplant, the diesel engine, which seemed to offer very substantial advantages to merchant ship owners, in particular in range. There was early naval interest in using diesels to extend the cruising range of ships, but the attempt initiated in 1911 in Germany to build a 12,000 hp battleship engine was too far ahead of the metallurgy of the period and resulted in failure (Landahn 1925).

3.1 Innovations in Oil-Fuelled Shipping

The innovation of oil fuels for ships was clearly revolutionary, but the technologically oriented response of the engineers was very different from the response of those concerned with geopolitics. The engineers were very clear on the technological advantages of oil fuel, but also on its ability to destabilize the shipping industry:

"The old stable situation, where the problems were mainly non-engineering in character, and competition depended on crew wages and size within a given size of ship, on shipping management, and possibly on government cooperation, has for the moment entirely disappeared, and while some of the old problems remain, new ones have come forward and are to-day dominant. The new ones are the result of (a) the use of oil as fuel in competition with coal; (b) the steam turbine; and (c) the internal combustion engine" (Lucke 1921: 14).

The policy responses focused more on the issue of the supply of oil. As demand rose there were fears that supply might falter, fears expressed through the "oil exhaustion" thesis. Delaisi (1922) and de la Tramerye (1924) both argued that the British had robbed France of possible oil rights in Mesopotamia she should have had under Sykes-Picot-Sazanoff but was maneuvered out of at Versailles and, later, at San Remo. Both authors linked hegemony explicitly to a nation's ability to control the oil supply and to the strength of its merchant marine.

> The nation which controls this precious fuel will soon see the wealth of the rest of the world flowing towards it. The ships of other nations will soon be unable to sail without recourse to

Table 2 Steam versus diesel power for merchant shipping (de la Tramerye 1924: 15)

Powerplant	Diesel	Steam
Horsepower	2,100	2,100
Weight of engine and accessories	1,000 t	3,400 t
Space required	5,300 m^3	10,000 m^3
Daily fuel consumption	100 t heavy oil	360 t coal
Consumption for a voyage of 15 days	1,500 t	5,400 t
Bunker space for a voyage of 15 days	1,700 m^3	7,000 m^3
Total space required for engine and fuel	7,000 m^3	17,000 m^3

> its stores of oil. . .[T]he nation which obtains the world's carrying trade takes toll from all those whose goods its carries, and so has abundant capital. . .At one stroke the controlling centre of the world's credit is displaced. This is what happened already in the eighteenth century when, with the development of British shipping, it passed from Amsterdam to London (de la Tramerye 1924: 10–11, based on Delaisi 1922: 2).

De la Tramerye's argument focused on the technological advantages of the diesel engined merchant ship, illustrated in Table 2. The horsepower rating given in the original as 21,000 is incorrect. No diesel that powerful had run by that time and steamships of that power would not be cargo liners. The two engines used in the first diesel engine motor ship, *Seelandia*, generated just over 2,000 hp in total, so one should assume de la Tramerye meant 2,100 hp. His other figures are consistent with other sources.

For oil firing of steam ships, there would be a reduction "by 70–80 %. of the stokers since a single man can look after several boilers. [For diesels the labor savings would be even greater]. The fuelling of the ship is effected cleanly and quietly in a few hours. Hundreds of tons of oil can be pumped into the cisterns in a negligible time, and that even at sea and in heavy weather. . .In the latest Cunard and White Star liners the economy of space thus realized has been as much as 33 %" (de la Tramerye 1924: 16–17).

The popular press had begun to point out the technical advantages of diesel engines much earlier than the "oil exhaustion" advocates. The first successful ocean-going motor ship was, appropriately, *Vulcanus*, an oil tanker built in 1910 for the Anglo-Saxon Petroleum Company. Such ships would use far less fuel for the same amount of work, have more usable cargo space, need less labor, and be much more rapidly refuelled (Carter 1911: 592–594). The first really notable motor ship was, however, *Seelandia,* which was built for the Far Eastern run where the range made possible between refueling mattered. *Seelandia* made her maiden voyage in 1912, and marked "the most striking advance at this time" (Smith 1938: 330).

3.2 *Oil and Naval Power*

As early as 1908, in *Oil Fuel and the Empire*, J. D. Henry, British author and founder of the journal *Petroleum World*, noted that "recognising that in swift ships oil saves time and labour, if it does not save money, the Admiralty has decided to use oil fuel in every type of warship, and it is this fact which gives the imperial aspects of the subject such tremendous importance" (Henry 1908: 16). In the run-up to World War One navalists such as Winston Churchill began to see oil-firing of steam turbine engine warships as a highly desirable technology that would substantially increase the fighting capacity of naval vessels by allowing higher speeds and radically easier refueling, plus lowered weights through lighter fuel and a substantial reduction in stokehold crew size, thus more armor plate and bigger guns. In 1913 Churchill oversaw the acquisition of APOC by the British state and it was such policies that pushed Britain into the acquisition of Mesopotamia as a source of oil that would be securely within the Empire (Yergin 1991).

3.3 Oil Firing of Merchant Ships

Before 1914 operators of merchant vessels began to follow a similar line regarding oil-firing for the new steam turbine powered ships coming to dominate the Atlantic shuttle by virtue of their speed. Such ships, like their warship cousins, used direct-drive turbines, which required a great deal of fuel. Smaller merchant ships, however, stuck with the more fuel efficient triple-expansion reciprocating steam engine and coal fuel until around the outbreak of World War One. One of the aftermaths of the war was, however, that constant strikes at the mines and the docks meant that from 1918 to 1920 coal prices rose faster than oil: "between 1919 and 1922 the tonnage of steamers fitted for oil-firing rose from 5.3 million to 14.5 million [tons]" (Henning and Trace 1975: 382).

The year 1914 was something of a watershed. "The geared turbine, the turbo-electric drive, and the Diesel engine made their appearance almost simultaneously around 1913 and by 1917 the triple-expansion and quadruple-expansion steam engine began to feel the competition of the Diesel engine and the geared turbine" (Chapman 1942: 110). The diesel-engined motorship burned only oil, the steam turbine ship mostly oil. American shipyards turned out a huge tonnage of basic merchant ships during World War One, the majority using reciprocating steam engines but increasingly burning oil. By 1920 56 % of the American merchant fleet burned coal and 44 % oil (USDC 1960: 444). When the war was over there was a steady shift to geared turbines and oil fuel. American shipbuilders, however, generally avoided diesels.

3.4 Diesel Engined Motor Ships

Before World War One steam engines were clearly important to the adoption of oil as a maritime fuel but in that same period the utility of the diesel engine also became apparent. Even in their early form diesels were ideally suited to powering merchant ships, particularly where long range was needed. *Seelandia's* sister ship, *Fionia*, re-named *Christian X*, was reported on in the *New York Times* upon its arrival in New York on its maiden voyage: "the steamship requires recoaling four times to circle the globe...A fuel consumption of between nine and ten tons for each day...makes it possible for the *Christian X* ...to make a continuous voyage around the world without replenishing her tanks" (Bogert 1912: 4). Motor ships did not offer high speed but they did offer substantial savings in fuel and stoke-hold labor costs as well as greater cargo capacity for a given tonnage (Table 2).

World War One proved to be a huge technology forcer in the area of diesel engines. Although the French navy was the first to use diesels in submarines the Germans quickly followed. During the war half Germany's "submarines ran on MAN's four-stroke diesels, mostly 1,200-hp units" (Smil 2010: 71). At the war's end the availability of a large number of such engines as war surplus helped speed the transition to an increasingly diesel powered merchant marine (Smith 1938, 337). By 1939 24.4 % of the world's merchant ships were powered by diesels. It was, however, the Scandinavian countries and Holland that most rapidly adopted diesels. By 1939 Norway led, with 62.2 % of its merchant fleet powered by diesels: Denmark was at 52.2, Sweden at 46.6, and Holland at 45.5 %. By contrast Britain was at 25.6 and Germany only at 26.2 %: even Japan surpassed those figures with 27.2 % of its merchant ships diesel powered (Henning and Trace 1975: 354).

America was much slower to adopt diesels than oil fuelling, in part because of the protectionism of the Jones Act. In 1940 69 % of American merchant ships were oil-burning steamships whereas less than 11 % used diesels (USDC 1960: 444). Adolphus Busch had licensed Diesel's engine as early as 1897 and in 1898 installed one to power his brewery in St Louis, MO (Krebs 1953). The Swiss company, Sulzer, licensed Diesel's engine in 1898 and joined forces with Busch in the American market. Initially Busch-Sulzer focused on the demand for diesels by local electricity generating stations. When America entered World War One the Navy had them develop submarine engines and after the war they moved into marine diesels. Even so, in the mid-1920s, out of fifteen manufacturers of diesel engines worldwide, only two, Bethlehem and Worthington, were American companies building American designed diesels, and neither was successful (Hardy 1926: 308).

A second reason for the slower American adoption of diesels was that the American economy of the early 1920s still lacked some of the skills needed to manufacture and maintain them. Diesels operated at much higher internal pressures and temperatures than even the most sophisticated steam engines, and thus required very high levels of manufacturing skill and excellent metallurgy that increased first cost. Other than Busch-Sulzer the first major American company to enter the market for diesels was General Motors, which did so in the early 1930s via its acquisition of the Electro-Motive Corporation (EMC) in 1930, re-named Electro-Motive Diesel (EMD) in 1941. By the late 1930s EMC had an efficient diesel engine at work in railroad locomotives, and in World War Two EMD turned its attention to marine diesels (Schramm 2010).

4 Conclusion

Commentators in Britain and America by the early 1920s were clear on the relationship between the developing struggles foreseen by Wilson in the arenas of marine transportation and petroleum. They argued that whoever controlled the right mix of marine transportation technology, in particular that of the merchant marine, and petroleum supply would control the emerging post-war international order. However, the ideas that were put around in the early 1920s with regard to the development of petroleum and shipping did not fully mesh with the realities that were apparent by1945. Four major technological and economic forces intervened to prevent clear American victory in this struggle, although in general America did better than Britain.

First, the "oil exhaustion" thesis of the early 1920s faded rapidly as new American oilfields came on line in the 1930s and oil production finally began in Persia and Iraq. Britain's policy advantages in the Middle East evaporated for three reasons. American companies such as SOCAL and The Texas Company developed practical experience with salt-dome oil deposits, while theoretical understanding of these was developed by geologists at Louisiana State University and at the University of Texas' Bureau of Economic Geology. British companies involved in the Middle East failed to understand the importance of salt-dome deposits and effectively denied their existence. Finally, sound diplomacy brought Saudi Arabia's massive oil reserves under American control by 1945. Despite fears in the 1920s that America might lose its grip on the global oil supply, from 1930 to 1945 that grip substantially tightened.

Second, although America remained the dominant global supplier of oil, America failed to capitalize on the remarkable increase in its ship building capacity that had been developed in World War One, failed to adopt the diesel engine for its merchant ships when the war was over, and never challenged British dominance of the world's merchant marine in the period between World Wars One and Two, however much that dominance slipped. In 1894 Britain built 79 % of the 1,323,538 t of merchant ships launched that year, America only 5 % and Germany just under 9. In 1919 7,144,549 t were launched, with America accounting for 57 % and Britain just under 23 %. Of the 2,117,924 t launched in 1936 Britain built just over 40 %, America just over 5 and Germany 18 %. Wilson's vision of an American merchant marine fell apart for three reasons. First, the ships built in a hurry during World War One were outdated technologically by the end of that war. Too many ships were coal-fired, triple expansion engine ships, and by 1919 oil fired, geared turbine engine ships and diesel engine motor ships were rapidly taking their place. Second, the fleet of passenger ships that Wilson hoped would compete for traffic on the lucrative Atlantic routes dominated by Britain were a mixed bag of largely German ships interned in American ports in 1914 and seized as war prizes when America entered the war in 1917. Such ships were stylistically and technologically outdated by 1920. Finally, few well-off passengers were willing to take passage on an American ship after Prohibition made American ships legally "dry" in 1919.

Third, the success of the motor ship encouraged by the 1930s significant improvements in the fuel efficiency of the oil-fired steamship that brought its specific fuel consumption much nearer to that of the diesel engine motor ship. The combination of superheated steam and higher boiler pressures allowed substantial increases in fuel efficiency in steam turbine engine ships. Whereas in 1919 the best such ships consumed one pound of fuel per horsepower hour and diesels 0.42 lbs., by 1940 those figures were 0.60 and 0.38 lbs. respectively (Chapman 1942: 249). The one strength of American merchant shipping in the 1919–1939 period was the refinement of the geared turbine, oil-fuelled steamship. Unfortunately this was ultimately a technological dead end, and an analysis such as Chapman's based only on engineering criteria was not realistic in the light of the real world of merchant shipping.

> The motorship's great advantage was in cargo carrying capacity. . .this additional cargo capacity accounts for the greater profit earned by motorships in the 1920s and early 1930s. Note also that whereas the proportion of the motorship's capacity required to be filled in order to break even varied between 67 and 91 %, there were years in which it was impossible, given freight rates ruling, for either the coal- or oil-fired steamships to break even (Henning and Trace 1975: 375).

Fourth, the price of refined diesel fuel moved against the price of the unrefined bunker C fuel that was burned in oil-fired steamships. By 1941, and corrected "for the differences in specific gravities of the two fuels. . .the ratio of fuel costs per ton. . . = 1.64" (Chapman 1942: 250). Given that the ratio of fuel use was 0.60 to 0.38, or 1.58, the oil fuelled turbine engine steam ship held a cost advantage. This, the Jones Act, and the short-haul geography of American shipping between American ports encouraged American shipping lines to stay with the geared turbine when they should not have. A diesel engine could run on bunker C fuel as well as the more refined and expensive diesel fuel as long as bunker C was pre-heated so it atomized properly. It was much harder to start and needed more maintenance, and American engineering texts of the 1940s were skeptical of its utility (Chapman 1942: 251).

The simple fact is that America maintained a huge advantage over Britain between 1919 and 1947 in the struggle over petroleum, and this despite the initial policy success of Britain in the Middle East. Where America failed was that it signally failed to capitalize on the advantages in international merchant shipping that it should have enjoyed as it emerged from World War One. In both cases technology played a major part. American success in Saudi Arabia was based on the development at SOCAL of the technology to exploit salt-dome deposits of oil. The failure in merchant shipping was the American failure to see the central importance of Diesel's radical new engine.

References

Ardagh, C. (1886). The Red Sea petroleum deposits. *Proceedings of the Royal Geographical Society, 8*(8), 502–507.

Bogert, J. J. (1912, September 26). The new oil engines. *The New York Times, 4*.
Bromley, S. (1991). *American hegemony and world oil: The industry, the state system and the world economy*. State College, PA: Pennsylvania State University Press.
Cadman, J. (1934). Middle east geography in relation to petroleum. *The Geographical Journal, 84* (3), 201–212.
Carter, C. F. (1911). The advent of the motor ship. *American Review of Reviews, 44*(5), 591–594.
Chapman, L. B. (1942). *The marine power plant*. New York: McGraw Hill.
de la Tramerye, P. L.'. E. (1924). *The world-struggle for oil*. New York: Knopf.
Deffeyes, K. S. (2006). *Beyond oil: The view from Hubbert's peak*. New York: Hill & Wang.
Delaisi, F. (1922). *Oil: Its influence on politics*. London: Allen & Unwin.
Denny, R. (1928). *We fight for oil*. New York: Knopf.
Denny, R. (1930). *America conquers Britain. A record of economic war*. New York: Knopf.
Earle, E. M. (1923). *Turkey, the great powers, and the Bagdad railway: A study in imperialism*. New York: Macmillan.
Earle, E. M. (1924). The Turkish Petroleum Company—A Study in oleaginous diplomacy. *Political Science Quarterly, 39*(2), 265–279.
Edgar, E. M. (1919, September 21). Sees oil control passing to Britain. *The New York Times*.
Fanning, L. M. (1947). *American oil operations abroad*. New York: McGraw-Hill.
Fischer, L. (1926). *Oil imperialism: The international struggle for petroleum*. New York: International.
Fitzgerald, E. P. (1991). The Iraq Petroleum Company, standard oil of California, and the contest for Eastern Arabia, 1930-1933. *International History Review, 13*(3), 441–465.
Fitzgerald, E. P. (1993). Business diplomacy: Walter Teagle, Jersey Standard, and the Anglo-French pipeline conflict in the Middle East, 1930-1931. *Business History Review, 67*(2), 207–245.
Fitzgerald, E. P. (1994). France's Middle East ambitions, the Sykes-Picot negotiations, and the oil-fields of Mosul, 1915-1918. *Journal of Modern History, 66*(4), 697–725.
Flis, J. E., & Price, R. C. (Eds.). (1990). *Permian basin oil and gas fields: Innovative ideas in exploration and development* (West Texas Geological Society Symposium, publication No. 90 -87). Midland: West Texas Geological Society.
Fox, E. W. (1971). *History in geographic perspective: The other France*. New York: Norton.
Fromkin, D. (1989). *A peace to end all peace. The fall of the Ottoman Empire and the creation of the Modern Middle East*. New York: Avon.
Goldstein, J. S. (1988). *Long cycles. Prosperity and war in the modern age*. New Haven, CT: Yale University Press.
Hanighen, F. C. (1934). *The secret war*. New York: John Day.
Hardy, A. C. (1926). *Motorships. An investigation into the character of mercantile vessels powered by internal combustion engines*. New York: Van Nostrand.
Henning, G. R., & Trace, K. (1975). Britain and the motorship: A case of the delayed adoption of new technology. *Journal of Economic History, 35*(2), 353–385.
Henry, J. D. (1908). *Oil fuel and the empire*. London: Bradbury, Agnew & Co.
Herald, F. A. (1957). *Occurrence of oil and gas in West Texas*. Austin, TX: University of Texas Bureau of Economic Geology.
Hugill, P. J. (1993). *World trade since 1431: Geography, technology, and capitalism*. Baltimore: Johns Hopkins University Press.
Hugill, P. J. (2003). Technology, its innovation and diffusion as the motor of capitalism. *Comparative Technology Transfer and Society, 1*(1), 89–113.
Hugill, P. J. (2005). Trading states, territorial states, and technology: Mackinder's unexplored contribution to the discourse on state types. In B. W. Blouet (Ed.), *Global geostrategy: Mackinder and the defence of the West* (pp. 108–125). London: Frank Cass.
Hugill, P. J. (2009a). The American challenge to British Hegemony, 1861-1947. *Geographical Review, 99*(2), 403–425.

Hugill, P. J. (2009b). Transitions in hegemony: A theory based on state type and technology. In W. Thompson (Ed.), *Systemic transitions: Past, present, and future* (pp. 33–54). London: Palgrave Macmillan.
Hugill, P. J., & Bachmann, V. (2005). The route to the techno-industrial world-economy and the transfer of German organic chemistry to America before, during, and immediately after World War One. *Comparative Technology Transfer and Society, 3*(2), 159–186.
Hugill, P. J., & Dickson, D. Bruce (Eds.). (1988). *The transfer and transformation of ideas and material culture*. College Station, TX: Texas A & M University Press.
Krebs, R. (1953). *Making friends is our business, 100 years of Anheuser-Busch* (written in collaboration with P. J. Orthwein). St. Louis: Anheuser-Busch, Inc.
Landahn, D. (1925). Krupp's big battleship diesel engine. *Journal of the American Society for Naval Engineers, 37*(1), 168–173.
Lucke, C. E. (1921). Internal combustion engines in marine service, parts I & II. *Journal of the Franklin Institute, 192(1)*, 11–45 and *192(2)*, 203–232
Mackinder, H. J. (1904). The geographical pivot of history. *The Geographical Journal, 23*, 421–444.
Mackinder, H. J. (1919). *Democratic ideals and reality: A study in the politics of reconstruction*. New York: Henry Holt.
MacMillan, M. (2001). *Paris 1919. Six months that changed the World*. New York: Random House.
Marcosson, I. F. (1924). *The Black Golconda. The romance of petroleum*. New York: Harper.
Mejcher, H. (2007). *The struggle for a New Middle East in the 20th century. studies in imperial design and national politics*. Berlin: Lit.
Modelski, G. (1978). The long cycle of global politics and the nation-state. *Comparative Studies in Society and History, 20*, 214–235.
Modelski, G., & Thompson, W. R. (1988). *Seapower in global politics, 1494–1993*. Seattle, WA: University of Washington Press.
Mohr, A. (1926). *The oil war*. New York: Harcourt, Brace.
Schramm, J. W. (2010). *Out of steam: Dieselization and American railroads, 1920–1960*. Bethlehem, PA: Lehigh University Press.
Smil, V. (2010). *Prime movers of globalization: The history and impact of diesel engines and gas turbines*. Cambridge, MA: MIT Press.
Smith, E. C. (1938). *A short history of marine engineering*. Cambridge: Cambridge University Press.
Stewart, C. C. E. (1886). The Herat Valley and the Persian Border, from the Hari-rud to Sistan. *Proceedings of the Royal Geographical Society, 8*(3), 137–156.
Stivers, W. (1982). *Supremacy and oil: Iraq, Turkey, and the Anglo-American World Order, 1918-1930*. Ithaca, NY: Cornell University Press.
Texas Railroad Commission. http://www.rrc.state.tx.us/about/history/chronological/chronhistory01.php. Accessed January 16, 2013.
Texas State Handbook. (2013). Entry on Permain Basin. http://www.tshaonline.org/handbook/online/articles/ryp02. Accessed January 16, 2013.
Tilly, C. (1990). *Coercion, capital, and European States, AD 990–1990*. Cambridge, MA: Basil Blackwell.
Tuma, E. H. (1987). Technology transfer and economic development: Lessons in history. *Journal of Developing Areas, 21*, 403–428.
U.S. Department of Commerce (USDC). (1960). *Historical statistics of the United States: Colonial times to 1957*. Washington, DC: U.S. Government Printing Office.
Venn, F. (1986). *Oil diplomacy in the twentieth century*. New York: St. Martin's.
Wallerstein, I. (1974). *The modern world-system* (Capitalist agriculture and the origins of the European World-Economy in the sixteenth century, Vol. 1). New York: Academic.

Winkler, M. F. (2003). Introduction to marine petroleum fuels. In G. E. Totten, S. R. Vestbrook, & R. H. Shah (Eds.), *Fuels and lubricants handbook: Technology, properties, performance, and testing* (pp. 145–168). West Conshohocken, PA: ASTM.

Yergin, D. (1991). *The prize: The epic quest for oil, money, and power*. New York: Simon & Schuster.

Zischka, A. (1933). *La Guerre secrete pour le Pétrole*. Paris: Payot.

Zischka, A. (1934). *Der Kampf um die Weltmacht Öl*. Leipzig: Wilhelm Goldmann.

Zischka, A. (1939). *Ölkrieg: Wandlung der Weltmacht*. Leipzig: Wilhelm Goldmann.

The Globalization of Intellectual Property Rights: Much Ado About Nothing?

Daniele Archibugi and Andrea Filippetti

Abstract It is hardly surprising that companies try to exploit their intellectual property rights (IPRs) globally. This has generated heated debates about the advantages and disadvantages associated with the global regime of IPRs. The aim of this chapter is to put the debate in the right context of what IP can actually do, and what they cannot do, in order to reward inventors and innovators and to prevent imitators. The generation, transmission and diffusion of knowledge are complex phenomena and both supporters and detractors of IPRs often tend to exaggerate the effects that IPRs have in the economy and society. We present two ideal-type models for and against IPRs and discuss their limitations. In our view, both models overemphasize the role of IPRs in the world economy.

Keywords Innovation • Intellectual Property Rights • Agreement on Trade Related Aspects of Intellectual Property Rights (TRIPS) • Technological appropriation • Patents • Copyright

D. Archibugi (✉)
Italian National Research Council, IRPPS, Rome, Italy

Birkbeck College, University of London, London, UK
e-mail: daniele.archibugi@cnr.it

A. Filippetti
London School of Economics and Political Science, Department of Geography and Environment, London, UK

Italian National Research Council, Institute of Regional Studies, Rome, Italy
e-mail: andrea.filippetti@cnr.it

M. Mayer et al. (eds.), *The Global Politics of Science and Technology - Vol. 1*, Global Power Shift, DOI 10.1007/978-3-642-55007-2_8,

1 Introduction

The international economic landscape is periodically rattled by controversies concerning intellectual property rights. To secure market shares, companies introduce new products and processes and this often leads to controversies for the real or supposed violations of patents, copyrights and trademarks. Very often, the companies involved in these battles are based in different countries. The so-called smart phones war (Filippetti 2012; Graham and Vishnubhakat 2013) has recently captured the attention of public opinion because the product at stake is in the hands of top managers and leading politicians as well as in the hands of school kids and housewives. Some companies have acted against governments that supposedly do not adequately protect their intellectual property, the most dramatic example represented by the coalition of large multinationals in the pharmaceutical industry, the so-called 'big pharma', against the South African governments for infringement of their IPRs on anti-HIV drugs (Fisher and Rigamonti 2005). In other occasions, the controversies do not directly involve competing companies, but rather the national governments themselves. This is, for example, the case with current controversies debated at the World Trade Organization (WTO) under the Trade Related Aspects of Intellectual Property Rights (TRIPS) agreement.

These controversies reflect the fact that in the current global economy, knowledge and intangibles have become increasingly important both as production factors and as consumption goods. While most of the controversies of the past were associated to trade in physical merchandise, today there is an increase in litigation associated to intellectual property rights (IPRs).[1] Companies should bear greater investments in Research and Development (R&D) and design in order to generate and bring to the market new products and services, but very often these core competitive assets can be imitated and replicated at costs that are substantially lower than the original cost. This is at the root of traditional tensions between innovators and imitators, a tension that for a long period mainly occurred within nations and that now has taken on a global dimension. The growth in international trade and foreign direct investment, associated to the rise of emerging markets, has thus increased the propensity of companies to search for profits related to their innovations at the global level.

It is often pointed out, and rightly so, that holders of IPRs are mostly a restricted number of gigantic multinational corporations.[2] The same companies are also responsible for a corresponding amount of expenditure in R&D, industrial design and investment in intangibles (Archibugi and Iammarino 2002; Archibugi and Pietrobelli 2003; Filippetti and Peyrache 2011). These companies have managed

[1] See among others: Archibugi and Filippetti (2010); Drahos and Mayne (2002); Henry and Stiglitz (2010); Maskus (2000).

[2] For an analysis of the patent offices in the new global context, see Drahos (2010). The concentration of patents in a few hundred multinational corporations has been analysed by Patel and Pavitt (1997) and Cantwell and Iammarino (2003).

to assure that basic IPRs are protected in their home country as well as with their most important economic partners, namely in the areas where they concentrate the bulk of their sales. So far, however, they have not managed to get an equally effective protection in emerging and developing countries. These markets to not represent at the moment part of their core business, but are expected to become steadily more important. This has led Western companies to demand greater international protection for their IPRs against unfair foreign imitators (Ryan 1998; Sell 2003).

The requests of multinational corporations to get stronger IRPs has generated passionate protests in developing countries since most inventions with commercial potential come from companies based in North America, Europe, Japan. It has been claimed that IPRs constitute a net transfer of resources from developing to developed countries, which retards growth in developing countries along with pushing up prices, and which, in their most insidious form, allow multinational corporations to claim monopoly rights over ancient forms of knowledge that were not actually developed by those corporations (Shiva 2001).

From the perspective of international political economy, this is an excellent case to explore the interplay of competition across companies *and* across states—an issue that Susan Strange and her followers address with great energy (see Strange 1988; Stopford and Strange 1991). This is because, on the one hand, national courts are often suspected to favour the interests of their own country and of home-based firms. Yet, on the other hand, agents in the WTO are governments, not companies, implying that the complaints of the business sector and of civil society will be discussed and taken into account if a government is willing to promote them.

This leads to the key question related to IPRs: what is the best architecture for the emerging global IPR regime? Can IPR policy alone really regulate the generation and diffusion of knowledge and innovation? The aim of this chapter is to reframe the debate regarding the globalization of IPRs so as to place it into the right context of what IP can actually do and what it cannot do to reward inventors and innovators, while also preventing imitators. As students of technological change, we have learnt that the generation, transmission and diffusion of knowledge are a complex phenomena and that both supporters and detractors of IPRs often tend to exaggerate the effects that they have in economy and society. The chapter is organized as follows. The next section outlines the general debate surrounding IPRs. We follow by describing the role of the US in promoting IPRs and how this has gathered ideational strength. In section three we discuss the attempt to generate a global IP regime through TRIPS. We then present two ideal typical models for and against IPRs and discuss their limitations. Finally, the last section discusses their policy implications.

2 What Are IPRs

Intellectual Property Rights are legal instruments designed to provide the holder of this right the exclusive use over certain creative activities. They include patents, copyrights, trade-marks, utility models, geographic indications and others. Each of these rights has a specific legislation and therefore covers a different domain. Each of these IPRs has a distinct economic and social impact and can be more or less relevant according to the nature of the industry, of the technology and of the geographical area. For example, patents are crucial for a few manufacturing industries, including pharmaceuticals, chemicals and electronics; copyright is the core instrument to protect audiovisual and literary products; trade-marks are relevant in the industries where brands represent an important competitive advantage; geographic indications are important for food and beverages (such as wine appellation). Often, companies use a combination of IPR instruments to increase the protection of their activities. For example, companies defend their software through a mixture of copyright and patents, while drugs combine patents with trade-marks.

The legislation on IPRs is *national* in scope. It is the responsibility of national governments to make decisions about the length and scope of each IPR. National legislation is also responsible for establishing the penalties associated with infringements. Finally, the enforcement of IPRs is also carried out by national authorities. As a result, police can be tougher or softer against violators while the courts which should take the final decision in controversies among economic agents are national. When the globalization of IPRs is discussed, it should be kept in mind that we are dealing with legal and policing practices that are predominantly national in scope.

But IPRs are also one of the areas where for more than a century governments have tried to reach a certain harmonization between states. The Paris Convention for the Protection of Industrial Property dates to 1883 and the Berne Convention for the Protection of Literary and Artistic Works to 1886. There were therefore important attempts to reach a uniform legislation, although this co-existed with the desire of each nation to get the IPR system congenial to its own economic and social advantage. Odagiri et al. (2011) shows how this has worked in practice in several countries.

As can be expected, countries that had more inventions and products to protect were keen to push the international system towards stronger protection. These countries saw IPRs as a method to increase their revenues and to strengthen the bargaining position of their national companies in host countries. Countries that relied more on knowledge and products generated elsewhere were more keen to have a permissive IRP system. Catching up countries interpreted a rigid IPR system as a further burden on their economy and their development strategy. The rules of the system co-evolved differently in each country. As shown by the country case studies reported in Odagiri et al. (2011), national institutions have evolved differently, and often they have tried to protect national industries through a variety of legal, institutional and customary devices.

Through complicated and long diplomatic negotiations—including the diffusion of experiences and attempts of companies and other economic agents to reach similar standards across countries—IPRs national legislation and practices have converged. The number of countries that joined the international conventions on IPRs has steadily increased and the legal norms are certainly less different across countries than they were 100, 50 or even 25 years ago. If we look at the expansion of the IPR system, it is clear that in the beginning it involved countries at a comparable level of economic development. Progressively, the IPR regime has incorporated countries at the periphery and, consequently, the members of the world IPRs system have become more heterogeneous.

One of the core principles of existing international conventions was to guarantee that public institutions in each country would not discriminate against foreigners. It was acceptable that some countries had longer or shorter validity for patents or copyright, stronger or weaker protection, but the various conventions tried to establish the principle that home and foreign inventors, authors and companies should be treated equally. The formal principle, however, is not easily enforced in practice. It is up to the police to identify if counterfeited CDs and DVDs are sold in an open market, it is a national patent examiner who should decide if the application for an invention is genuinely original, and it is the remit of national courts to settle business controversies. In spite of all the principles stated in the various conventions, treaties and national legislation related to non-discrimination, there have always been allegations that national institutions tend to favor national interests and that the police, patents and trademarks offices and courts are inherently biased. The fact that recently Apple took Samsung in a Californian court, while Samsung responded taking Apple in a Korean court seem to be consistent with this view (Filippetti 2012).

It can be questioned if it is advantageous for developing and catching-up countries to implement the IPR system as designed by developed countries (see, for example, the analysis of Drahos 2010). It has also been argued that weak economies have also bad negotiators or have underestimated the costs associated to IPRs agreements (May 2002; Sell 2003; Heller 2008). Others, on the contrary, have argued that countries with Western-like IPR systems have facilitated industrial development (Branstetter et al. 2010), induced foreign investment by multinational corporations (Dinopoulos and Segerstrom 2010) and induced technology transfers (Mansfield 1994, 1995). In the next sections, we will discuss how the situation has changed over the last two decades and if such a change has obtained its desired outcome.

3 The Silent IPR Revolution in the United States

The United States has always been the country where violations of IPRs have been more vocally denounced. For the whole twentieth century the most spectacular controversies about intellectual property have occurred in the American theatre and the twenty-first century looks largely the same. It is difficult to explain why the United States economy, society and legal system are so keen to consider somebody a villain when they use another's intellectual property without authorization. In Europe and Japan, there are daily controversies on IPRs infringements, but intellectual property is not surrounded by the aura that it has in the United States. The unauthorized download of a song, the cracking of software, the replication of a patented product are much less disapproved in Europe than in the United States. Even more, in developing countries hackers and imitators are often admired as sorts of modern Robin Hoods that takes knowledge from those who have it and provides it to those that do not have it (for a contemporary praise of the hacker, see Himanen 2011).

The reason why intellectual property in the United States is much more appreciated than in other countries has both material and ideological foundations. The United States is the largest R&D spender and patent generator in the world, its companies have invested massively in trademarks and brands and it hosts the largest number of large multinationals with operations across the five continents. Old companies, such as General Motors, Disney and IBM, and new companies, such as Microsoft, Google and Amazon, have business lines heavily dependent on IP protection. Moreover, the United States government has also the economic, political and diplomatic muscle to protect the property rights of its companies both at home and abroad, instruments that are often lacking for other countries.

But there are perhaps also other cultural reasons that make the US society praise and reward individual creativity rather than the societal context in which knowledge is generated. The Anglo-American ideology strongly supports individual values and freedom even when they are at the expenses of the public interest. In continental Europe, on the contrary, the public interest generally prevails over individual rights. This US vision spans across a variety of human rights including property rights and it extends to *intellectual* property rights. European societies are generally keen to consider limits to property and to intellectual property especially. It is therefore understandable that obtaining strong protection of IPRs has been a lesser priority in Europe than in the United States. The Asian vision is even more likely to praise collective effort over individual enterprise. In most of the Asian countries, the introduction of IP legislation has been directed more by the need to please the United States and the other Western parties rather than by a genuine endogenous sentiment to reward individual creativity and company investment in immaterial goods.

It is true that Asian countries such as Japan and China have also been *catching up* countries, the former in the 1950s and 1960s and the second since the 1990s. As catching up countries, they had a clear interest to adopt technologies developer abroad and weak IPRs system could serve such a purpose. But it is also true that since the 1980s Japanese companies started to be major producers of innovation and yet they have still been reluctant to use IPRs as principal instrument to seize the return of their innovations.

The attempt to strengthen the IPRs regime and to obtain greater enforcement started in the United States more than 30 years ago. A "silent revolution" (Jaffe and

Lerner 2004) took place in the form of four interconnected changes. The first was the Bayh-Dole Act, which allowed Universities and other research centres to commercialize and profit from the innovations generated with public money—a legal transformation that has been later imitated by several OECD countries. Second, the scope of the patent system has progressively grown, allowing patenting in new areas, such as software, which were previously covered by other forms of intellectual property or not covered at all. Third, the US Patent and Trademarks Office (USPTO) started to be funded through a fee charged on applicants rather than by the government. This apparently innocuous change made the USPTO keen to grant patents to as many applications as possible, even when the novelty and utility is not self-evident and when the invention is in areas traditionally excluded from the range of patents. Fourth, the courts have become increasingly tougher with violators of IPRs (see also Coriat and Orsi 2002).

The effects of the silent revolution in the United States are controversial and many critics argue that this has been harmful to the US economy since it has reduced the rate of innovation (Heller 2008; Heller and Eisenberg 1998; Jaffe and Lerner 2004), it has generated excessive litigation and increased costs (Bessen and Meurer 2008) and it has increased monopoly power (Boldrin and Levine 2008). So, within the United States there have been many voices that have argued that the "silent revolution" has discouraged innovation and distorted resources from innovation to patent protection.[3] But when the hegemonic country moves in one direction, it is very likely that other countries will somehow follow the path and attempt to create a stronger regime of IPRs—which as will be discussed below, soon reached the international scene.

4 From the American Silent Revolution to the Vociferous Global Scene

The "silent revolution" considerably increased the strength of IPRs within the United States and has been able to better protect US companies within their own internal market against foreign competitors. However, in a globalizing economy, this was only part of what US corporations desired, because supposed and real IPRs infringements continued to occur outside the United States. The problem was that US companies had the possibility to use legal devices to block violations in their own country but had little possibility to retaliate when violations occurred elsewhere. The traditional legislation on intellectual property inaugurated by existing conventions required individual states to prevent discrimination and to protect foreign intellectual property rights. As with many other international covenants,

[3] On 4th July 2013, President Barack Obama himself expressed concern that there are too much patents in the USA that do not reflect genuine inventive activity and to reduce "frivolous litigations". White House 2013.

there was no guarantee that states actually introduced homogeneous legislation and, even when they did, national institutions were obliged to enforce it. Many governments had little interest to use their authority to protect the IPRs of foreign companies.

Since the end of the Second World War, the United States has been a generous supplier of knowledge, technology and technical assistance to its allies. But at the end of the 1970s, US corporations and their government started to take a different turn. The undisputed technological leadership of the US started to be eroded by its economic partners, productivity gaps narrowed, and competing countries continuously improved their innovative potential (see Nelson and Wright 1992; Pianta 1988). Japan constantly increased its export share in high-technology products and Europe was progressively performing better. The US trade deficit was no longer associated to the import of raw materials and traditional products, but also to high technology products, and this was enough to shock a public which for most of the twentieth century was proud of its leadership in technological innovation. A growing concern emerged in the United States and the culprit was easy to be found: Japanese companies (and, to a lower extent, European companies) had better performance because they were exploiting commercially the knowledge generated in the United States, often infringing their IPRs. Less attention was devoted to look at how companies based in Japan, Germany, Switzerland, Sweden and other countries were investing massively in R&D and design, and even less to the fact that these companies generated an increasing number of inventions and innovations for which they claimed patent protection in the United States and elsewhere.

The stories heard in the 1980s soon re-emerged with reference to the Asian tigers. These countries followed patterns very similar to those of Japan. Comparable catching up processes have occurred for more than a decade in China, a country with a population size much higher than Japan and the other East Asian tigers. The widespread feeling in the United States continued and still continues to be that the national investment in knowledge and in other intangible assets is not adequately rewarded and that its competitors in the international market also make unauthorised use of technologies, design and other intangible assets generated by US corporations. This position holds that, as long as the international market would not provide adequate remuneration for its innovations and intangibles, and institutions do not punish infringements, the United States would continue to be penalized.

It took a while for the US corporations to find a method that was able to better guarantee their IP outside the United States. The retaliations available ranged from diplomatic reproach to sanctions to military interventions; none of them was particularly effective to protect companies' economic interests. As is well documented by Susan Sells (2003) and Michael Ryan (1998), a group of Chief Executive Officers of leading corporations joined forces through the creation of the Intellectual Property Committee (IPC) and the International Intellectual Property Alliance (IIPA). The IPC explored the available opportunities and eventually agreed that the most effective and perhaps sole way to penalize countries not adequately protecting IPRs was retaliating on trade. Since negotiations to reduce trade barriers were already on the agenda, the United States government effectively pushed with all its political, diplomatic and economic muscle to make IPRs a crucial pillar of the GATTs replacement—the new World Trade Organization (WTO)—which is where the Trade Related Aspects of Intellectual Property Rights (TRIPS) were born.

Some commentators were surprised that so many countries signed TRIPS: after all, if these countries were using knowledge and intangibles without paying a proper remuneration they should not have an interest to sign an agreement that implied greater transfer of resources from the imitators to the innovators (Wade 2003). Developing countries had an interest to liberalize the international trade in textile apparel—the Multifiber Agenda—and agricultural products in the US market, along with the European market, was the most attractive for them. If the price to be paid to access the American and European markets was to promise to introduce a tougher IPRs legislation, it seemed that this was an acceptable deal. TRIPS was born as a bargain: developing countries accepted the risk of retaliation if they were not able to introduce adequate legislation for IPRs in their own country in exchange for the opportunity to enter into the American and European markets. The powerless conventions of the twentieth century were therefore integrated with a potentially vigorous tool: trade retaliations.

It has been argued that TRIPS was accepted by developing countries because the US government was a much better negotiator, and because it had the instruments to convince other governments to sign the agreement even if it was against their interest (Sell 2003). This is certainly part of the story, but it does not also mean that the US made a good deal. The deal was cash against a promise: developing countries did not get much, but they obtained access to the market of developed countries. In turn, developed countries got the promise that all countries that signed the treaty would have accurately introduced IPRs in the future. But, as we will argue, the US did not properly consider three core aspects: (1) *Legislation does not necessarily imply enforcement*: even if developing countries were induced to apply more rigid IPR norms, this did not necessarily imply their willingness and capacity to enforce them; (2) The reasons why some of its multinational corporations focussed so much on TRIPS were not necessarily beneficial to the American economy, especially, as we will argue, *their strategy was to decentralize production*, including knowledge-based production, *rather than protecting home-base activity*, abroad; (3) *The real economic effectiveness of IP is not as important as it was assumed.* Next section presents two alternative models on the importance of IPRs.

5 Two Models on IPRs

Since the beginning, IP had enthusiastic supporters and fierce enemies (May and Sell 2006, trace the genealogy of these arguments). Yet, much has changed in the generation, imitation and diffusion of invention and creativity over the last several centuries, but the arguments in favour and against IPs have somehow been repeated. It is possible to identify two opposite models in the contemporary debate, the first that recommend a strong regime of IPRs for economic development, and a second that argues that a weak regime or even the absence of IPRs is conducive to the diffusion of knowledge and economic growth.

5.1 Defending Strong IPRs

The model that supports the strong protection of IPRs both at the national and global level argues that in the long-term strong IPRs will not only be beneficial for the producers of knowledge but also for users and therefore for society at large. This model applies the same logic at the international level, assuming that IPRs will benefit developing as much as developed countries.

The pro-IPRs model stresses that incentives and rewards to inventive and innovative activity are crucial to generate further investment. It is not denied that *statically* the existence of IPRs may reduce the diffusion of innovations to those that cannot afford to pay its price, but *dynamically* this will be an incentive to invest more in the future. This is supported by four arguments:

- Invention, innovation and more generally creativity is generally costly to generate, especially since it involves a great degree of uncertainty. But they can be imitated or replicated at substantially lower costs. The absence of protection will discourage profit-seeking agents to invest in these activities.
- IPRs protection encourages full disclosure of inventions that eventually will become part of the public domain. In absence of disclosure, knowledge may be kept secret and this will reduce its dissemination.
- The revenues generated by inventions are one of the core resources to finance further projects. It is therefore relevant that the current inventive activities are able to generate profits to keep the system going.
- New entrants have to face the alternative of imitating existing devices or investing in generating fresh solutions, and strong IPRs will induce them to opt for the latter strategy that sparks inventiveness.

These reasons are used to justify IPRs within nations with knowledge that IPRs will have some advantages for individuals and companies that generate knowledge and some disadvantages for consumers or supplier-dominated companies. But it is assumed that the advantages for the economic system are greater than the disadvantages and well-tuned legislation could balance the interests of both producers and users of knowledge (for example by regulating the length and scope of IPRs and even by using compulsory licensing in cases of palpable public interest).

The pro-IPRs model advances additional reasons to suggest that developing countries will also benefit from a global IRPs regime. It is a fact that the distribution of scientific and technological capabilities is extremely polarized across the globe. This means that the net recipients of IPRs royalties and fees are based in the North and the net payers are based in the South. The advocates of IPRs argue that Southern countries with a strong regime of IPRs will benefit from greater inflows of technology transfer. Companies in the North may be reluctant to establish production facilities, to build R&D labs, to licence knowhow and to engage in strategic technology agreements in countries that do not properly guarantee IPRs. On the contrary, if IPRs are secure, companies may be willing not only to licence the knowledge generated in their home country, but they may also decentralize in emerging and developing countries by moving some of their R&D and innovative facilities and to collaborate with local companies on common projects. In a world where developing and emerging countries still need to acquire the knowledge generated in the North, a well functioning IPR system is the best guarantee that Southern companies would not be excluded.

This view has particularly emphasized the role of multinational firms and foreign direct investment. For instance, Dinopoulos and Segerstrom (2010) developed a model of North-South trade and find that stronger IPR protection in the South leads to: (1) a permanent increase in the rate of technology transfer to the South within multinational firms and (2) a permanent increase in R&D employment by Southern affiliates of Northern multinationals. Along similar lines it has been shown that improvements in IPR protection has led US based multinational firms to increase technology transfer to their affiliates and to shift toward more technologically based products abroad (Branstetter et al. 2004, 2010). Other scholars point to the indirect benefits for the countries hosting multinational firms, in terms of growth of local suppliers (Javorcik 2004a), and transfer of advanced knowledge and skills to the local workforce (Görg and Strobl 2005; Poole 2010). Also, the establishment of reliable and harmonized IPRs systems lead to the creation of 'markets for technology' that facilitate and encourage knowledge diffusion trough formal transactions of technology (Arora et al. 2001; Athreye and Cantwell 2007).

5.2 In Praise of Weak IRPs

The pro-IPR model has been contrasted with fierce arguments. The first argument against IPRs is that they are harmful because, by creating a legal monopoly, they obstruct and reduce the diffusion of knowledge (Andersen 2006; Boldrin and Levine 2008). Since knowledge drives the generation of further knowledge, IPRs may create a vicious circle that stops inventive activity. This is particularly true within industries in which innovation activity is based on sequential inventions and complementary technologies. In these cases, imitation may promote further innovations, while strong patents might actually inhibit it (Bessen and Maskin 2009; Merges and Nelson 1990). By becoming tighter, IPRs increase legal costs more than investment in innovation, leading to a scientific and technological system based on litigation rather than on research (Bessen and Meurer 2008). Eventually, lawyers are the main beneficiaries of a strong system of IPRs.

Second, IPRs may also distort the investment for knowledge since this is likely to be directed towards the areas that promise greater profits or that can be better protected rather than towards those that are more likely to generate socially useful results or where there are more technological opportunities. For instance it has been shown that the type and strength of the patent regime will influence not only the rate of innovation activity but also the direction of technical change (Moser 2005). It is therefore not in the public interest to get a strong IPRs system (Macmillan 2006).

These beliefs are reinforced when the specific needs of developing countries are taken into account. In these nations, knowledge generating institutions are still in their infancy and they will be the most affected by a strong international regime where there is a price to be paid for any technology transfer. In a North-South perspective, the IPR regime may hamper or impede catching up (Chang 2002). It has been claimed that transferring the IPR practices of the developed world to developing countries leads to reduced knowledge flows, less imitation activity and increased prices (Helpman 1993; Lai 1998; Parello 2008). Thus, a system prior to TRIPS, with a strong IPR regime in developed countries and a weaker one in developing countries, is more congenial so as to allow the lower to middle income countries to catch up by benefitting from the knowledge developed elsewhere. The consequence of TRIPS is therefore damaging development and it will make it more difficult to allow emerging and developing countries to build solid innovation systems. Moreover, it will extract resources from developing to developed economies and this will delay further catching up.

From a historical perspective, it is often pointed out that most countries managed to catch up through copying and imitating from more developed countries (Boldrin and Levine 2008). It was only later that tighter IPRs regimes were established *as an effect* of the development of the country (Lerner 2002; Mokyr 2002; for a detailed list of national case studies, see Odagiri et al. 2011). By making international technology transfers more onerous through a strong regime of IPRs, developed countries "kick the ladder away" and make it more difficult for laggard countries to catch up (Chang 2002).

Some critics of IPRs also focus on selected key products protected by IPRs (see, for example, Correa 2000; Shiva 2001). A paradigmatic case is the anti-retroviral drugs patented by leading US and European pharmaceutical corporations (the so-called 'Big Pharma'), but marketed in South Africa through locally owned generic and un-authorized manufacturers. This led to a legal battle between Big Pharma and the South African government. The government argued that the vast majority of South Africans affected by HIV could not afford to pay the price charged by Big Pharma (Fisher and Rigamonti 2005). The campaign was sufficiently powerful enough to induce Big Pharma to withdraw from the case. In this domain, intellectual criticism overlaps with social and political activism. Due to the public good related to certain products, new social movements have started to contest the pro-IPR policies carried out by multinational corporations, national governments and the World Trade Organization. Big Pharma, Microsoft and other companies have become the frequent targets of many campaigns against intellectual monopoly.

6 A Different View: How Powerful IPRs Are?

The models for and against IPR both describe real aspects of the generation, transmission and diffusion of innovation. But, surprisingly, both models are based on a common textbook assumption: *they assume that a legal system of IPRs is much more powerful than it actually is*. Both of them take for granted the fact that strong IPRs can guarantee the protection of invention, innovation and intangibles and that weak IPRs will on the contrary allow imitators to acquire the related knowledge. This is not the case and there is old and new empirical evidence that illustrates that IPRs are much less effective than generally assumed by both models.

From the perspective of the producer of invention, innovation and intangibles, it should be clarified that its main economic interest is not to secure IP per se but rather to profit from it. In order to appropriate the returns from their inventions, innovations and intangibles, companies have to develop complex strategies that include R&D, design, lead-time, ability to deliver products to market, and to combine effectively industrial secrecy with IPRs. IPRs are just one element in this strategy and certainly not the most important one. Surveys carried out for US and European manufacturing companies have indicated that patents and other legal methods are, in fact, the two least important appropriability factors, while companies have ranked as more important lead time, industrial secrecy, complementary manufacturing and complementarily sales and services (Cohen et al. 2000; Arundel 2001).

From the perspective of potential users, would-be imitators cannot manage to acquire knowledge for production facilities just by getting knowledge unprotected by IPRs. The use of knowledge for production is associated with a much larger variety of factors. IPRs may, at most, report and protect some codified knowledge, but there is an equally important component represented by tacit knowledge that is not, nor can be, reported in patents, handbooks, software, blueprints, and other codes (Pavitt 1987; Nelson 1992). Imitators will need to acquire this knowledge to properly use it. A musical score and a violin are not sufficient to play Beethoven's violin concert, likewise the free use of relevant patents will not immediately result in a company manufacturing a good car. Patent rights last for not more than 20 years: after this period, all the knowledge protected by patents is publicly available. It is, however, difficult to imagine that, if the protection accorded by patents will be immediately abolished worldwide, developing countries will be able to assimilate and put into practice the knowledge of the most developed countries. And the reason is very simple: the acquisition of knowledge is not mainly blocked by legal devices but rather by the lack of competences of prospective imitators and complementary assets. In a nutshell, IPRs are less important than assumed for both generators and users of knowledge.

Of course, it is difficult to make generalizations regarding a complex economy where there are many products and industries with radically different characteristics. In fact, the available empirical research has shown that most industries and products are affected differently by IPRs. Within the manufacturing industry, pharmaceuticals and, to a much smaller degree, chemicals, are heavily dependent on patents, and respondents argue that as many as two thirds of innovation in pharmaceuticals, and one third in chemicals, would not have been introduced in the absence of patent protection (Mansfield 1986). Likewise, copyright is a crucial factor for the audiovisual and software industries where the final consumers have the possibility to copy the products directly.[4] In most other high-tech industries, including computers, electronics, aerospace, automobiles, mechanical engineering, IPRs are overwhelmed and/or complemented by other methods of appropriation of innovation.

The fact that IPRs are effective only in a few industries is reflected in the composition of foreign direct investment. It has been shown that the importance of IPR protection varies between industries (Mansfield 1994, 1995) and that weak IPR discourages investors only in sensitive sectors (Javorcik 2004b). In a study examining the drivers of the surge of patents in China, it is shown that foreign direct investment is mostly limited to electric machinery, transportation equipment, and chemical industries (Hu and Jefferson 2009). Also, it has been argued that far from being automatic, adoption of foreign technologies from developing countries is contingent on the development of an adequate level of skills and technological capabilities (Benhabib and Spiegel 2005; Parello 2008). Other research that has tried to explain how Western companies have increased their patent applications in countries with weak appropriability regimes, have found that preventing imitation or securing royalties are partial reasons (Keupp et al. 2012).

Innovating companies know well that in developing countries they cannot sell their products for the same price as they sell it for in developed countries. Many products have a substantial difference between the average and the marginal cost, and in products such as drugs, software and audiovisuals the difference is enormous. Companies apply price discrimination in order to maximise the revenues from the same product innovation across different markets. What these companies are most worried about is the possibility that the same products are re-imported in the Western markets. For example, one main concern of the Big Pharma when they sued the South African government for its unwillingness to stop the diffusion of generic versions of the retro-viral drugs against HIV/AIDs, was the concern that the generic version could also reach the much more lucrative Western markets (Muzaka 2011).

To sum up, IPRs on their own cannot guarantee or reduce returns to invention, innovation and intangibles. Profits for innovations are obtained through a variety of channels and, if companies are asked to rank the relative importance of them, it emerges that most companies think patents to prevent imitators and patents to secure royalties are not among the most important methods to appropriate returns from innovation. This leads to a logical question: *if IPRs have so little relevance,*

---

[4] If products such as CDs, DVDs and software could not be technically copied, they would not even need to rely on *intellectual* property rights since the standard property rights will be more than sufficient to protect the producer.

how come companies, governments, lobbies and social movements are so concerned about them? A first tentative answer is to suggest that IPRs are somehow readily visible and can be more easily modified by institutions. But what they represent—the system of incentives to generate, transmit and diffuse knowledge and creativity—are much more complex and often less visible. As a result, in some sense, the debate is seemingly concentrating on the finger rather than on the moon it's pointing to.

In addition, most discussion has concerned the issue of drugs. This confirms the importance of IPRs, namely patents, to the pharmaceutical industry, and also explains the attention of media and social movements, given their substantial and often immediate interest in health. True, on a moral level, there have been convincing claims that every citizen in the world should have a right to medications she needs regardless her income and in order to live a minimally decent life (Brown and Paremoer 2014). However, it is difficult to argue that, should TRIPS be abolished tomorrow, the pharmaceutical industry would suddenly become the engine of technological development and growth in developing countries. Thus one has to make a distinction between the two problems related to IPRs in the pharmaceutical sector. The former concerns the right of access to drugs, while the latter regards the capacity of economic and technological development in low to middle income countries. Our conviction is that TRIPS has more to do with providing incentives for the former rather than just focusing on the latter (Bhalotra and Pogge 2012).

7 How Effective TRIPS Are?

As is well known, a passionate debate emerged when the WTO integrated IPRs into its core competences. As discussed above, the critics of IPRs argued that the governments of developing countries made a fundamental mistake in signing the TRIPS agreement since they will get no advantages and will bear all disadvantages. The critics argue that TRIPS could seriously hamper developing countries from catching technological opportunities. This, in turn, is based on the assumption that TRIPS are effective in preventing imitators to use the protected knowledge unless this is properly licensed.

After more than 15 years since the implementation of TRIPS, an overall assessment of their economic and social impact is much needed. In practice, the disputes brought at the WTO concerning IPRs are not many. From 1995 to 2011, the WTO Dispute Settlement Process (DSP) machinery has been activated 29 times for IRPs related issues (Lee 2011). It is true that in 17 of these cases the United States is the complainant country, i.e. the economic dominant country. But only seven of these complaints are directed toward developing countries, while ten are towards other OECD countries. Of course, this does not imply that the most serious infringements of IP occurs in the OECD area, but rather that real and perceivable economic damages largely occur in OECD countries. The US government, the most active in using the WTO machinery, did not bother to use the DSP when the markets in which violations take place are not particularly attractive. Moreover, the DSP has some clear limitations:

1. The DSP can be activated against unfair legislation but much less against the lack of effective enforcement of IPRs, which continues to be the prerogative of sovereign states;
2. The DSP process is lengthy (it takes up to 3 years) and the remedies that a country agrees to implement may take up to a couple of years. For areas of rapid technological change, this means that a decision may be implemented when it is no longer relevant;
3. The parties to the WTO and its DSP are the states and not the companies. Governments often act to pursue the interests of the companies based in their own country and are often solicited by the companies themselves. But when dealing with multinational corporations, the national interests are more difficult to assess and governmental take-up will be weighed alongside other strategic interests.
4. Trade retaliations have so far seldom been authorized.

TRIPS has at most established a general framework that provides the possibility to link two things that are not clearly associated, namely the regime of IPRs and trade retaliations. But TRIPS does not provide effective machinery to punish countries that do not have a strong IPRs regime. That said, as said before, they do provide a legal basis to associate trade to IPRs which gives some precedence for developed countries (most notably the United States) to link the two issues together in bilateral agreements (although these agreements are still open to negotiated outcomes that promote interests on both sides).

If we move a step beyond and look at litigation and enforcement, the role of national institutions is even more important. IPRs controversies need to be interpreted by national courts, and infringements need to be policed. Take the spectacular litigation between Apple and Samsung concerning smartphone (the most updated and detailed source is the Wikipedia (2013) entry "Apple Inc. v. Samsung Electronics Co., Ltd." and, more generally "Smartphones Wars", see also Filippetti 2012; Graham and Vishnubhakat 2013). The competing companies have headquarters in nations with rather uniform norms regarding IPRs and, as expected, litigation has already occurred in at least nine different countries. These litigations are already post-TRIPS, since what TRIPS can do is to ask nations to introduce comparable limitations, yet it cannot interfere with any decisions taken by a national court. In light of that, the rulings of national courts have been, so far, very different, and often reaching opposite legal opinions. And, with no surprise, the rulings of national courts are generally favourable to national firms. As a result, the basic principle of the international covenants of non-discrimination holds in theory much more than in practice.

8 What Is the Future of IPRs in a Global Economy?

We have argued that IPRs are much less effective than generally believed in guaranteeing the appropriation of innovation. It is quite natural to ask why companies bother to patent at all: to spend an increasing amount of resources to develop patent applications, to extend them across countries, to engage in litigations and so on. The real problem is that IPRs are just one of the tools that companies use in a competitive environment and there are multifarious reasons why companies take patents. Furthermore, this goes beyond conjecture, since all the empirical evidence on technological appropriation confirms this multidimensionality (see Cohen et al. 2000; Arundel 2001; Blind et al. 2006). The prevention of imitators is, of course, a crucial motivation, but often this is associated to a defensive strategy to prevent competing companies to patent first. Patents are also very important to allow companies to engage in negotiations, i.e. to cross-licence patent rights with competitors. Other reasons exist and are equally important, such as a company's wish to set new standards, build a reputation, and use patents as an indicator of internal efficiency (Shapiro 2001; Blind et al. 2006). Furthermore, the cases where IPRs are operationally crucial are strongly confined to a limited number of industries and, within these industries, mainly to specific products.

We have also seen that there has been a consistent attempt to create a stronger IPRs regime in the United States and globally. In part, this is precisely what has happened. But as soon as the costs associated to acquire innovations and knowledge intensive products have risen, new players were ready to take it as an opportunity to enter into profitable markets. When software started to be policed seriously, many companies introduced formal policies to prevent the use of cracked programmes by their employees. In turn, this has led software companies to increase the price of their products. But the consequence is that consumers started to search for other viable alternatives. This led to an unexpected diffusion of 'open source software' and non proprietary forms of operating systems such as Linux. Extra profits associated with protected and copyrighted programs have been reduced. When the United States complained against the illegal distribution of Hollywood films in India and obtained greater protection, this provided an impetus to scale-up new Bollywood productions, an industry that is now a potential treat to Hollywood (Sunder 2011). Companies are often aware that the best way to profit from their inventions and innovations is to leave IPRs relatively flexible since this is the best way to increase the diffusion of their products. This was the policy carried out by Microsoft in the 1980s.

It is often argued that Western countries should make a net profit from stronger IPRs. But, again, this conclusion is not substantiated from the evidence. Or, more precisely, greater profits for the companies are not necessarily a gain for the home-country of the very same companies. As already indicated by the Lieberman Report (Lieberman 2004), the main aim of multinational corporations in obtaining a stronger global IPRs regime was to off-shore production facilities and knowledge-intensive jobs in countries with lower wages. This can hardly be called an advantage for the United States and other advanced economies. In fact, it opens up questions about whether the US government has actually pursued the interests of its people, or whether it pursued the interests of MNCs in pushing a harmonized global IPRs regime.

So, should the IPRs regime be fixed? The economic practice is very different from what is often suggested as the institutional ideal. So far, even the advocates of TRIPS recognize that the agreement did not have much impact, either positive or negative, on the least developed countries (Hold and Mercurio 2012). There are clearly contesting forces at work, with net knowledge producers trying to appropriate the returns of their investments. But neither strong nor weak IPRs can guarantee technology transfer. What is needed for a successful strategy of technology transfer and innovation-based growth is an overall strategy of developing countries to assimilate, apply, transform and improve the knowledge generated elsewhere. IPRs may block imitators in some specific areas, or they can make it more onerous to access information, but in general IPRs cannot impede the use of knowledge to those that have the competences to use it. Once again, we are in the classical situation in which the main policy implication is "if it ain't broke, don't fix it" (Winter 1989). And the debate around the issue IRP is some real sense recalls the classic Shakespearian notion of much ado about nothing.

Acknowledgment Part of this research has been supported by a Marie Curie Intra-European Fellowship project FP7-PEOPLE-2011-IEF-298167-REGIO_SPIN, under the EC Grant Agreement No: PIEF-GA-2011-298167.

References

Andersen, B. (2006). *Intellectual property rights. Innovation, governance and the institutional environment*. Cheltenham: Edward Elgar.

Archibugi, D., & Filippetti, A. (2010). The globalization of intellectual property rights: Four learned lessons and four thesis. *Journal of Global Policy, 1*, 137–149.

Archibugi, D., & Iammarino, S. (2002). The globalization of technological innovation: Definition and evidence. *Review of International Political Economy, 9*, 98–122.

Archibugi, D., & Pietrobelli, C. (2003). The globalisation of technology and its implications for developing countries. Windows of opportunity or further burden? *Technological Forecasting and Social Change, 70*, 861–884.

Arora, A., Fosfuri, A., & Gambardella, A. (2001). *Markets for technology: The economics of innovation and corporate strategy*. Cambridge, MA: The MIT Press.

Arundel, A. (2001). The relative effectiveness of patents and secrecy for appropriation. *Research Policy, 30*, 611–624.

Athreye, S., & Cantwell, J. (2007). Creating competition? Globalisation and the emergence of new technology producers. *Research Policy, 36*, 209–226.

Benhabib, J., & Spiegel, M. M. (2005). Human capital and technology diffusion. In P. Aghion & S. N. Durlauf (Eds.), *Handbook of economic growth* (pp. 935–66). North-Holland: Elsevier B.V.

Bessen, J., & Maskin, E. (2009). Sequential innovation, patents, and imitation. *The RAND Journal of Economics, 40*, 611–635.

Bessen, J., & Meurer, M. J. (2008). *Patent failure. How judges, bureaucrats, and lawyers put innovators at risk*. Princeton, NJ: Princeton University Press.
Bhalotra, S., & Pogge, T. (2012). Ethical and economic perspectives on global health interventions. In G. W. Brown, G. Yamey, & S. Wamala (Eds.), *The handbook of global health policy*. Cambridge: Wiley-Blackwell.
Blind, K., Edler, J., Frietsch, R., & Schmoch, U. (2006). Motives to patent: Empirical evidence from Germany. *Research Policy, 35*, 655–672.
Boldrin, M., & Levine, D. K. (2008). *Against intellectual monopoly*. Cambridge: Cambridge University Press.
Branstetter, L. G., Fishman, R., & Fritz Foley, C. (2004). Do stronger intellectual property rights increase international technology transfer? Empirical Evidence from U.S. firm-level panel data. *Quarterly Journal of Economics, 121*(1), 321–349.
Branstetter, L. G., Ray Fisman, C., Foley, F., & Saggi, K. (2010). Does intellectual property rights reform spur industrial development? *Journal of International Economics, 83*, 27–36.
Brown, G. W., & Paremoer, L. (2014). Global health justice and the right to health. In G. W. Brown, G. Yamey, & S. Wamala (Eds.), *The handbook of global health policy*. Cambridge: Wiley-Blackwell.
Cantwell, J., & Iammarino, S. (2003). *Multinational corporations and European regional systems of innovation*. London: Routledge.
Chang, H.-J. (2002). *kicking away the ladder: How the economic and intellectual histories of capitalism have Been Re-Written to justify neo-liberal capitalism*. London: Anthem Press.
Cohen, W. M., Nelson, R. R., & Walsh, J. P. (2000). *Protecting their intellectual assets: Appropriability conditions and why U.S. Manufacturing Firms Patent (or Not)* (NBER Working Paper 7552).
Coriat, B., & Orsi, F. (2002). Establishing a new intellectual property rights regime in the United States: Origins, content and problems. *Research Policy, 31*, 1491–1507.
Correa, C. M. (2000). *Intellectual property rights, the WTO and developing countries. The TRIPS agreement and policy options*. London: Zed Books.
Dinopoulos, E., & Segerstrom, P. (2010). Intellectual property rights, multinational firms and economic growth. *Journal of Development Economics, 92*, 13–27.
Drahos, P. (2010). *The global governance of knowledge. Patent offices and their clients*. Cambridge: Cambridge University Press.
Drahos, P., & Mayne, R. (2002). *Global intellectual property rights. Knowledge, access and development*. New York: Palgrave MacMillan.
Filippetti, A. (2012, September 18). Apple vs. Samsung: Something doesn't quite fit! *Open Democracy*. http://www.opendemocracy.net/openeconomy/andrea-filippetti/apple-vs-samsung-something-doesn%E2%80%99t-quite-fit. Accessed January 29, 2014.
Filippetti, A., & Peyrache, A. (2011). The patterns of technological capabilities of countries: A dual approach using composite indicator and data envelopment analysis. *World Development, 39*, 1108–1121.
Fisher, W. W., & Rigamonti, C. P. (2005). *The South Africa AIDS controversy. A case study in patent law and policy*. Cambridge, MA: Harvard Law School. The Law and Business of Patents. http://cyber.law.harvard.edu/people/tfisher/South%20Africa.pdf. Accessed January 29, 2014.
Görg, H., & Strobl, E. A. (2005). Spillovers from foreign firms through worker mobility: An empirical investigation. *Scandinavian Journal of Economics, 107*, 693–709.
Graham, S., & Vishnubhakat, S. (2013). Of smart phone wars and software patents. *Journal of Economic Perspectives, 27*, 67–86.
Heller, M. (2008). *The Gridlock economy. How too much ownership Wreck Markets, stops innovation, and costs lives*. New York: Basic Books.
Heller, M. A., & Eisenberg, R. S. (1998). Can patents deter innovation? The anticommons in biomedical research. *Science, 280*, 698–701.

Helpman, E. (1993). Innovation, imitation, and intellectual property rights. *Econometrica, 61*, 1247–1280.

Henry, C., & Stiglitz, J. E. (2010). Intellectual property, dissemination of innovation and sustainable development. *Journal of Global Policy, 1*, 237–251.

Himanen, P. (2011). *The Hacker Ethic: And the spirit of the information age*. New York: Random House.

Hold, A., & Mercurio, B. C. (2012). *Transitioning to intellectual property: How can the WTO integrate least-developed countries into TRIPS?* (NCCR Working Paper 2012/37) http://papers.ssrn.com/sol3/papers.cfm?abstract_id=2169352. Accessed January 29, 2014.

Hu, A. G., & Jefferson, G. H. (2009). A great wall of patents: What is behind China's recent patent explosion? *Journal of Development Economics, 90*, 57–68.

Jaffe, A. B., & Lerner, J. (2004). *Innovation and its discontents. How our broken patent system is endangering innovation and progress, and what to do about it*. Princeton, NJ: Princeton University Press.

Javorcik, B. S. (2004a). Does foreign direct investment increase the productivity of domestic firms? In search of spillovers trough backward linkages. *The American Economic Review, 94*, 605–627.

Javorcik, B. S. (2004b). The composition of foreign direct investment and protection of intellectual property rights: Evidence from transition economies. *European Economic Review, 48*, 39–62.

Keupp, M. M., Friesike, S., & von Zedtwitz, M. (2012). How do foreign firms patent in emerging economies with weak appropriability regimes? Archetypes and motives. *Research Policy, 41*, 1422–1439.

Lai, E. L.-C. (1998). International property right protection and the rate of product innovation. *Journal of Development Economics, 55*, 133–153.

Lee, E. (2011). Measuring TRIPs compliance and defiance: The WTO compliance scorecard. *Journal of Intellectual Property Law, 18*, 414–445.

Lerner, J. (2002). 150 years of patent protection. *American Economic Review Papers and Proceeding, 92*, 221–225.

Lieberman, J. I. (2004). *Offshore outsourcing and America competitive edge*. Washington, DC: United States Senate.

Macmillan, F. (2006). Public interest and the public domain in an era of corporate dominance. In B. Andersen (Ed.), *Intellectual property rights: Innovation, governance and the institutional environment* (pp. 46–69). Cheltenham: Edward Elgar.

Mansfield, E. (1986). Patents and innovation: An empirical study. *Management Science, 32*, 173–181.

Mansfield, E. (1994). *Intellectual property protection, foreign direct investment, and technology transfer* (World Bank Discussion Paper No. 19). http://www-wds.worldbank.org/servlet/WDSContentServer/WDSP/IB/1994/02/01/000009265_3970311123634/Rendered/PDF/multi_page.pdf. Accessed January 29, 2014.

Mansfield, E. (1995). *Intellectual property protection, foreign direct investment, and technology transfer. Germany, Japan and the United States* (World Bank Discussion Paper No. 27). http://elibrary.worldbank.org/doi/pdf/10.1596/0-8213-3442-5. Accessed January 29, 2014.

Maskus, K. E. (2000). *Intellectual property rights in the global economy*. Washington, DC: Institute for International Economics.

May, C. (2002). *A global political economy of intellectual property rights*. London: Routledge.

May, C., & Sell, S. K. (2006). *Intellectual property rights: A critical history*. London: Lynne Rienner Publisher.

Merges, R. P., & Nelson, R. R. (1990). On the complex economy of patent scope. *Columbia Law Review, 90*, 839–916.

Mokyr, J. (2002). *The gifts of Athena. Historical origins of the knowledge economy*. Princeton, NJ: Princeton University Press.

Moser, P. (2005). How do patent laws influence innovation? Evidence from nineteenth-century world's fairs. *The American Economic Review, 95*, 1214–1236.
Muzaka, V. (2011). *The politics of intellectual property rights and access to medicines*. Basingstoke, GB: Palgrave Macmillan.
Nelson, R. R. (1992). What is 'commercial' and what is 'public' about technology, and what should be? In N. Rosenberg, R. Landau, & D. C. Mowery (Eds.), *Technology and the wealth of nations* (pp. 57–71). Stanford: Stanford University Press.
Nelson, R. R., & Wright, G. (1992). The rise and fall of American technological leadership: The post-war era in an historical perspective. *Journal of Economic Literature, 30*, 1931–1964.
Odagiri, H., Goto, A., Sunami, A., & Nelson, R. R. (Eds.). (2011). *Intellectual property rights, development and catch-up. An international comparative study*. Oxford: Oxford University Press.
Parello, C. P. (2008). A north-south model of intellectual property rights protection and skill accumulation. *Journal of Development Economics, 85*, 253–281.
Patel, P., & Pavitt, K. (1997). The technological competences of the world's larges firms: Complex and path-dependent, but not much variety. *Research Policy, 26*, 141–156.
Pavitt, K. (1987). On the nature of technology. Reprinted (1999). In K. Pavitt (ed.) *Technology, management, and systems of innovation* (pp 3–14), Cheltenham: Edward Elgar
Pianta, M., (1988). *New technologies across the Atlantic*. Hemel Hempstead: Harvester & Wheatsheaf.
Poole, J. P. (2010). *Knowledge transfer from multinational to domestic firms: Evidence from worker mobility* (Working Paper). Department of Economics, Santa Cruz: University of California
Ryan, M. P. (1998). *Knowledge diplomacy. global competition and the politics of intellectual property*. Washington, DC: Brooking Institution Press.
Sell, S. K. (2003). *Private power, public law. The globalization of intellectual property rights*. Cambridge: Cambridge University Press.
Shapiro, C. (2001). Navigating the patent thicket: Cross licenses, patent pools, and standard-setting. In A. B. Jaffe, J. Lerner, & S. Stern (Eds.), *Innovation policy and the economy* (Vol. I, pp. 119–150). Cambridge: MIT Press.
Shiva, V. (2001). *Protect or plunder?: Understanding intellectual property rights*. London: Zed Books Ltd.
Stopford, J., & Strange, S. (1991). *Rival states, rival firms. Competition for world market shares*. Cambridge: Cambridge University Press.
Strange, S. (1988). *States and markets*. London: Pinter Publishers.
Sunder, M. (2011). Bollywood/Hollywood. *Theoretical Inquiries in Law, 12*, 275–308.
Wade, R. H. (2003). What strategies are viable for developing countries today? The World Trade Organization and the shrinking of "development space". *Review of International Political Economy, 10*, 621–644.
White House. (2013). Fact sheet: White House tasks force on high-tech patent issues. http://www.whitehouse.gov/the-press-office/2013/06/04/fact-sheet-white-house-task-force-high-tech-patent-issues. Accessed January 29, 2014.
Wikipedia. (2013). *Entries Smartphone patent Wars and Apple Inc. v. Samsung Electronics Co., Ltd.* Accessed January 29, 2014.
Winter, S. G. (1989). Patents in a complex context: Incentives and effectiveness. In V. Weil & J. W. Snapper (Eds.), *Owning scientific and technical information: Value and ethical issues* (pp. 41–60). New Brunswick, NJ: Rutgers University Press.

Telegraph Technology and Administrative Internationalism in the Nineteenth Century

Douglas Howland

Abstract This essay examines telegraph technology in the context of the establishment of the International Telegraph Union (ITU), and Japan's membership in that union as an aspect of Japan's introduction to international society. The ITU represented a novel internationalism in the nineteenth century—an "administrative internationalism" open to states, semi-sovereigns, and colonies. To a state such as Japan, membership in the ITU offered linkages to the global network of telegraph cables and represented a version of international society in which Japan might be the equal of the treaty powers that had forced Japan in the 1850s into the European order of great powers and their domination.

Keywords Japan • International administrative law • International Administrative Union • International Telegraph Union • Global cable network • Territoriality

The invention of the electric telegraph, patented in 1837, and its expansion across the globe invited visionaries in the nineteenth century to celebrate a "new internationalism." Two commentators cited most often, Paul Reinsch and Leonard Woolf, observed that the new internationalism was based, not on an idealism looking to eternal peace, but on a practical realization among people across the world: the mutual advantages to be secured by their common interests justified some collective administration of their mutual activities. The world was achieving a new stage of integration, and a central fact of this new world order was the new international administrative union inaugurated by the International Telegraph Union in 1865 (Reinsch 1911: 1–3; Woolf 1916: 141–142; see also Krajewski 2006: 23–29).

This essay examines the telegraph in two international contexts: first, the establishment of the first international administrative union, the International Telegraph Union (ITU), and particularly its rules of membership and representation; and

D. Howland (✉)
University of Wisconsin-Milwaukee, Milwaukee, WI, USA
e-mail: dhowland@uwm.edu

M. Mayer et al. (eds.), *The Global Politics of Science and Technology - Vol. 1*, Global Power Shift, DOI 10.1007/978-3-642-55007-2_9,

second, Japan's membership in that union as an aspect of Japan's introduction to international society. What links these two contexts is my argument that international administrative unions represented an alternative version of internationalism in the nineteenth century—what I call "administrative internationalism." This was an internationalism that differed from our idea of international society as a family of sovereign states dominated by the great powers. To the contrary, international unions such as the ITU welcomed semi-sovereigns, dependents, and colonies—an existing telegraph administration was the fundamental criterion of membership. Hence, for a state such as Japan, membership in the ITU represented an alternative arena of international society where Japan might be the equal of the treaty powers that had forced Japan in the 1850s into the European order of great powers. This essay offers evidence that the Meiji government worked successfully to roll back foreign domination through technical means—a point missing from the historical analyses of the English school of international relations (Gong 1984: 164–200; Suginami 1984; Suzuki 2009).

This essay thus expands upon Madeleine Herren's important analysis of "governmental internationalism." But where Herren emphasizes this new internationalism in the nineteenth century as the cooperation of states for purposes of standardizing international norms in technical and scientific matters, this essay is much more interested in the political issues surrounding the ITU as an administrative union—particularly the politics of territory and membership. Herren has argued that international organizations in the nineteenth century provided a "back door to power" for states such as Belgium, Switzerland, and the U.S., who were not among the great powers of Europe (Herren 2000: 5–11, 18–19, 26–28, 40–43, 62–64; Herren 2001). For Japan, too, the story of its engagement with the ITU demonstrates that participation in international organizations was an alternative means of asserting Japanese state power against the colonial powers on its territory, as well as a new arena in which members of the international unions treated each other as equals.

To be sure, global perspectives of nineteenth-century world integration stress the fact that international unions had a prominent commercial impact: transportation, communications, weights and measures, patents and copyright. So much so that many commentators upon international unions at the time described the ITU and the Universal Postal Union (UPU) as economic or commercial organizations. Where Reinsch and others argued that the spread of democratic societies encouraged the internationalization of interests, Woolf agreed but added that such "national" interests of democratic societies are actually the interests of powerful minorities—especially the capitalist class (Reinsch 1911: 135; Woolf 1916: 353–356). As Marxists would argue then and now, the political economy of capitalism structured social relations in the world of the nineteenth century, particularly with the advance of imperialism, and a primary question that engages the best of historians of international organizations is the priority of the economic base.[1]

[1] (Pashukanis 2002: 73–84; Chimni 1999; Miéville 2005: 103–110; Murphy 1994: 82–118).

In contrast, both Reinsch and Pieter Eijkman, writing just before the first world war, as well as Herren and Japanese scholar Mogami Toshiki today, would insist that the multilateral cooperation which we observe in the nineteenth century was not merely a strategy to ensure the functioning of a developing global economy; it was the beginning of a new world order (Eijkman 1911: i–xii; Herren 2001: 143; Mogami 2006: 31–32; Reinsch 1911: v, 9–13, 68–73). Efficiency and practicality are not only economic values, and the integration of telegraph administrations in the ITU was not solely a phase of imperialism. In addition to diplomatic and commercial correspondence, press syndicates were keeping people abreast of current news and friends might send messages by telegraph. This essay argues first, that telegraph technology served both universal and national ends as it integrated sovereign territories; second, that the ITU realized an integrative version of internationalism through its administrative unification of those territorial telegraph administrations; and third, that a state such as Japan, in joining the ITU and other administrative unions, made use of those universal and national goals which were supported by both the technology and the novelty of administrative internationalism.

1 Technology and Territoriality

The growth of the nation-state in the nineteenth century was closely linked to the territorialization of state power. In the German lands after 1848, for example, legal theorists sought to depersonalize the state in order to move it away from monarchy and, at the same time, to rely upon the territoriality of state sovereignty in order both to assert the legal personality of the state as sovereign and to pursue a uniform practice of administration over the society within that territory (Lindenfeld 1997: 76–91, 125–128; Stolleis 2001: 207–218, 292; Wieacker 1995: 347–349). Many of these developments fall within the range of policies that Michel Foucault and his students have analyzed in terms of "governmentality"—sanitation, streetlights, water supply, and hygiene—which represent the efforts of the territorial sovereign state to manage its domestic population (Barry et al. 1996; Burchell et al. 1991). Yet much of the early international administrative law, from the 1860s to the end of the century, was especially based in technology—creating standards for telegraph linkages, standard time zones and railway gauges for train transportation, and standards for postal rates and transport. These new technologies problematized territoriality as an abstract principle and, at the same time, transform our view of a global order of territorially independent sovereign states.

From the outset, the ITU planned an international network of telegraph lines (*réseau international*) that would link the cities of all members of the union. Assisting the realization of this network of territories was the fact that administrative unions such as the ITU deemphasized sovereign status, in that any state, government, or administration was welcome to join the union. Agreement with the foundational treaty linked different territories into the larger space, defined by uniform service and increasingly by standard time zones (Kern 1983: 12–14; Krajewski 2006: 28–37). Technology transcended territory in theory and accordingly, the international administrative union that managed the technology and its services was prepared to represent that universal space uniformly across territorial divisions reflecting sovereign or national differences. Some media scholars have identified a technological determinism implicit in the "liberal internationalism" of the nineteenth century. That is, the promise of new communication technologies was to liberate their users from the potentially hostile interests of business and politics.[2]

But this was only part of the situation. For a technology such as the telegraph, a number of national interests were implicit in the usefulness of the technology: it integrated territory in a new manner when, in conjunction with the railroad, armies could be sent to suppress rebellion or food supplies could be sent to relieve a local population in distress. The telegraph was not simply a neutral technology of universal progress, but played a role in developmental and military projects that were key to the normalization of the territorial state and the integration of its subject or citizen populations. And the ITU honored the security conceits of sovereign states, by including in its convention a member's right to suspend telegraph transmissions that it deemed hostile or otherwise contrary to its national interests.

Hence, contrary to universalist histories, a historian of technology such as Pascal Griset argues that the international telegraph cable network was subject to geopolitical strategies of the great powers (Griset 1991: 2–5, 25; 1999: 67). Because British companies controlled most of the global cable network, the British approach in particular "was based on a systematic and physical control of space, involving the control of islands, strategic points and straits, the patient weaving of a cable network, year after year extending an ever more dense net over planetary space." In that regard, the British quest for geostrategic superiority resembled "a war of position where superiority could only be maintained by ceaselessly increasing one's fortifications."[3]

This technical-strategic aspect of the international network has informed different interpretations of the territoriality of the ITU. Where Russian legal scholar M.P. Kasansky argued that the universality of the ITU lay not so much in the geographical spread of the union as in the integration of relations among all the member countries and administrations, others emphasized the fact that public and private associations constituted the very integration of the international network—particularly in light of the private ownership of undersea cables. Otto Kunz, for example, argued that the territory of the ITU was a manifestation not so much of the force associated with the territory of the state, as the arrangements made by the great number of private companies that operated within and between states

[2] Winseck and Pike (2007: 159); Briggs and Burke (2009: 88, 132–141); Matthewman (2011: 15–20); Murphy (1994: 7–10, 13–18).

[3] Griset (1999: 70). See also Griset (1991: 82–85); Boyce (2001); Headrick (1991: 28–49); Hugill (1999: 25–51); Winseck and Pike (2007: 100–103, 110–111, 184–187); Zacher (2002).

(Kasansky 1897: 182; Kunz 1924: 52–53). These public-private arrangements of telegraphy not only reflected the reality of an international society and its needs and services, but also pointed to the awkward presence of non-sovereign interests that lay within international law—those of private companies (Codding 1952: 55; Rolland 1901: 16, 82).

Private companies were in fact the complicating factor in the nineteenth century, particularly because the telegraph lay in private hands in Britain and the U.S. Although Griset identifies two opposed conceptions of the national operation of the telegraph—one grounded in liberalism and private enterprise, and the other in monopoly and state administration—such a binary distinction is not tenable. In the U.S., so-called private enterprise produced the near monopoly of the Western Union Company and, according to Jorma Ahvenainen, monopoly and government subsidy characterized the private British telegraph industry (Griset 1991: 87–88; Briggs and Burke 2009: 139; McDowell 2002: 222–224; Ahvenainen 2009). Because most international lines were in British hands and Britain was not a member of the ITU, the private telegraph companies generally had a free hand with rates. Governments may have encouraged private companies to adopt the regulations of the ITU, but companies did not necessarily go along with the preferred tariffs, and many scholars point to the high costs of using the telegraph in order to explain its limited use in the nineteenth century. Businesses and the press dominated the lines, while private persons found the telegraph too expensive for frequent use. Only with the introduction of competition, after multiple telegraph cables had been installed by rival states, did rates decrease and usage expand—and that on the eve of the Great War (Ahvenainen 2009: 68–77).

The international territory of the telegraph, in other words, was composed of multiple interests, including the sovereign claims of governments and the commercial interests of private companies. These may have been competing for relative advantages, but the novelty of administrative internationalism—through the institution of the ITU—managed to integrate the technology into a global network. This new if abstract territoriality of the network was a product of neither technological nor political determinism. Rather, it was a universally conceived space which member administrations supported through their commitment to the union.

2 The International Telegraph Union

Although the commissions for European rivers such as the Rhine and the Danube were the earliest international associations in the nineteenth century, these were not strictly "universal" unions, as the members of these river commissions were largely limited to the states through which a river flowed and which, accordingly, had direct interests in river traffic. By contrast, the ITU was an international union with membership open to all state governments and telegraph administrations, including semi-sovereigns, dependencies, and colonies.

The ITU and all other international unions—most prominently the Universal Postal Union (UPU)—were created by a treaty convention that bound all signatory members of the union to the common goals specified in the convention. This aspect of the convention and union—this commitment to the common interests of the members—distinguished the administrative union as a novelty in the nineteenth century. Legal scholar Heinrich Triepel described such a union convention as a *Vereinbarung*, or "law-making treaty," which sought to unite common ends. Because the signatories of a law-making treaty judged themselves bound by the international legal principle of *pacta sunt servanda*—that one must honor one's agreements—the ITU convention arguably assumed the status of international administrative law (Triepel 1899: 49–74; Kasansky 1897: 180; Kunz 1924: 40–42, 125–126; Vec 2006: 112–123).

In addition to the convention, however, another aspect of the novelty of administrative unions was the addition of a set of regulations or *règlement* that was separate from the convention. Members signed both the permanent convention and the *règlement*; the ITU convention included a third part as well, the comprehensive tariff. Because the ITU understood that the details of telegraphic services would need constant updating and renegotiating, the regulations were kept separate from the foundational convention, which registered the original commitment of a member to the mission of the ITU (Boisson 1932: 4–5, 16–20).

The ITU was created in 1865; its primary goal was to ensure international telegraph service under the control of state governments or telegraph administrations. The ITU sought to guarantee both the security of transmissions and the regularity and reliability of transmissions—in return for which members paid a membership fee that supported the work of the union. Because the material installations of telegraphs were on state territories, governments were in a position to encourage private telegraph companies to adhere to ITU guidelines; and, to facilitate that goal, private companies were welcomed to attend ITU conferences as non-voting participants from 1872 on. Because the telegraph required a great deal of capital for the initial installation of lines and only gradually recovered that investment and became profitable, ITU conferences witnessed an ongoing debate over international rates, for users of the telegraph wanted them lowered but the installers of telegraph lines—particularly the expanding network of undersea cables—wanted rates higher. In the wisdom of the founders of the union, members were always free to enter into separate, special treaties between themselves in order to contract arrangements that differed from ITU guidelines.[4] For example, when the ITU declined to reduce rates for news services, a group of five (Germany, France, England, Switzerland, and Holland) planned a special agreement to that end (Kunz 1924: 63).

[4] Boisson (1932: 11–21); Codding (1952: 20–30); International Telecommunications Union (1965: 57–76); Kunz (1924: 55–73); Lyall (2011: 23–37); Rolland (1901: 141–145, 151–159, 166–174).

The organizational structure created by the ITU convention was likewise a novelty in the nineteenth century. The ITU was a simple organization. Its periodic and independent collective Conferences of members—which assembled every 5 or 6 years—were the legal and legitimate substance of the ITU, and only these bodies could make decisions and create policy for the ITU. To assist the periodic Conferences of the ITU in their work, the ITU created the International Bureau of the ITU, which was located in Berne, Switzerland. The International Bureau had exclusively administrative functions; its purposes were to gather statistics of use and other relevant data, to monitor member input between the periodic Conferences, and to keep members informed of new developments in telegraphy. Occasionally, it might assist the Conferences of the ITU in organizing a special Conference or commission to investigate some problem or to arbitrate a disagreement between members.[5]

What distinguished the ITU as a form of administrative internationalism, however, was its membership rules. As already noted, states and other telegraph administrations could become members, but the ITU convention introduced a striking measure of ambiguity because it referred to all members as "governments," "contracting states" and "high contracting parties." So, for example, Britain did not initially become a member (since its telegraph system was in private hands), but British India was an early member of the Union. Because, like India, they constituted telegraph administrations, a number of semi-sovereigns, vassals, colonies and dependencies became members of the ITU: hence Egypt, Persia, Bulgaria and Tunis were early members. Yet for practical purposes, they were all simply referred to as "governments," "states" and "high contracting parties."[6]

At the start of the ITU, all members—both states and administrations—possessed equal voting rights. But problems arose when colonial powers sought to assert control over their dependents and to accrue the voting rights of their dependent telegraph administrations. A primary cause of this situation was a discrepancy between the ITU Convention and its periodic Conferences. Although all members had signed the Convention, delegates to the periodic Conferences were increasingly telegraph officials and experts who negotiated and voted upon the occasionally contentious changes to the regulations. An early ITU policy was to insist that a government that represented more than one telegraph administration could have only one vote, unless a separate delegation representing one of these telegraph administrations specifically requested a separate vote. In the absence of Britain as a member of the ITU, Britain's colonial administrations voted separately, but after Britain joined the ITU in 1876, it eventually garnered 13 votes—a situation which drew the ire of Russia, which had only one vote. Of course, an important administration such as India could vote differently from its colonial master, but the ITU never solved this problem. Unlike the Universal Postal Union, which tied postal

[5] Boisson (1932: 19–21); Clark (1931: 96–98, 105–108); Codding (1952: 48–52, 57–59); Kunz (1924: 74–82, 85–89, 107–115); Moynier (1892: 13–36).

[6] *Convention télégraphique internationale* (1969: 202). See also Clark (1931: 102–103); Kunz (1924: 45–48); Rolland (1901: 206–210).

rates and membership fees to quantities of mail transferred and distances traversed, and thereby equalized the representation of member administrations, the great colonial powers came to dominate the ITU by century's end (Codding 1952: 39–40; Kasansky 1897: 181–183; Kunz 1924: 93–99).

If members of the ITU were not necessarily states, and the identities of states and members were asymmetric, a related aspect of this administrative internationalism is the fact that the ITU and other international unions followed the model of classical diplomatic relations: Delegates to ITU Conferences were treated as diplomatic personnel, for diplomatic *plénipotentiaires* had signed the original ITU Convention, Regulations, and Tariff, and attended the periodic Conferences. But as the ITU developed, technical experts dominated the ranks of representatives to the ITU Conferences (Lyall 2011: 8–11). The ITU reasoned that, because such representatives were "under instruction" by their governments, they were deemed to have the "full powers" of diplomats and able to sign new versions of documents to which their governments were already committed. When this procedure was questioned at the 1885 Conference in Berlin, the ITU officially confirmed this arrangement, following the example of the UPU (Kunz 1924: 56–57, 60, 102–107).

Moreover, the effects of the union took place within the domain of international law. The ITU Convention constrained members as public administrations in their territorial homelands, for members had to account for the international oversight of the ITU. To describe this as international administrative law is to point out how the ITU treaty extended the international sphere of action to domestic administration. Structurally, an international administrative union linked administration to international law through the organizational requirement that members had to fulfill the *règlement*, and through the distinct organ which supervised that requirement on behalf of the union members—the Bureau. Informing both of these organizational requirements was a confidence in the collective wills of the members.[7]

The key issue for early commentators regarding the status of state sovereignty within unions like the ITU was the voting method, for disagreements over state representation and voting had arguably undermined the success of the second Hague Peace Conference in 1907 (Simpson 2004: 132–164). As many contemporary scholars insisted, organizations that punctiliously respect the sovereignty of member states require unanimous decisions, so as to guarantee the sovereign independence of each state. The consent that binds a state's will must be expressed through explicit agreement, and such practices reinforce a mutual recognition of the sovereignty of each member. The ITU began on such terms, with unanimous decisions required for any change to the convention, the regulations, or the tariff. Because unanimity was hard to achieve, particularly over alterations to the tariff, the ITU introduced a right of "formal refusal," which allowed a member to announce that it would not be bound by a new rule but continue to adhere to the previous standing rule. Future negotiations presumably would work out the

[7] von Kirchenheim (1880: 465–77); Neumeyer (1924: 343–45, and 1925: 104–108); Dendias (1938: 249–56, 290); Kazansky (1902: 360–63); Rapisardi-Mirabelli (1925: 348–58, 361–62).

disagreement. Although this right was not often used, it potentially undermined the unity of the ITU, for the regulations remained in place in spite of a formal opponent among the members. Otto Kunz suspected that this situation put increasing pressure on the ITU to turn to majority voting—a policy formally adopted in 1908.[8]

The ITU remained, in sum, a complex but largely public international network of fragmented territories endowed with numerous different kinds of political status. It asserted three general principles in its 1875 revised treaty: that telegrams be secret; that telegraph administrations be subject to common regulations in order to insure uniformity of service; and that tariffs be fixed on the basis of two regions—telegrams within Europe and those outside Europe (International Telegraph Treaty 1969: 418–421). Although some commentators interpreted this set of requirements as a potential infringement upon the sovereignty of its member countries, the union did not require any partial delegation of sovereignty on the part of a member to the ITU; rather, the ITU Convention solicited an agreement to maintain common standards for the telegraph in the interests of a unified system and in solidarity with fellow members. The ITU was not an all-powerful federation of states with a political will superior to the combined wills of its members, and compulsion doesn't appear to have been necessary or possible. The Bureau was without power to take any initiative, and the Conference usually negotiated a consensus on the *règlement*. So rather than describe the Union's effect as an infringement on sovereignty, it would be more accurate to state that voting marked a member's assent to whatever self-imposed restriction or alteration of domestic practice that the ITU required.

3 The Case of Japan

Let us consider Japan's construction of telegraph lines and it subsequent membership in the ITU, for the Japanese example demonstrates not only the technological linkage of national to international territory but also the utility of multilateral cooperation represented in administrative internationalism. For comparative purposes, we will also briefly look at Japan's membership in the UPU.

The leisurely manner in which Japan joined the ITU contrasts strikingly with the Japanese urgency in establishing a telegraph network within Japan in the 1870s. The revolutionary Meiji government began work on the telegraph promptly in 1869, when it installed a wire between Yokohama and Tokyo. A Danish firm, the Great Northern Telegraph Company, contacted the Japanese government in 1870 and negotiated the construction of two international lines on Japan's behalf, one

[8] Codding (1952: 45–46); Kasansky (1897: 183); and Kunz (1924: 99–102). See also Mangone (1954: 4). At p. 76, Mangone argues that a major shift took place with an 1872 amendment that specified that if a contracting administration did not reply within 4 months to a technical question or regulation to which it had been referred, its agreement would be considered as accepted.

linking Nagasaki and Vladiavostok and a second linking Nagasaki and Shanghai—much to the chagrin of the British minister in Japan, Harry S. Parkes, who had intended that a British firm would secure the work. The lines opened in January 1872 and, in order to ensure that domestic telegraph lines stayed under Japanese control, the Meiji government quickly built a Tokyo-Nagasaki line, which opened in November 1872. In order to accomplish this work, the Japanese government imported 50-some British telegraph experts, quickly moved from importing equipment to manufacturing its own, and established a training institute for telegraph engineers and operators.[9] This was one of the most rapid and successful pieces of Japanese westernization in the 1870s, one result of which, Daqing Yang has observed, was that "Japan is probably the only nation to have had a complete domestic telegraphic network in place from the outset of its industrialization" (Yang 2010: 24).

At the same time, Danish diplomats and representatives of the Great Northern Telegraph Company encouraged Japan to join the ITU. The Austro-Hungarian chargé d'affaires in Shanghai and the Danish Minister in Tokyo first invited the Japanese government to send an observer to the 1872 ITU Conference in Rome—it was in Japan's interest, they argued, to learn about the ITU and the way it could contribute to Japan's development and progress.[10] Additional pressure came with the agreement between the Great Northern Company and Japan renewed in March 1872, which proposed that the two rely on ITU regulations to govern Japan's international correspondence. Although Japan sent an observer in 1872 and to the St. Petersburg Conference in 1875, it did not join the ITU until March 1878, its membership taking effect in January 1879.[11]

One factor that encouraged Japan finally to integrate itself into the global telegraph network was the pressure from Great Northern, but Japan joined the ITU only in 1878 especially on account of two domestic concerns. One was a need for military intelligence. Both the Taiwan incident in 1874 (in which Taiwanese aborigines attacked Japanese fishermen) and then the Satsuma rebellion in 1877 (in which former samurai fought against the new state) impressed upon the government the significance of rapid domestic communications. And as the Japanese state looked to the Asian continent, with trepidation at the unrest in both Korea and China, it perceived the need for faster international intelligence. Second was a concern for national security. The Japanese government began to question its commitment to the Great Northern Telegraph Company and to worry about Japan's vulnerability in light of its complete dependence on a foreign corporation for its international connections. Both the pending negotiations over Great Northern's contract renewal in 1882 and Japan's wish for its own undersea cables encouraged

[9] Westney (1990: 41–54); Takahashi (1969: 30–35, 39, 59–63, 76–86); Takahashi (1986: 160–63); Yūseishō (1971: 49–55, 113–121); Jakobsen (2002: 331–333); Checkland (1989: 51–53); Ahvenainen (1981: 38–48, 186–188).

[10] Japan, Gaimushō, *Nihon gaikō monjo*, repr. ed. (Tokyo: Nihon kokusai rengō kyōkai 1950–1963), vol. 7, part 2 (1871): 943f. Hereafter cited *NGM*.

[11] *NGM*, vol. 14 (1878): 203–205; *NGM*, vol 8 (1872): 606–610; *NGM*, vol. 7, part 2 (1871): 953. See also Jakobsen (2002: 333)

the Japanese government to pursue both of these matters as a member of the ITU, which offered precedents for government relations with private cable companies and negotiable international rates. However, largely for want of the financial and technical wherewithal, the Japanese government continued to renew its contract with Great Northern—until imperialist expansion offered alternatives. In 1896, Japan laid its own cable from Japan to its new colony, Taiwan, with a newly purchased cable-laying ship; and in 1906, the Japanese Army laid a cable from Japan to Manchuria. At the same time, a new U.S.-Japan cable connected the two countries across the Pacific Ocean, so that Japan gradually undermined the monopoly that it had granted Great Northern in 1870 (Yang 2010: 27–28, 33–39; Jakobsen 2002: 333–338; Ahvenainen 1981: 189–191).[12]

Consider, for the sake of comparison, Japan's quite different membership in the UPU. Japan's interest in joining the UPU began in 1875, just after the creation of the UPU. This effort was undertaken in conjunction with negotiations with Britain over Japan's desire to assume delivery of British—and all foreign—mail in Japan. As Japanese officials explained to British minister Parkes in Tokyo, it was most inconvenient that overseas Japanese—particularly diplomats—had to mail their letters back home to a treaty port in Japan, and that Japanese post going overseas had to leave likewise from a foreign treaty port.[13]

Aoki Shūzō, Japan's Minister in Berlin, managed negotiations between Japan and the UPU. Aoki had reported the creation of the UPU in 1874 to the Foreign Ministry, and he explained that, since the UPU attempted to integrate mail services throughout the world, he hoped that membership in the UPU might well provide a means of eliminating British and French post offices from Japan. Aoki found promising the fact that within the UPU, members retained their rights over their own post offices and in fact, as members of the UPU, Egypt had managed to defend its postal autonomy and Turkey had successfully recovered its sovereignty over the post in Turkey. Hence, Aoki encouraged his government to seek membership (*NGM*, vol. 11 (1875): 689–691, 695, 697).

From 1875 through 1876, three major developments ensued in tandem with Japan's application to the UPU. First, Japan's interest in the UPU coincided with negotiations with the U.S. over revisions to the U.S.-Japan Postal Agreement of 1874. Although the agreement had come into effect only in January 1875, new revisions were ratified in 1875 and 1876—some of which had direct bearing on Japan's application to the UPU.[14] Most importantly, the U.S-Japan Postal Agreement gave Japan valuable preparatory experience with postal negotiations: Japan

[12] The first undersea cable that Japan installed was an experimental line in 1876 from Okayama to Shikoku through the Inland Sea; see Takahashi (1969: 59–63). For an account of Japan's relations with the Great Northern Telegraph Company, see Ishihara (1999: 3–8, 47–72, 109–121).

[13] Foreign Office.11.19., in Great Britain, National Archives, Foreign Office Archives, file FO 46/168: [299-300]. Hereafter cited FO. See also Maejima (1936: 41).

[14] For the 1874 Agreement, see *NGM*, vol. 9 (1873): 766–795; and *NGM*, vol. 10 (1874): 705–736. For the 1875 and 1876 revisions, see *NGM*, vol. 12 (1876): 783–796. See also Maejima (1936: 49–54); Yabuuchi (1975: 227–232).

became familiar with postal fees, the types and weight standards of postal materials, and so on. Japanese officials remained grateful to the U.S. for its sympathy to Japanese wishes, and the U.S. fully supported Japan's effort to join the UPU (Takahashi 1969: 91–95).

A second development at this time—alarming to Britain—was Japanese and U.S. cooperation over Japan's new shipping lines. In the wake of the initial agreement of 1874, the U.S. had graciously provided the Japanese post office with free transportation between Yokohama and Shanghai (Hunter 1976: 21–22; Kamikawa 1958: 85; Yūseishō 1971: 88). At the end of 1875, Japan's Mitsubishi company established its own shipping line on that route, and both the creation and expansion of a Japanese system were being planned—a Yokohama-Hong Kong route and a Yokohama-San Francisco route. In addition, Japan hoped to increase its range by taking advantage of the San Francisco-New York railway line in the U.S. Although Japanese officials noted that Britain was more likely to assist Japan with a postal agreement if Japan were to avail itself of Britain's India routes, British minister Parkes so disparaged Japanese plans and progress and so constantly slighted the Postmaster General of Japan, Maejima Hisoka, that Japan shunned such cooperation, and British officials remained unhappy with the U.S. and Japan in making arrangements that were certain to compete with British shipping (Parkes to Derby 1877.3.31: FO 46/269: [43-46]).

The key to the establishment of a Japanese postal network was the third project under way in 1875: Japan opened its own post office in the foreign settlement of Shanghai in April 1876 so as to manage Japanese mails to and from Shanghai. With the direct link to Shanghai, Japan could take advantage of preexisting shipping routes to other parts of the world as it developed the new Japanese shipping industry. Moreover, the Japanese post office in Shanghai was central to U.S.-Japan negotiations in 1875. The U.S. facilitated the project because Japan agreed to lower by 3/5 the postage levied against newspapers and other printed matter, as well as samples of merchandise. And the U.S. agreed to turn over the shipping of U.S. post to Japan from Shanghai at the start of 1877; the Japanese post office would carry U.S. post from Shanghai to Yokohama on Mitsubishi's new Mail Steamship Company [*NGM*, vol. 11 (1875): 706–725; *NGM*, vol. 12 (1876): 789–791; Takahashi (1969): 112–114].

Hence, Japan successfully applied for admission to the UPU. In June 1876, Aoki made a formal request to Swiss postal officials to admit Japan to the UPU. The U.S., as the only member country with which Japan had full postal relations, welcomed Japan's admission; and U.S. officials testified that Japan and the U.S. had just agreed to lower postal rates in accord with UPU standards. Contrary to Japan's apprehensions regarding Britain, the latter made no objections. Thus Japan was admitted to the UPU and its membership took effect on 1 June 1877. (*NGM*, vol. 12 (1876): 801–806; vol. 13 (1877): 78f., 84–91, 97–99, 102–104. See also Schröter 1900: 68–69). By participating within the administrative internationalism of the ITU and UPU, Japan developed its national programs in the contexts of international networks of telegraph and postal administrations. Yet Japan's memberships in the ITU and UPU reveal that Japan sought to bring domestic power to bear against the colonial powers already on her shores—to assert control over the telegraph and postal administrations and eventually eliminate its colonialist rivals.

4 Conclusions

How do we understand the expansion of the telegraph, the advent of the International Telegraph Union, and Japan's memberships in international administrative unions such as the ITU as an historical development? In the first place, we have a variety of arguments put forward by historians of international unions and international law, and historians of the telegraph. F. S. L. Lyons and Miloš Vec, for example, argue that the international administrative unions were especially effective in the dissemination of technical norms (Lyons 1963: 15–16, 223–237; Vec 2006: 126–129; 2009). By comparison, historians such as Jorma Ahvenainen and Rolf Walter emphasize the economic role of the telegraph in changing commercial practices by the late nineteenth century: business transactions shifted to real-time exchanges and the rapid circulation of price indices hastened the fluctuations of commodities markets (Ahvenainen 2001; Walter 2001). Japan was certainly engaged in the former process: its development of telegraph technology, with the assistance of British experts and a Danish partner, encouraged European norms of development; furthermore, its membership in the ITU committed Japan to international standards and expectations. By contrast, the commercial consequences of the telegraph were problematic for Japan and initially resisted by the Japanese government, primarily because of Japan's subordinate economic position within the treaty regime. Because the foreign powers had determined Japanese import and export tariffs with the unfair treaties of 1858–1869, the Japanese government was especially interested in protecting, for example, its domestic silk producers from the disturbances of global market prices. Telegraph linkages to international markets threatened its ability to insulate the Japanese market from international trade fluctuations in raw silk products.

A second line of theoretical arguments is clearly reflected in the history of Japan in the late nineteenth century, supporting those historians of geopolitics who emphasize the "space-time compression" produced by technologies such as the railroad, steamship, and telegraph. With distances reduced by steam and rail and communications made nearly instantaneous by the telegraph, states could rule not only their national territories more effectively but also their colonial outposts more thoroughly (Bell 2005; Kern 1983: 206, 214, 220, 223; Matthewman 2011: 11, 13; Tuathail 1992). As has been noted, Japan's interest in the telegraph and membership in the ITU was very much motivated by national security and a concern for international geopolitics. National integration and the suppression of both samurai and peasant dissent, combined with linkages to Japan's outpost in Shanghai and colonial communities in Taiwan, Korea and Manchuria, guaranteed that the telegraph would assist the development of Japanese diplomacy and military prowess, particularly as Japan readied itself for battle in the Sino-Japanese and Russo-Japanese wars.

But this essay argues instead that technologies such as the telegraph encouraged the development of an alternative internationalism in the nineteenth century—an internationalism open to all administrations of technology, whether sovereigns, semi-sovereigns, dependents, or colonies. The order more familiar to us today—

that of sovereign states under the leadership of the great powers—was a competing order under formation in the nineteenth century. The conundrum that notions of international society faced at the end of the century is perhaps most pointed in Louis Rolland's quite forward-looking statement of 1901. On the one hand, Rolland asserted confidently that "the society of states is closed to those that are not sovereign states," sovereignty being the sole and competent authority informing the right to make laws for oneself. It was for this reason that states had developed the custom of the mutual recognition of sovereignty (Rolland 1901: 187–189). But on the other hand, Rolland was aware that private telegraph companies had interests different from those of the states whose shores their cables linked, and that administrative unions such as the ITU and UPU had—for practical considerations—included members that were not sovereign states.

Rolland's tentative solution was to differentiate the proper international society of states from that amorphous group of simple or collective entities (outside of any one state) which conduct some international activity (Rolland 1901: 182–183). Drawing on Pasquale Fiore's concept of "interest," such that an international interest creates international legal personality, Rolland rather presciently concluded that private telegraph companies, international administrative unions, and even "non-sovereign states" had international interests and therefore international legal personality (Rolland 1901: 197–200; Fiore 1918: 116). Hence, it was reasonable that the ITU and the UPU had admitted these non-sovereign entities that were best understood as analogous to states. But the administrative unions had confused the right to be represented in such a union with the right to engage in international acts. Because the ITU and UPU conventions were treaties, only sovereign states should have signed them—the treaty was, after all, a fundamental representation of international society. By contrast, because the regulations concerned international activities of interest to concerned parties, both states and these other "analogous bodies" were competent to sign those documents (Rolland 1901: 226). Rolland's solution, however, was not workable. His categorical definition of international society as composed of sovereign states was threatened by this alternative internationalism institutionalized in the ITU and UPU and, at best, his solution would undo and remake the international administration developed by those international unions.

For so-called semi-civilized nations such as Japan, Egypt, or Persia, or colonies such as India or Hong Kong, membership in the ITU offered an arena open to all governments on the basis of equality. For the fundamental fact was that all members were equal signatories of a convention that committed each to its principles and regulations. In this regard, Herren's argument that international unions provided a "back door to power" helps to illuminate Japan's motivation to become a member of the ITU—and even more the UPU—as a path to power. But this was not the sort of power sought by Herren's examples—Switzerland, Belgium, and the U.S.—who were neutrals on the European front and sought an alternative path to the power that might influence the great powers. Rather, Japan sought to augment the domestic power which it could bring against the colonial powers already on her shores—to assert control over the telegraph and postal administrations and, later, eliminate great power rivals. Japan wanted to assert domestic sovereignty over its own territory and technology, but it did so nonetheless by participating within this larger administrative internationalism that linked sovereign territories as it established international networks of telegraph and postal administrations. Telegraph technology, in other words, and its organization within the ITU supported an alternative model of international society, which sovereign states nonetheless struggled to employ for their own ends.

Acknowledgment I thank Patrice Petro and Maximilian Mayer for their encouragement in developing this essay, and the editors of the volume for their comments on a previous draft. I remain grateful to the University of Chicago Center for East Asian Studies for support of the research informing this work.

References

Ahvenainen, J. (1981). *The far eastern telegraphs*. Helsinki: Suomalainen Tiedeakatemia.

Ahvenainen, J. (2001). The role of telegraphs in the 19th century revolution of communications. In M. North (Ed.), *Kommunikationsrevolutionen: Die neuen Medien des 16. Und 19. Jahrhunderts* (2nd ed., pp. 73–80). Köln: Böhlau.

Ahvenainen, J. (2009). The international telegraph union: The cable companies and the governments. In B. Finn & D. Yang (Eds.), *Communications under the seas: The evolving cable network and its implications* (pp. 61–79). Cambridge: MIT Press.

Barry, A., Osborne, T., & Rose, N. (Eds.). (1996). *Foucault and political reason: Liberalism, neo-liberalism and rationalities of government*. Chicago: University of Chicago Press.

Bell, D. S. A. (2005). Dissolving distance: Technology, space, and empire in British political thought, 1770-1900. *The Journal of Modern History, 77*, 523–562.

Boisson, H. (1932). *La Société des Nations et les Bureaux Internationaux des Unions Universelles Postale et Télégraphique*. Paris: Pedone.

Boyce, R. (2001). Submarine cables as a factor in Britain's ascendancy as a World Power, 1850-1914. In M. North (Ed.), *Kommunikationsrevolutionen: Die neuen Medien des 16. Und 19. Jahrhunderts* (2nd ed., pp. 81–99). Köln: Böhlau.

Briggs, A., & Burke, P. (2009). *A social history of the media* (3rd ed.). Cambridge: Polity.

Burchell, G., Gordon, C., & Miller, P. (Eds.). (1991). *The foucault effect: Studies in governmentality*. Chicago: University of Chicago Press.

Checkland, O. (1989). *Britain's encounter with Meiji Japan, 1868-1912*. London: Macmillan.

Chimni, B. S. (1999). Marxism and international law: A contemporary analysis. *Economic and Political Weekly, 34*(6), 337–349.

Clark, K. (1931). *International communications: The American attitude*. New York: Columbia University Press.

Codding, G. A., Jr. (1952). *The International Telecommunication Union: An experiment in international cooperation*. Leiden: Brill.

Convention télégraphique internationale conclue à Paris, le 17 mai 1865 (1969). In C. Parry (Ed.) *Comprehensive Treaty Series*, Vol. 130 (pp. 198–225). Dobbs Ferry: Oceana.

Dendias, M. (1938). Les principaux services internationaux administratifs. *Recueil des Cours, 63 (1938, pt. 1),* 243–366.

Eijkman, P. H. (1911). *L'Internationalisme scientifique (sciences pures et lettres)*. La Haye: W.P. van Stockum & fils.

Fiore, P. (1918). *International law codified* (From the fifth Italian edition by E. Borchard, Trans.). New York: Baker, Voorhis and Co.

Foreign Office (FO). Great Britain, National Archives, Kew.
Gong, G. W. (1984). *The standard of 'Civilization' in International Society*. Oxford: Clarendon.
Griset, P. (1991). *Les révolutions de la communication: XIXe-XXe siècle*. Paris: Hachette.
Griset, P. (1999). Technical systems and strategy: Intercontinental telecommunications in the first quarter of the twentieth century. In O. Coutard (Ed.), *The governance of large technical systems* (pp. 58–72). London: Routledge.
Headrick, D. R. (1991). *The invisible weapon: Telecommunications and international politics, 1851-1945*. New York: Oxford University Press.
Herren, M. (2000). *Hintertüren zur Macht: Internationalismus und modernisierungsorientierte Außenpolitik in Belgien, der Schweiz, und den USA, 1865-1914*. München: Oldenbourg.
Herren, M. (2001). Governmental internationalism and the beginning of a New World Order in the late nineteenth century. In M. H. Geyer & J. Paulmann (Eds.), *The mechanics of internationalism: Culture, society, and politics from the 1840s to the First World War* (pp. 121–144). London: German Historical Institute.
Hugill, P. J. (1999). *Global communications since 1844: Geopolitics and technology*. Baltimore: Johns Hopkins University Press.
Hunter, J. (1976). The abolition of extraterritoriality in the Japanese post office, 1873-1880. *Proceedings of the British Association for Japanese Studies, 1*(1), 17–37.
International Telecommunications Union. (1965). *From semaphore to satellite*. Geneva: International Telecommunications Union.
International Telegraph Treaty signed at St. Petersburgh July 10 [22] 1875 (1969). In C. Parry (Ed.) *Comprehensive treaty series*, Vol. 148 (pp. 416–484). Dobbs Ferry: Oceana.
Ishihara, F. (1999). *Kokusai tsūshin no Nihon shi*. Tokyo: Tōkai daigaku shuppankai.
Jakobsen, K. (2002). In struggle for control over the Far Eastern Telegraphs: The Great Northern Telegraph Company and Japan (1870-1943). In H. Bonin et al. (Eds.), *Transnational companies (19th-20th centuries)* (pp. 331–345). Paris: PLAGE.
Kamikawa, H. (Ed.). (1958). *Japan-American diplomatic relations in the Meiji-Taisho era*. Tokyo: Pan-Pacific Press.
Kasansky, P. (1897). L'Union télégraphique internationale. *Journal Télégraphique, 21*(8), 180–185.
Kazansky, P. (1902). Théorie de l'administration internationale. *Revue Générale de Droit International Public, 9*, 353–367.
Kern, S. (1983). *The culture of time and space, 1880-1918*. Cambridge: Harvard University Press.
Krajewski, M. (2006). *Restlosigkeit: Weltprojekte um 1900*. Frankfurt am Main: Fischer.
Kunz, O. (1924). *Die internationalen Telegraphen-Unionen*. Stuttgart: F. Enke.
Lindenfeld, D. F. (1997). *The practical imagination: The German sciences of state in the nineteenth century*. Chicago: University of Chicago Press.
Lyall, F. (2011). *International communications: The International Telecommunications Union and the Universal Postal Union*. Farnham: Ashgate.
Lyons, F. S. L. (1963). *Internationalism in Europe, 1815-1914*. Leyden: A.W. Sythoff.
Maejima, H. (1936). *Yūbin sōgyō den*. Tokyo: Teishin kyōkai.
Mangone, G. J. (1954). *A short history of international organization*. New York: McGraw-Hill.
Matthewman, S. (2011). *Technology and social theory*. London: Palgrave Macmillan.
McDowell, S. D. (2002). Understanding shifts in the form and scope of communications governance. In J. N. Rosenau & J. P. Singh (Eds.), *Information technologies and global politics* (pp. 211–237). Albany: SUNY Press.
Miéville, C. (2005). *Between equal rights: A marxist theory of international law*. Leiden: Brill.
Mogami, T. (2006). *Kokusai kikō ron* (2nd ed.). Tokyo: Tokyo daigaku shuppankai.
Moynier, G. (1892). *Les bureaux internationaux des unions universelles*. Genève: A. Cherbuliez.
Murphy, C. N. (1994). *International organization and industrial change: Global governance since 1850*. Cambridge: Polity.
Neumeyer, K. (1924). Les unions internationales. *Revue de Droit International de Science Diplomatiques, Politiques, et Sociales, 2*, 18–41, 139–144, 343–362.

Neumeyer, K. (1925). Les unions internationales. *Revue de Droit International de Science Diplomatiques, Politiques, et Sociales, 3*, 20–25, 102–108.

Nihon Gaikō Monjo (NGM). (1950–1963). Gaimushō (Japan. Foreign Ministry) (Repr. ed.). Tokyo: Nihon kokusai rengō kyōkai.

Pashukanis, E. (2002). *The general theory of law and marxism*. New Brunswick: Transaction.

Rapisardi-Mirabelli, A. (1925). Théorie générale des unions internationales. *Recueil des Cours, 7* (Part II), 331–393.

Reinsch, P. S. (1911). *Public international unions: Their work and organization; A Study in International Administrative Law*. Boston: Ginn & Co.

Rolland, L. (1901). *De la correspondance postale et télégraphique dans les relations internationales*. Paris: Pedone.

Schröter, C. (1900). *Der Weltpostverein: Geschichte seiner Gründung und Entwicklung in 25 Jahren*. Bern: K. J. Wyss.

Simpson, G. (2004). *Great powers and outlaw states: Unequal sovereigns in the international legal order*. Cambridge: Cambridge University Press.

Stolleis, M. (2001). *Public law in Germany, 1800-1914*. New York: Berghahn.

Suginami, H. (1984). Japan's entry into international society. In H. Bull & A. Watson (Eds.), *The expansion of international society* (pp. 185–199). Oxford: Clarendon.

Suzuki, S. (2009). *Civilization and empire: China and Japan's encounter with European International Society*. London: Routledge.

Takahashi, Z. (1969). *Oyatoi gaikokujin: Tsūshin*. [*Oyatoi gaikokujin*, U. Noboru (Ed.), Vol. 7] Tokyo: Kashima kenkyūjo shuppankai.

Takahashi, Z. (1986). *Tsūshin [Nihonshi shōhyakka]* (Vol. 23). Tokyo: Kondō Shuppansha.

Triepel, H. (1899). *Völkerrecht und Landesrecht*. Leipzig: Hirschfeld.

Tuathail, G. Ó. (1992). Putting Mackinder in his place: Material transformations and myth. *Political Geography, 11*(2), 100–118.

Vec, M. (2006). *Recht und Normierung in der industriellen revolution*. Frankfurt am Main: Klostermann.

Vec, M. (2009). Die Bindungswirkung von Standards aus rechtsgeschichtlicher Perspektive: Globale Normsetzung und Normimplementation am Beispiel des Weltpostvereins von 1878. In T. M. J. Möllers (Ed.), *Geltung und Faktizität von Standards* (pp. 221–251). Baden-Baden: Nomos.

von Kirchenheim, A. (1880). Les congrès internationaux de la poste et du télégraphe. *Revue de Droit International et de Législation Comparée, 12*, 465–477.

Walter, R. (2001). Die Kommunikationsrevolution im 19. Jahrhundert und ihre Effekte auf Märkte und Preise. In M. North (Ed.), *Kommunikationsrevolutionen: Die neuen Medien des 16. und 19. Jahrhunderts* (2nd ed., pp. 179–190). Köln: Böhlau.

Westney, D. E. (1990). Building the National Communications System. In J. W. White, M. Umegaki, & T. R. H. Havens (Eds.), *The ambivalence of nationalism: Modern Japan between East and West* (pp. 39–59). Lanham: University Press of America.

Wieacker, F. (1995). *A history of private law in Europe, with particular reference to Germany* (T. Weir, Trans.). Oxford: Clarendon.

Winseck, D. R., & Pike, R. M. (2007). *Communication and Empire: Media, markets, and globalization, 1860-1930*. Durham: Duke University Press.

Woolf, L. S. (1916). *International government: Two reports*. New York: Brentano's.

Yabuuchi, Y. (1975). *Nihon yūbin sōgyōshi*. Tokyo: Yūzankaku.

Yang, D. (2010). *Technology of empire: Telecommunications and Japanese expansion in Asia, 1883-1945*. Cambridge: Harvard University Asia Center.

Yūseishō (ed.). (1971). Yūsei hyakunen shi. Tokyo: Yoshikawa kōbunkan.

Zacher, M. W. (2002). Capital, technology, and liberalization: The International Telecommunications Regime, 1865-1998. In J. N. Rosenau & J. P. Singh (Eds.), *Information technologies and global politics* (pp. 189–210). Albany: SUNY Press.

Biotechnology and Empire: The Global Power of Seeds and Science

Sheila Jasanoff

Abstract Following the cold war, interest grew in the possible rise of new forms of imperial rule and in the likely role of science and technology in processes of global governance. In particular, just as the life sciences advanced the interests of bygone empires, so modern biotechnology is poised to support today's transboundary exercises of political, economic, and cultural power. Drawing on analyses of large-scale political and technological systems, this paper suggests that contemporary biotechnology may be enrolled into empire-making in several different modes, including bottom-up resistance, top-down ideological imposition, administrative standardization, and consensual constitutionalism. At present, biotechnology seems more likely to increase the power of metropolitan centers of science and technology than that of people at the periphery. Institutional innovations will be needed to bring global biosciences and biotechnologies under effective democratic control.

Keywords Biotechnology • Empire • Agriculture • Power • Transnational enterprises

Books about imperialism, many sporting the word "empire" in their titles, appeared by the dozen at the turn of the twenty-first century (influential contributions include: Johnson 2000; Cannadine 2001; Hardt and Negri 2000, 2004; Hall 2002; Colley 2003; Harvey 2003; Ferguson 2003, 2004; Khalidi 2004; and Slaughter 2004). Through them, and through endless journalistic commentaries (c.f.: Ignatieff 2003; Maier 2009), the attention of much of the reading world turned to a particular instance of imperial expansion: the post–cold war United States, driven by what

I am grateful to the Universities of Wageningen, Netherlands, and Halle, Germany, for invitations to present earlier versions of this paper. This chapter is an updated and revised version of "Biotechnology and Empire: The Global Power of Seeds and Science," *Osiris* (2006) 21: 273–292.

S. Jasanoff (✉)
Harvard University, Cambridge, MA, USA
e-mail: sjasan@fas.harvard.edu

M. Mayer et al. (eds.), *The Global Politics of Science and Technology - Vol. 1*, Global Power Shift, DOI 10.1007/978-3-642-55007-2_10,

many saw as a runaway ambition to impose military dominance, ideological conformity, and cultural homogeneity on the rest of the world.[1] It was as if a potential left implicit by Ronald Reagan's famous appellation for the Soviet Union—"the evil empire"—came to fruition in George W. Bush's Manichaean vision, which pitted an imagined, divinely blessed, "good" America against its "evil" enemies, the states that harbor terror. Global power struggles were recast as a fight to the finish between the imperial forces of light and of darkness. Presidential rhetoric reprised popular culture: George Lucas's hugely successful trilogy of intergalactic conflict, *Star Wars*,[2] provided not only the template for dividing the world into two vast opposing armed camps but also the visual and metaphorical resources for reducing warfare between them to the starkness of black and white.

Empires, however, are patchier constructs than the simple dualisms of presidential imaginations, shaped by Hollywood imagery, would have us believe (c.f. Judt 2004). Neither culturally nor normatively homogeneous, they invite analysis as spaces in which power is exercised through complex, often subterranean means. From the Roman imperium to the territories ruled by Britain at the height of its Victorian expansion, diversity rather than homogeneity has been the characteristic look of empire. Possibly the most successful empires have been those that allowed multiple divergences in language, religion, dress, diet, and customs to flourish, within an envelope held together by various consolidating moves that coordinated, but did not erase, difference. For insights into these processes, we may turn to scholars of colonialism and postcolonialism, who have pointed out the disparate moves made to differentiate, as well as integrate, the populations under the ruling regime's control. On the one hand were steps that clarified and firmed territorial boundaries, imposed common linguistic and educational standards, and produced shared categories to reason and rule with (c.f. Anderson 1991; Radcliffe 2001). On the other hand were strategies for preserving hierarchies of power, including rules of cohabitation allowing or disallowing mixing between the rulers and the ruled (Stoler 1989, 2002). Empires then were places of hybrid identities, with all the tensions for regularity and order that hybridity entails.[3] The wonder is that they nonetheless held and that similar formations may yet hold in other times and places. In this respect, empires can be seen as analogous to large technological systems,

---

[1] American progressives would like to detach what many see as the illegitimate path of unilateral militarism from the legitimate, indeed desirable, path of economic and social globalization driven by the "soft power" of culture and markets, see Joseph S. Nye (2004). Celebrations of America's role in leading the world to free-market democracy include Friedman (1999).

[2] Directed by George Lucas, the trilogy opened in 1977 with *Star Wars*, the film that gave its title to the series. It was succeeded by *The Empire Strikes Back* (1980) and *Return of the Jedi* (1983). Appearing in the waning years of the cold war, the films exercised a particular influence on Ronald Reagan, America's first Hollywood president. The idea of a satellite-based missile defense shield was initially broached in the Reagan era, and the project, which remained mired in conflict during his presidency, was nicknamed Star Wars.

[3] See, e.g., the account of collectors and collecting in the eighteenth-century British and French proto-empires, Maya Jasanoff (2005).

like electric power grids (Hughes 1983) or civil aviation: so complex, heterogeneous, loosely pinned together, even jerry-built on close inspection that their stability is the thing that needs explanation. By contrast, as illustrated by the terrorist attacks of September 11, 2001, in the United States, mundane technological systems such as high-rise buildings, regarded as not seriously vulnerable to external threats, can reveal deep structural faults under unexpected attack (9/11 Commission 2004).

Viewing empires as social technologies, that is, as human-made assemblages that enable power to extend beyond its original spatial and cultural locations[4] raises a critically important set of questions for international relations. What is the role of conventional technological systems, those built around material components such as guns, butter, and newspapers, in the production and maintenance of new forms of transnational rule? How, in particular, might the human capacity to instrumentalize nature influence the possibilities for cross-border politics in a globalizing world? Will the major technological revolutions of our time—in the life sciences, information and communication technologies, computers and weaponry, energy systems, and most recently nanotechnology—favor emancipation or recolonization? Will they make people around the world more or less connected, more or less free, more or less comfortable, and most important for our purposes, more or less democratic? Will the radically unequal distribution of wealth and privilege in the contemporary world reinscribe itself through technological means, continuing older forms of hegemony and dominance? If that danger exists even in principle, are there institutions or processes through which a global citizenry can assert the right to shape the technologies that may, if widely deployed, shore up global regimes of control?[5]

I approach these questions in this chapter through the lens of modern agricultural biotechnology. Still in its infancy more than three decades after its first experimental successes in western laboratories, so-called green biotechnology has rapidly become a global industry promising enormous benefits to the world's poor. Its proponents claim it has the capacity to overcome nature, making plants that can resist drought, ward off insects, and with the ability to produce micronutrients engineered into their genes, even transcend the "normal" dividing line between food and pharmaceuticals.

Biotechnology by some definitions is as old as "second nature," the first successful prehistoric attempts by human societies to harness nature's growth to serve their basic needs for food, fuel, clothing, and shelter. Under another definition, the one I use here, biotechnology is much newer. It is the name given to an array of manipulative techniques based on alterations of the cellular and subcellular

[4] This way of thinking about empires is consistent with contemporary work in science and technology studies—see, in particular: Jasanoff (2004), Latour (1990). Richard Drayton (1993) adopts a similar perspective when he speaks of empire as "an ecological system," stressing the interconnections among politics, economy, and nature that define empires.

[5] For an argument that such demands are already being expressed through a tacit and unwritten form of global constitution-making, see Jasanoff (2003).

structures of living things enabled by the 1953 discovery of the structure of DNA (Bud 1993). These techniques include, most notably, not only genetic engineering, or gene splicing, but also operations such as cell fusion and cell culturing carried out at levels of structure significantly smaller than the whole organism. How will these technological developments, heralding what some have called a second Green Revolution,[6] affect flows of power and opportunities for self-determination around the world?

In looking for answers, I begin in effect with a typology of empire, based on the diverse ways in which the extension of imperial power has been conceptualized by analysts of large-scale political, as well as technological systems. The life sciences, as much research has shown, have long been implicated in serving the designs of empire builders. Modern biotechnology, I suggest, can similarly be drawn into the service of possible imperial constructions, and I ask in what ways this particular global production system is likely to influence today's transboundary exercises of political, economic, and cultural power. This analysis suggests that, without institutional innovations, biotechnology as currently governed may increase the power of so-called metropolitan centers of science and technology in relation to people regarded as being at the periphery. In conclusion, I reflect on the prospects for democratic governance of technological systems such as agricultural biotechnology that are centrally involved in contemporary processes of political and economic globalization.

1 Imperial Constructions

How are empires held together? Not mainly, as I have suggested, through homogenized identities and uniform allegiances that make the residents of imperial territories as if carbon copies of one another. Clues may be found in those areas of the social sciences that occupy themselves with the stability of heterogeneous constructs, in such fields as international relations and law, science and technology studies, colonial and postcolonial history, and cultural anthropology. Work in all these domains suggests that the fabrication of empire proceeds not through any single grand gesture of unification, nor by a revolutionary process of mass struggle as suggested by two theorists of the Left, Michael Hardt and Antonio Negri (2000) but through a series of contingent, overlapping, altogether human practices that build coherence and cohesion while staving off dispersal.

As shown in Table 1, we can discern five distinct modes of imperial governance—that is, five mechanisms, not mutually exclusive, through which the unruly

[6] The first Green Revolution was the introduction worldwide of high-yielding grain varieties pioneered by Nobel laureate Norman Borlaug and other plant biologists. Their work was sponsored in part by the Rockefeller Foundation. For accounts of the scientific and social dimensions of the Green Revolution, see Anderson et al. (1988), Hazell and Ramasamy (1991), Kay (1993).

Table 1 Modes of imperial governance

Empires of resistance	Emergent, agentless form of rule, constituted in possibly violent opposition between global ruling institutions and resisting citizens ("the multitude")
Empires of ideology and force	Communal norms and beliefs imposed through force, persuasion, surveillance, and sanctions
Empires of legibility	Communal standards imposed through administrative simplification and efficiency (Weberian)
	Communal standards achieved through classification, normalization, and erasure (Foucauldian)
Empires of identity	Imagined communities built through mass media, official representations, political and cultural symbols
Empires of law and constitutions	Rule of law under constitutional principles, enabling liberal individualism and free movement of goods and people

heterogeneity of empires can be made more orderly and therefore more tractable to transborder rule.

The vision of empire put forward by Hardt and Negri stands in a somewhat anomalous relation to the others in Table 1, partly because the empire they envision is a global formation lacking any particular sovereign at the head, and partly because of the authors' disregard for the micro-processes of agency and governance that have loomed large in the work of other theorists of national and imperial power.[7] The empire whose emergence Hardt and Negri ambitiously prophesy is a revolutionary construct, propelled in part by the consolidation of a global multitude whose demands nation-states are no longer able to satisfy. Bottom-up political action in an inchoate agonistic field, mediated through the Internet, is seldom strategic or coordinated, but, through repeated, decentralized gestures, it can achieve something of the character of continuous mass protest. Hardt and Negri's account has drawn vigorous criticism for its lack of clarity, inattention to specifics, denial of agency, and leftist nostalgia for violence as a means of radical social change. At the same time, it provides a vision of uncoordinated, multicentric, populist, political, and normative action—propelled by ideas and beliefs—that is, in some ways, more appealing than the tight, and equally faceless, administrative networking of the world contemplated by analysts grounded in positive international law or international relations theory (Slaughter 2004). Something resembling the dynamics of the multitude, as we will see below, is not altogether absent in the contemporary global politics of biotechnology.

Turning to more conventional articulations of empire, those constituted by (or *as*) an identifiable sovereign state, we note that the processes and practices that sustain imperial rule do not have to be consensual or responsive to the popular will and that violence remains very much an instrument of top-down domination. This is clearest in the case of empires of ideology and force, such as the former

[7] Contrast in this respect Hardt and Negri (2004), with Slaughter (2004). See also Hale and Slaughter (2005).

Soviet Union and perhaps the American empire, powered by weapons such as drones, in which adherence to a common ideology (socialism and market capitalism, respectively) has been achieved through the forceful subordination of countervailing belief systems and forms of life. Technology, historically, played a central role in the effectuation of such extended ideological dominion: not only military technologies, though these were of course essential, but also technologies of surveillance, punishment, and mass communication. Built to control hybridity, such control technologies are themselves hybrid, marrying the hardware of computers or cameras, for instance, with social supports from law and administration, and increasingly the mass media.[8] In this way, technologies of force shade into technologies of legibility and standardization, which are tools of imperial construction in their own right.

When we speak of standardization as a form of political control, we begin inevitably with Max Weber. Empires were, in the first instance, vast bureaucracies. They were administered domains, and their management called forth the production and diffusion of professionals of every stripe: scientists, engineers, surveyors, physicians, lawyers, linguists, archaeologists, and archivists, among others. Their task was to make government more efficient and rational, enable communication and exchange, and—in the more beneficent of imperial imaginations—extend the virtues of knowledge, reason, and productivity equally throughout the empire. Less clear in Weber's time, however, was the extent to which the imposition of administrative rule altered, or even created, the identities of the subjects being governed. Nor did Weber put front and center how the administration of natural resources, such as oil, or attempts to conquer threats from nature, such as vector-borne disease, could interact with administering people, and so produce new topographies of power and subservience (Mitchell 2002).

Michel Foucault's work on "governmentality" fills that gap, and it provides another indispensable starting point for understanding imperial standardizations today (Foucault 1986). Foucault represented governmentality as a specific form of rule that emerged with European modernity, coincident with the waning of absolute monarchical power and the rise of science. In this social order, the governors and the subjects to be governed became part of the same enterprise, linked through their allegiance to new truth regimes, grounded in technical disciplines (preeminently the human sciences) that provide the means for authoritatively characterizing both social bodies and social problems. Experts trained in professional discourses can identify populations and, through clinical work, their individual members as healthy or sick, sane or mad, normal or deviant, racially pure or impure, criminal or socially responsible. These definitions become essential, not only to those who exercise power to keep illness, insanity, deviance, racial commingling, and criminality at bay, but also to their subjects, who, as what the philosopher Ian Hacking has called "interactive kinds," (Hacking 1999) come to see and recognize each

[8] Consider, for example, the U.S. military's practice of "embedding" journalists with ground forces during the conduct of the 2003 Iraq war.

other in terms of the dominant classification systems of their time and place. Bureaucracies fitted out with elaborate expert support systems develop norms and regulations based on the experts' classifying knowledge (Bowker and Star 1999; see also: Jasanoff 1990). Government (the project of the rulers) and mentality (the state of mind of the ruled) then fuse, as both begin to perceive the world in identical conceptual terms and reinforce each other's perceptual frames.

Governmentality, despite its pretensions of neatness, seldom divides the world into cleanly defined categories. It takes work of a special sort—specifically, boundary work—to smooth out the messy spaces between classes and to create the appearance of sharp divisions, or bright lines as lawyers call them (Jasanoff 1990; Gieryn 1999). In the process of classification, problematic hybrids and hard-to-fit entities or communities may be erased, either through forcible elimination or through administrative and symbolic moves, such as selective mapping or listing, that take the unclassifiable things out of the ruler's visual space. Thus, unproductive citizens may be cleared out of slums and city streets (Collins and Bromley 2003),[9] wildernesses replaced by planned forests, last names substituted for patronymics, and medieval streets overlaid with the familiar grid pattern of the surveyable and policeable modern city.[10] The political theorist James Scott refers to these simplifications as a process of creating legibility, a concept lying somewhere between Weberian administrative efficiency and Foucauldian governmentality. Modern statecraft, Scott argues, has consisted in the main of taking "exceptionally complex, illegible, and local social practices" and creating "a standard grid whereby it could be centrally recorded and monitored" (Scott 1998: 2).

While Scott and, to some extent, Foucault stress the role of the state and its docile experts in making knowledge and order, others have asked (as indeed Foucault did in connection with the "mentality" component of governmentality) how subjects buy into the imperial projects of which they are a part. James Morris's splendid popular account of the British Empire at what he calls the moment of its climax in 1897, the diamond jubilee of Queen Victoria (Morris 1979), provides one illustration on an imperial scale of the argument advanced by Benedict Anderson in his influential treatment of nationhood. A nation, Anderson suggested, is best regarded as "an imagined political community—and imagined as both inherently limited and sovereign." (Anderson 1991: 6) Characterizing what holds a nation, or, in Morris's case, an empire together then becomes a task for history and ethnography, for the definition orients our attention to the practices through which the state

[9] Under India's prime minister Indira Gandhi, in close association with her son Sanjay Gandhi, the slogan *garibi hatao* (eradicate poverty) became equated with a program of forcible slum clearance—in other words, eradicating not poverty but the visibly poor.

[10] Laid out on modern lines in the 1950s by the French-Swiss architect Le Corbusier, at the behest of Prime Minister Jawaharlal Nehru, the city of Chandigarh, the capital of Punjab and Haryana, accommodates a degree of traffic surveillance that I have not encountered in other Indian cities. Just over a hundred years before Chandigarh was inaugurated, Baron Georges-Eugène Haussmann substantially rebuilt Paris for Napoleon III, razing many old districts and replacing winding streets with broad boulevards so that the state could better control potential revolutionaries.

and its minions train the collective imagination of a national or imperial community. Morris's imperial moment called forth an unprecedented outpouring of celebration and circulation, of people, goods, vessels, language, profits, and plants, that criss-crossed the empire on which, famously, the sun never set. But what of the work that was needed to produce such a worldwide convergence? To see this, we need more disciplined histories.

Anderson, his own imagination challenged by the unlikely agglomerate of the Indonesian nation-state, stressed the role of structuring élites, in particular the unifying work of the print media and, in an elaboration of his original argument, also of the state-sponsored census, map, and museum. Whereas Anderson looks primarily to the public spaces and instruments of national identity-making, Ann Stoler, the feminist anthropologist, provides a Foucauldian account of the intrusions into private life undertaken by colonial regimes for the sake of creating and maintaining relations of dominance. In the Dutch East Indian colonies, she argues, carefully constructed rules governing sexual relations among Europeans and between whites and natives preserved necessary demarcations between the governors and the governed (Stoler 1989, 2002). For both Anderson and Stoler, making empires is an active, creative, and dynamic process of ordering, centering on producing and, especially for Stoler, reproducing a vision of the thing being made.

The emergence of the European Union (EU) as an autonomous political force in the late twentieth century illustrates one more modality of imperial construction, based in part on constitutional principles and the rule of law, and designed to further the free flow of goods and services in an open market. The EU's tightening integration through successive treaties, the adoption of a common currency (the euro), the admission of ten new member states in May 2004, and the signing of a constitution in Rome on October 29 of the same year marked the production of a new kind of empire, but one allegedly founded on the democratic consent of its citizens.[11] Declining turnout in EU parliamentary elections, widespread popular disenchantment with Brussels, the stinging rejection of the EU constitution in French and Dutch referenda in 2005, and rumblings over a EU referendum in Britain were indicators that constitutionalism on such a scale carries huge risks of alienation along with the promise of enhanced economic and political integration. The Eurozone crisis beginning in 2009 revealed added vulnerabilities in a union whose currency system had arguably been prematurely standardized, without due attention to the social and political dimensions of member states' spending policies. What matters for our discussion, however, is the very availability of a constitutional process, with all of its positive connotations for democracy, in validating the EU's supranational authority; even the continuing debates on membership and the hotly contested bailout programs may be seen, at one level, as supporting the idea of a

[11] The ten new members met the so-called Copenhagen criteria, according to which they had to "be a stable democracy, respecting human rights, the rule of law, and the protection of minorities; have a functioning market economy; and adopt the common rules, standards and policies that make up the body of EU law." See http://ec.europa.eu/enlargement/ (accessed Feb. 2014).

common European project. We will return below to the implications of the constitutional model of imperialism for governing biotechnology globally.

2 Biology in the Service of Empire

As if echoing the explosion of historical and political writing about empire, there has been an explosion of writing on the uses of science in the cause of imperial expansion, with the scientific management of nature commanding center stage. Colonial historians have observed that the human and biological sciences came into their own to serve imperial needs from the eighteenth century onward, in much the same way that Scott's twentieth-century planning states used engineering and social sciences to achieve legibility. Anthropology, botany, ecology, geography, linguistics, and even early forensic sciences have deep colonial roots: to rule effectively, occupying governments had to map their territories, classify populations into identifiable groups, and catalog flora, fauna, languages, and cultural practices.[12]

Making things grow, often under unfavorable natural conditions in nonnative habitats, gave a push to imperial ecology, conservation biology, and agricultural science (MacKenzie 1990; Rajan 1996; Anker 2001). Sometimes the motives were crassly extractive and exploitative, as in the harvesting of wild rubber in King Leopold II's Belgian Congo, where violence and force were the notorious instruments of colonial rule (Hochschild 1999). Elsewhere, colonists heedlessly harvested tropical timber or took commercially useful plants such as cinchona (from which quinine is derived) or breadfruit for cultivation in new territories (see: Philip 1995; Drayton 2000). Sometimes otherwise well-intentioned migrations had disastrous results. For instance, rabbits transported to Australia for hunting became an uncontrollable pest, as Morris colorfully records (Morris 1979: 77f.). Yet more altruistic motives also prevailed. Richard Grove traces the roots of western environmentalism to early modern European encounters with tropical islands (Grove 1995). As self-contained and containable spaces, these islands appealed to voyagers' Edenic and Romantic sensibilities, as well as to their protective instincts. Lush islands brought to life idyllic conceptions of the gardens of paradise; at the same time, in those bounded preserves, travelers could easily observe the destructive effects of resource depletion and environmental degradation. The island of Mauritius, in Grove's account, became the site of some of the world's earliest systematic efforts at nature conservation and scientific forest management. These practices, in turn, provided practical models for conservation efforts in India and elsewhere from the 1830s onward (Grove 1995 9f., 168–263).

[12] On colonial histories of the human and natural sciences, see Cohn (1996), Edney (1997), Philip (2004). On the colonial origins of fingerprinting, see Cole (2001).

Colonial enterprise also laid the basis for western ideologies of development. Along with concerns for the moral and religious education of the strangers they went to live among, the rulers of empires exhibited a compelling desire to improve the new territories under their command. British engineers laid roads and railways, built irrigation systems, and left indelible architectural imprints throughout India. Just as pervasive was Britain's (and in other regions, France's) engagement with botany and agriculture. Already in the early nineteenth century, a coalition of professional scientists and administrators had converted the Royal Botanic Gardens at Kew into a publicly run center of knowledge for the productive management of nature (Drayton 2000). Problems of sugar cane cultivation in the West Indies led to the formation of the Imperial Department of Agriculture at the end of the nineteenth century. A source of scientific expertise for West Indian sugar cane growers, the department also became, under the leadership of Joseph Chamberlain, the Liberal secretary of state for the colonies, a breeding ground for early discourses of development (Storey 2004). Like enlightened estate managers back home, those entrusted with the welfare of colonial "properties" felt a need to ameliorate the conditions of life for the local poor. Promoting development abroad, they also thought, would transform the colonies into more advantageous trading partners, thereby producing useful returns for domestic constituencies. Improving agricultural production was a favored route to achieving these goals, although access to metropolitan knowledge remained stratified, with native farmers, in many cases, continuing to cultivate their lands without the benefits of modern science (Storey 1997).

The first half of the twentieth century cast the imperial project of biology in a darker light as the improvers' attention turned toward standardization for control, and broadened to include humans in addition to plants and animals. The enthusiasm of progressive social reformers for eugenics at the turn of the century led to decades of discrimination in the United States, including the exclusionary Immigration Act of 1924, numerous state sterilization laws, and *Buck v. Bell,* the infamous 1927 Supreme Court decision upholding the sterilization of a Virginia woman, Carrie Buck, on the ground that "[t]hree generations of imbeciles are enough."[13] The eugenicists' concern for selective breeding and race purity was carried to pathological extremes in the Nazi period, when millions of humans deemed undesirable by German race theorists—Jews, gays, Gypsies—were uprooted and eliminated throughout the Third Reich. For the sociologist Zygmunt Bauman, these atrocities were the natural descendants of the same enlightenment ideals that had led Frederick the Great of Prussia to exclaim, "It annoys me to see how much trouble is taken to cultivate pineapples, bananas and other exotic plants in this rough climate, when so little care is given to the human race." (Bauman 1991: 27) The modern "gardening state," Bauman argues, turned Frederick's metaphor into crude reality

[13] Justice Oliver Wendell Holmes Jr., an enthusiast for eugenics, wrote the majority opinion in *Buck v. Bell,* 274 US 200 (1927).

by ruthlessly weeding out everything that its planners saw as standing in the way of reason, order, and progress.

In spite of these midcentury turmoils and disruptions, the alliance between biology and power has only grown more intimate and pervasive in subsequent decades. Foucault saw biopower and biopolitics as essential technologies with which modern states must control their populations—by assuming responsibility for the health, safety, and stability of citizens' collective lives (Foucault 1978). Central to the exercise of biopower, then, is the state's ability to characterize human bodies and behavior in ways that rationalize and, in democratic societies, publicly justify that state's policies. Increasingly, the state asserts itself under the umbrella of epidemiology: as the master diagnostician of ills that threaten groups of people in society. The polarizing debates on gay marriage before and during the 2004 U.S. presidential campaign may be seen in this light as part of a more general discourse on sexuality and the family, with competing political factions claiming citizens' allegiance by defining what counts as deviance in sexual behavior and family mores. In the culturally heterogeneous United States, as in Stoler's East Indian colonies, the rules of sexual conduct serve as powerful instruments for building social cohesion, by decreeing who falls inside and who outside the accepted forms of domestic order. A decade later, as the United States began converging toward a liberal national policy on gay marriage, the rising salience of homosexuality as a political issue in other regions evidenced a resistance to the hegemonic extension of U.S. values and the relevance of biopower as an instrument of international relations in a globalizing world.

Today as before, moreover, biopower extends into all of life on the planet, not only the lives of humans but also the natural worlds with which humans live in close symbiosis. Sick and failing plants, no less than sick and failing people, fall within the biopolitical imagination of the neoliberal state and its corporate partners, whose innovative capacity is as essential to underwriting state action as is the capacity of expert professionals to define and apply the technical criteria of governmentality.[14] Governing *bodies*, after all, proceeds not only through exclusion, or weeding out, but also through therapeutic processes of making whole and bringing the previously sick back into the community of viable beings. The ordering state is most powerful when it is at the same time, demonstrably, a healing state, and such a state engages science for therapeutic, as well as diagnostic, ends. Let us return, then, to agricultural biotechnology as a field of contemporary biopower that continues the historical partnership of the life sciences with the state and, in so doing, intersects with each of the modes of empire-building described above.

[14] For an account of the changing social contract among science, state, and industry with respect to the life sciences, see Jasanoff (2005a).

3 Plants for the Planet: The Empires of Biotechnology

Apart from occasional radical social misfits such as the so-called Unabomber, Theodore Kaczynski,[15] few any longer question the vital role of science and technology in human development. Even opponents of particular technological projects—large dams (Khagram 2004), for example, or genetically modified (GM) foods[16]—rarely dismiss technology outright; rather they favor smaller, more transparent, or more locally governable technological systems. The question that preoccupies students of science and technology, then, is not whether, but how, to integrate innovation into people's lives so as to make a positive difference. Years of research in the social psychology of risk perception (c.f. Slovic 2000) and public understanding of science (Wynne 1995) have established that popular fear or rejection of new technology often rests, at bottom, on an uneasiness about the ways in which technology is managed or, more accurately, governed. What do these observations imply for an industry with global ambitions, like agricultural biotechnology? How, more specifically, does biotechnology contribute to ways of political worldmaking beyond the nation-state, and what implications do the engagements between biotechnology and global politics have for democratic governance?

In reaching for answers, it is useful to think of biotechnology operating politically in several different registers. It is, of course, most plainly a material technology: it makes new instruments for warding off harm and disorder, such as plants that resist insects, weeds, or drought, and it redesigns pieces of nature, such as genes, to perform new tasks in new environments. In this respect, biotechnology is, concurrently, a metaphysical and a normative device; it brings new entities into the world and through that process reorders our sense of rightness in both nature and society (Jasanoff 2005b). At the same time, biotechnology is a discourse: to some, of progress and improvement, beneficence and utility; to others, of risk, invasiveness, and domination from afar. Proponents of agricultural biotechnology tell particular stories about a world in which plant genetic modification is possible, and these stories carry political and cultural weight. Lastly, biotechnology is an institution of governance; it shapes forms of social life by influencing how people choose to, or are able to, live with the products of bio-industry. Each of these registers, as we see below, has been activated in the global politics of biotechnology.

[15] Theodore Kaczynski, a mathematician educated at Harvard and the University of Michigan, conducted a single-handed letter-bombing campaign against representatives of various industries from his cabin in Montana between 1978 and 1996. These attacks killed three people and injured many others. He was caught when his brother recognized as his work a long letter he had sent to the *New York Times*, see Kaczynski (1995).

[16] On transatlantic divisions over genetically modified crops and food, see Bernauer (2003).

3.1 The Resisting Multitude

In May 2004, a scientific journal reported that German researchers were keeping secret the locations of some thirty sites planted with GM corn for fear that anti-GM activists would destroy the crops, as they previously had elsewhere in Germany (Stafford 2004). Failure to disclose these locations was contrary to the EU Directive 2001/18, which requires GM crop sites to be publicly registered. Noncompliance with European law in traditionally law-abiding Germany may have been newsworthy, but the threat to GM crops was anything but novel. From the late 1990s onward, attacks on field trial sites began evolving into a form of international protest that seemed to epitomize Hardt and Negri's thesis about an emerging, assertive, global multitude: in Britain, hundreds of demonstrators dressed in decontamination suits uprooted GM plants in test fields in 1999; in India and Brazil, farmers' unions organized similar protests; in France, José Bové, the charismatic head of the radical Confédération Paysanne (Peasant Confederation), became a folk hero by orchestrating the destruction of thousands of GM plants, as well as a partially built McDonald's outlet, in 1999. His subsequent trial, fine, and terms of imprisonment left him and his supporters undaunted, indeed ready to resume battle as much as 5 years after their initial transgressions. The movement continued well into the second decade of the twenty-first century, with farmers in the Philippines destroying field trials of nutrient-enriched "golden rice" while scientists deplored what they saw as irrational prejudice and latter-day Luddism.

Field trial sites were not the only theater of protest against GM agriculture. Antiglobalization activists early identified biotechnology as a symbol of the environmental, economic, and cultural homogenization they wished to resist. Demonstrations against Monsanto and GM corn (or maize), together with evocations of risks to nontarget species such as the monarch butterfly, were part of the repertoire of street protest during the Third Ministerial Conference of the World Trade Organization (WTO) in Seattle in 1999. In this and similar episodes, representatives of a loosely networked global citizenry asserted their right to debate technological futures in terms other than those conventionally used by nation-states and their expert advisers: the formal discourses of law, molecular biology, economics, risk assessment, and bioethics. At stake was who had power to determine how much global harmonization there should be and which scientific, technological, and economic innovations should be allowed to diffuse throughout the world. Those opting for more local, bottom-up visions won a salient victory when Monsanto decided, under rising public pressure, to withdraw its plans to develop sterile seed technology, through use of the so-called Terminator gene[17]; later, citing a drop in

[17] The Terminator gene would have disabled grain seeds from sprouting in consecutive years. Farmers who had routinely planted seed stored from the previous year's harvest would then have been forced to buy new seed each year. The coalition that forced Monsanto to abandon this technology, at least for a time, included both indigenous organizations and the influential Rockefeller Foundation (Jasanoff 2003: 171).

global demand, the company also announced that it would put on hold its plans to market genetically modified Roundup Ready wheat.[18]

3.2 *Ideology and Enforcement*

Not everyone saw the antiglobalization movement as the promising vanguard of planetary resistance against an outmoded, corporate-dominated, neoliberal world order. Using the classical ordering machinery of science and the law, proponents of agricultural biotechnology sought to promote their visions of social and technological progress, stifling opposition and dissent.

At the February 2000 annual meeting of the American Association for the Advancement of Science, Senator Christopher "Kit" Bond, Republican from Missouri, Monsanto's home state, was openly dismissive of the Seattle protest. He represented it as a struggle between scientific expertise and the misguided, if exuberant, ignorance of youth: "The scientific debate is not being controlled by Ph.D.s but apparently by young people with a proclivity for street theater. (. . .) It's coming to the point that scientists are going to have to get dressed up as corncobs to get the attention of the media."[19] At the same meeting, Madeleine Albright, a Democrat and President Clinton's secretary of state, also cast the conflict as one between reason and unreason. "But science," she said, "does not support the 'Frankenfood' fears of some, particularly outside the United States, that biotech foods or other products will harm human health."[20] Both speakers, from different political parties, enlisted science as their ally in defending biotechnology against its critics. This invocation of scientific authority in support of technological innovation is a marker of America's commitment to a particular ideology of technoscientific progress (Jasanoff 2005a: chapter 5).

A look across the ocean at contemporaneous UK debates on biotechnology helps bring into relief the ideological dimensions of the American position. The term "Frankenfood" was widely used in the British tabloid press to reflect and, some said, reinforce public anxieties. But concerns were not restricted to the media and the ignorant public. The British scientific community had all along expressed greater uncertainty about the safety of GM crops than its American counterpart, particularly with respect to the environmental consequences of commercial use

[18] Roundup is a popular weed killer marketed by Monsanto, and Roundup Ready plants are genetically modified to withstand the use of that product. Many observers thought Monsanto's decision was motivated by opposition to GM crops in Europe and Japan (New Scientist 2004).

[19] Senator Christopher Bond, Annual Meeting, American Association for the Advancement of Science, Washington, DC, Feb. 23, 2000.

[20] Secretary of State Madeleine Albright, Annual Meeting, American Association for the Advancement of Science, Washington, DC, Feb. 21, 2000.

(Jasanoff 2005a: chapter 2). These doubts led British experts to reject the official U.S. position that the process of genetic modification carries no special risks; all that matters for regulatory purposes is the end product. Scientific and public opinion in Britain united behind a more cautious approach, demanding more experimentation—for example, through farm-scale trials (AEBC 2001)—before authorizing the commercialization of GM crops. As doubts intensified, Tony Blair's government decided on a highly unusual three-pronged review of the science, economics, and public acceptability of these products to reevaluate the case for their introduction.[21] The immediate outcome of this process was a decision to approve the commercialization of only one variety of GM corn, at least to start. Thus, while American neoliberalism treated biotechnology as just another stream of products, adequately controlled by the market except for assessments of their safety to human health and the environment, Britain's more cautious and communitarian political culture granted the public some say in deciding which products they wanted to allow into the market.

Whereas consultative procedures such as Britain's GM debate and referenda in countries such as Denmark and Switzerland sought to defuse public opposition, elsewhere legal sanctions were employed to beat down what biotechnology promoters saw as unacceptable acts of intransigence. Thus, demonstrators such as José Bové who destroyed GM crops were prosecuted for damaging property in several countries. At the international level, the United States brought a case against the EU at the WTO for imposing an allegedly illegal moratorium on the importation of GM crops and foods. Foundational to the U.S. case was the argument that there were no good scientific reasons for keeping these products off the European market, and that the moratorium therefore amounted to illegal protectionism.[22]

Intellectual property law, too, has been invoked in safeguarding the investments made by multinationals such as Monsanto in GM crops. Particularly interesting were the prosecutions brought against farmers in the United States and Canada who were found to be growing GM crops patented by Monsanto without a license. In one of the best known of these cases, a 73-year-old Saskatchewan farmer named Percy Schmeiser was sued for growing genetically modified Roundup Ready canola, which he claimed had blown into his fields from neighboring farms. A 5-4 decision of the Supreme Court of Canada upheld Monsanto's patent infringement claim, saying that Schmeiser's unlicensed use of seed containing Monsanto's patented gene was sufficient to constitute infringement.[23] In a Solomonic turn, though, the Court awarded no damages to Monsanto, on the ground that Schmeiser had not benefited economically from his unlawful act; equally, Schmeiser was not required to pay Monsanto's court costs. The case warned GM crop producers that, under

[21] The most unprecedented feature of this process was a nationwide public consultation known as *GM Nation?* See http://www.gmnation.org.uk/ (accessed Feb. 2014).

[22] For details of the case, as well as an argument against the U.S. positions on science and risk assessment, see Winickoff et al. (2005).

[23] *Monsanto Canada Inc. v. Schmeiser,* [2004] 1S.C.R. 902, 2004 SCC 34.

Canadian law, they would have a difficult time collecting damages for patent infringement; at the same time, they could be subject to potentially unlimited liability if their seeds accidentally contaminated, and thus damaged, the products of certified GM-free organic farms.

The point to note for the politics of international relations is not simply that intellectual property law protects property rights in biological materials across geopolitical boundaries, though that itself is a potent mode of extending dominance. Equally noteworthy is the displacement of ownership from one form of production to another: from the farmer's practices of cultivating whole organisms to the biotechnology company's practices of extracting, inserting, and circulating traits defined at more basic levels of biological organization. That displacement in transnational understandings of who owns the stuff of life, and how, redraws the relations between global centers and peripheries.

3.3 Legibility

Advertisements for agricultural biotechnology frequently show fields of grain laid out in neat parallel lines, illustrating both the fertility and the increased control that genetic modification allegedly delivers. One could hardly find more compelling images of the "legibility" described by Scott. Intrusive weeds, barren patches, unruly growth have all been eliminated in favor of healthy, predictable, quantifiable yields—achieved through the precision of genetic control. However, just as the mid-twentieth century grand planners' dreams of legibility were achieved at a cost, so legibility in modern GM agriculture demands unseen labors of standardization, and consequent elimination of ambiguity, to achieve its surface regularity. Four dimensions of standardization are worth noting: ontologies, epistemologies, socioecologies, and forms of life. All four maintain traditional relations of power between center and periphery, and all can be illustrated through the case of "golden rice," the poster crop for a new generation of nutrient-enriched GM crops to feed the developing world (Jasanoff 2007). The name was given to a strain of rice bioengineered to produce beta-carotene, which colors the grain a pale gold; when ingested, it converts to vitamin A in the body and protects consumers against vitamin deficiency leading to possible blindness. Developers claimed that, if widely marketed and eaten, golden rice might prevent hundreds of thousands of cases of blindness annually induced by vitamin A deficiency around the world.

For the products of GM agriculture to locate themselves securely in global markets, there has to be broad agreement on what these entities actually *are*. This ontological question may seem straightforward at first—proponents of golden rice, for instance, claim that it is nothing more than a more nutritious plant variety—but food crops straddle too many categorical boundaries for their identity in the political domain to be anything but hybrid. There are, to begin with, regulatory classifications. Should a crop engineered to produce ingredients of medicinal value be considered a food or a drug? Even if such issues can be settled by formal administrative definitions, the North-South debate surrounding GM crops shows how difficult it is to achieve ontological closure around a commodity that is at once a natural kind (a plant with specific genes and traits) and a social kind (a product of particular economic and political orderings, and thereby also a potential reorganizer of society).[24]

How one should know the properties of GM crops is similarly open to question. U.S. authorities have insisted that the only proper basis on which to evaluate the impacts of these novel entities is through science-based risk assessment. Yet, as the dispute between the United States and Europe at the WTO graphically illustrates, vast disagreements persist about the epistemological status of risk assessment. Is it a "science" at all, in the sense of being a well-demarcated, uncontroversial, paradigmatic (in a Kuhnian sense) method of representing the world; or is it instead a patently political and culturally constructed instrument for managing the uncertainties that inevitably accompany large projects of reconfiguring nature or society? (Winickoff et al. 2005) To accept producers' contention that crops such as golden rice are "safe," one has to buy the former, not the latter, characterization. If, however, risk assessment is an expression of political culture by other means, then one should not be surprised if that form of analysis does not travel friction-free across political and cultural boundaries (on the relationship of risk assessment to political culture, see Jasanoff 2005a).

GM crops are developed in the laboratory, usually in science-rich Western nations, tested in the field, and transported thence for commercial propagation in both naturally and socially variable environments. Monsanto, in this respect, is like the Kew Gardens of the nineteenth century: a metropolitan "center of calculation" (Latour 1990) from which standardized products flow out to take root in the world's economic and political peripheries. Key to sustaining this mode of production is the assumption that socioecologies are as standard as the crops grown within them—put differently, that social and ecological circumstances at the periphery are not so radically different from those at the metropolitan center as to defeat the project of global technology transfer. Yet accidents occurring even within the boundaries of single nation-states show that transfers from the laboratory to the field can bring unpleasant surprises. For example, in one costly U.S. episode, ProdiGene, a GM corn variety containing an insulin precursor, trypsin (Hord 2003), was planted in an unmarked field in rural Iowa. The manufacturer agreed with the U.S. Department of Agriculture, which approved the field trials, that the field would be quarantined the following year so as to remove any volunteer plants.[25] In fact, the fields were not properly isolated and an undetermined quantity of the GM crop was harvested along with about 500,000 bushels of soybeans during the following season. Similar

[24] Such ontological hybridity is taken as part of the order of things in the work of many science studies scholars, see, in particular Callon (1986), Latour (1993). Hybrids complicate the clean separation that philosophers such as Ian Hacking (1999) have sought to draw between natural ("indifferent") and social ("interactive") kinds.

[25] "Volunteer" plants are those that emerge spontaneously, usually from a previous season's growth, in places where they were not intentionally planted.

failures resulting from unforeseen couplings of technology, environment, and human behavior are all the more probable when transfers occur across disparate cultures of farming and of hazard control.

Expanding on this point, it has become clear that complex technological systems are forms of life, uniting human and nonhuman components in a common purposive framework, as much as they are targeted attempts to improve upon aspects of human life by physical or biological means. Thus, transportation systems do not only move people about from place to place. They remake social structures and self-understandings. A car culture, for example, gives rise to different visions, and valuations, of time, distance, autonomy, community, environmental quality, and the cost of life than does a culture dependent chiefly on bicycles or public transportation. Similarly, industrial agriculture is organized and managed on different principles from small family farms; the two systems of production rest on different economic, social, and technological infrastructures, and their impacts on human solidarity and on the environment are correspondingly divergent. Conventional risk assessment methods take little or no account of the social and ethical ramifications of technological systems, including the threats they pose to long-settled patterns of living. This blindness to technology's disruption of established forms of life, underwritten by the allegedly scientific power of risk assessment, has fueled much of the criticism of agricultural biotechnology in the global South (see, particularly, the arguments on this topic by the well-known Indian author and activist: Vandana Shiva 1993, 1997, 2001).

3.4 Identity and Community

Empires, no less than nation-states, engender and depend on feelings of belonging. Devices for producing imperial imagined communities have included, besides the grand, polarizing, ideological discourses of the cold war, mundane practices such as performing national celebrations,[26] teaching a common language, training administrative and judicial élites, and building infrastructures for commerce and communication. Science and technology, we have seen, have long served as agents of imperial governmentality, helping to produce the mission consciousness and the associated forms of knowledge and skill that serve as instruments for extending power. Modern biotechnology, similarly, provides a discourse of development that continues colonial traditions, although the agents, recipients, and specific mechanisms of the development project have been partially reconfigured in modern times.

The discovery of Africa as a site for biotechnological development, through the propagation of crops such as golden rice, offers perhaps the clearest illustration. In

[26] Morris (1979: 21–34), for example, describes Victoria's jubilee celebrations in London as a crystallizing moment for the British Empire in 1897. See also the account of the Imperial Assemblage of 1877 in Delhi by Cohn (1983).

the rhetoric of development specialists, and the scientific and industrial institutions that serve them, Africa is represented through tropes of crisis and charity that render the continent's condition as dire and the offers of scientific and technological solutions as salvationary (Jasanoff 2007: 190–194). In one instructive example, Gordon Conway, former president of the Rockefeller Foundation, and a colleague wrote an article in the prestigious journal *Science* on biotechnology's capacity to help Africans. Though presented as scientific, the article merged the empiricist register of science with a narrative register that was little short of missionary. At the center of the discussion was a fictional African housewife, "Mrs. Namurunda," who the authors said was not a real person but "a composite of situations existing in Africa." (Conway and Toenniessen 2003) The story begins with Mrs. Namurunda, a farmer and single mother, eking out a hard-scrabble existence on fields infested with every form of insect blight, under adverse conditions of drought and soil degradation. It ends with scientific biotechnology solving her problems, enabling her to turn a profit and secure a brighter, better educated, more enlightened future for her children.

This script follows Foucault's delineation of biopower with uncanny precision. An entire continent becomes a medicalized body, requiring urgent therapeutic intervention, both as a collective and for its individual members. The fictional person of Mrs. Namurunda, unveiled in the pages of one of the world's leading scientific journals, becomes a symbol for Africa's "composite" ailments. Advanced societies' power to develop and deliver the requisite treatments offers them the right, indeed the obligation, to engage in a new *mission civilisatrice*—built on a biomedical ethic of cure rather than, as in earlier times, a religious model of grace. But, this time, eschewing the forceful, state-led constellations of power that undergirded colonial rule, the neoliberal state works through a lightly regulated global industry and a largely selfregulating scientific community. *Their* expansion into new territories carries the promise of better jobs and higher incomes back in the home country, thereby allowing the economically more powerful state to justify itself where votes are counted, in its own national community of citizens. The sick and incapacitated recipient, however, has little or no say in either the diagnosis or the treatment of the alleged pathology.

3.5 *The Constitutional Turn*

We turn now to the fifth modality of empire-making identified above—the constitutional approach, which relies for its robustness on the formal consent of citizens. The European Union at the turn of the twenty-first century represents perhaps the most ambitious working out of this approach. With 25 member states as of May 2004, the EU brought within a single constitutional regime one of the most linguistically and culturally heterogeneous political assemblages ever created. In contrast to the institutionally inchoate, emergent empire discerned by Hardt and Negri, the EU is very much an orthodox space of governance, circumscribed by law and accountable to its members and (as illustrated by the French and Dutch "no" votes on the EU Constitution) to the particularities of their domestic politics.[27] On its Web pages, the public face it presents to the electronically plugged-in world, the EU takes considerable pains to explain itself: why it exists, how it was formed, its past achievements, and its hopes for the future. At one level, the talk is highly Weberian, a matter of official institutions and considered policies, justified in terms of an overall mission of peace, safety, solidarity, and a European model of society. At this discursive level, *Europe* very much exists; the question is only how to realize, through concerted, practical action, its already formed sense of collective identity.

At another level, however, Europe's identity is still very much in the making, and its constitutional union is but a cover for working out varying conceptions of what it means to be European; domains in which European-ness remains an open question, subject to multiple interpretations, include the development and deployment of the life sciences to advance communal interests in the EU. Looking at European engagements with biotechnology, both in Brussels and in the member states, one gets some sense of the issues in this debate, as well as some of the ways in which Europe has approached the problem of coordinating differences among its members without erasing them. The European example offers, in this respect, an intriguing alternative to the totalizing, disciplining vision of global biopower.

To be sure, European policy for biotechnology has followed to some extent familiar modernist impulses toward standardization and central control. Brussels has sought for decades to foster technological innovation and create new jobs, partly for the sake of continued European economic growth and partly in response to perceived threats from U.S., and now Chinese and Indian, innovation. Older discourses on international competitiveness (Gottweis 1998) have been joined of late to new worries about labor mobility within Europe, the out-sourcing of jobs to developing countries, and concomitant pressures to lower regulatory and ethical barriers to the free flow of scientists within the European research area. Since 1990, the EU has issued directives on research with genetically modified organisms (GMOs), release of GMOs into the environment, labeling of foods containing GMOs, and patenting of the products of biotechnology. In its efforts to counter popular resistance, the EU has also sponsored research on the public understanding of science—constituting in the process a citizenry whose needs the European state can characterize and cater to with aggressive programs of science and risk communication (Jasanoff 2005a).

These centralizing initiatives from Brussels, however, have run up against resistance from members states and their polities, showing that—at least in Europe—the prerogative of imagining technological futures no longer rests with

[27] This system of distributed accountability has resulted in a union whose members have not equally bought into all aspects of the EU vision. Thus, Sweden, Denmark, and Britain have not adopted the single currency (euro); Ireland and Britain are not parties to the Schengen agreement on frontier controls; and Britain thus far has not adopted the Community Charter of Fundamental Social Rights for Workers.

governments alone but must be shared with increasingly knowledgeable publics. Those publics, moreover, approach the promises of biotechnology with significantly different ethical sensibilities toward nature and different attitudes toward uncertainty and responsibility from the industries wishing to commercialize the new technologies (Marris et al. 2001). While public perceptions converge in important respects across Europe, the means through which people express their concerns and seek reassurance remain different, conditioned by national political culture and traditions. Thus, the nationwide public debate on GM crops held in Britain had no exact parallels anywhere else in the EU; other states conducted their own consultative exercises, in the form of citizen juries, consensus conferences, and referenda. The results, too, have varied, with member states disagreeing about how to establish the adequacy of data bearing on risk, as well as in the actions they have taken with respect to specific GM crops. Particularly noteworthy in this light are the EU's policy initiatives to foster coexistence between organic (non-GM) and GM-friendly modes of agricultural production, through standards that preserve the integrity of the former against contamination by the latter. In effect, such attempted coexistence authorizes a form of "epistemic subsidiarity," allowing citizens of member states to maintain their own belief systems about best practices in agriculture (Jasanoff 2013). Unification under the banner of scientific risk assessment is no longer seen as essential to the preservation of international cooperation and order.

In sum, European experience with the governance of biotechnology indicates that, in an empire built on constitutional principles, there may be broad agreement in public attitudes toward technology and on the rulers' willingness to take account of public views and values while actively pursuing the agenda of technological development. At the same time, democratic consultation pursued with genuine respect for diversity may produce locally specific accommodations that bear little resemblance to the global legibility sought by some twenty-first-century multinational corporations, or striven for in vain by Scott's over-ambitious twentieth-century planning states.

4 Conclusion

Imperial projects, as many are arguing today, did not end with the end of colonialism but may be resurfacing in new guises with the passage of time. Since early modernity, these projects have benefited from the enterprises of science and technology, and the biological sciences in particular have been caught up for centuries in the spread of imperial forms of governance. It is no surprise, then, to find contemporary biotechnology enrolled in various modalities of empire-making, whether through bottom-up resistance, top-down ideological imposition, administrative standardization, or consensual constitutionalism. In particular, as shown above, the capacity to engineer the genetic characteristics of plants has blended seamlessly with state and corporate projects of managing human populations so as to legitimate the exercise of power. Both nation-states and, in an era of neoliberalism, the multinational corporations that states are in league with have displayed their readiness to deploy agricultural biotechnology in advancing their interests on a global scale.

Struggles over the governance of biotechnology complicate any easy, linear narrative of progress. Instead, the nexus of globalization and technological innovation emerges on closer inspection as a politically contested site, where opposing conceptions of how human societies should live, and what other life forms should sustain them, remain very much at play. The example of European integration around biotechnology strongly suggests that there is considerable cross-cultural variation in the lines that human societies, even closely similar ones, choose to draw between nature and culture and the extent to which they are willing to tolerate line-crossings between those two domains. Given a chance to express themselves democratically, moreover, stable societies often opt to retain old boundaries and forms of life, preferring gradual, internally motivated change to imported, alien visions of progress, no matter how glittering the offerings presented to them.

These observations should not be taken as closing the door on the global promises of agricultural biotechnology, which may be considerable, even if not immediately on the horizon. The genie of genetic manipulation is with us in any case: there are not many precedents for turning the clock back on what human inquiry has revealed of the workings of the natural world, although highly developed techniques have occasionally been lost or gone into long periods of recession. Nor *should* we seek refuge in regress from innovation. The challenge, rather, is to constitute in tandem with global advances in technology the institutional capacities that might permit citizens to participate meaningfully in debating the implications of the new technologies. This essay speaks for more enlightened uses of our knowledge and capacity, preferably employed within constitutionally governed systems—keeping in mind that enlightenment flows not only from ingenious ways of tinkering with the material world but also as much, or more, from reflecting on how we should deploy for the good our profoundly human ingenuity.

References

9/11 Commission. (2004). *Final report of the Commission on Terrorist Attacks upon the United States*. New York: GPO.

Agriculture and Environment Biotechnology Commission (AEBC). (2001, September). *Crops on Trial*. London: AEBC.

Anderson, B. (1991). *Imagined communities* (2nd ed., rev. and exp.). London: Verso Books.

Anderson, J. R., Herdt, R. W., & Scobie, G. M. (1988). *Science and food*. Washington, DC: World Bank.

Anker, P. (2001). *Imperial ecology: Environmental order in the British Empire, 1895–1945*. Cambridge, MA: Harvard University Press.

Bauman, Z. (1991). *Modernity and ambivalence*. Ithaca, NY: Cornell University Press.

Bernauer, T. (2003). *Genes, trade, and regulation: The seeds of conflict in food biotechnology*. Princeton, NJ: Princeton University Press.

Bowker, G. C., & Star, S. L. (1999). *Sorting things out: Classification and its consequences*. Cambridge, MA: MIT Press.
Bud, R. (1993). *The uses of life: A history of biotechnology*. Cambridge, MA: Cambridge University Press.
Callon, M. (1986). Some elements of a sociology of translation: Domestication of the scallops and the fishermen of St. Brieuc Bay. In J. Law (Ed.), *Power, action, and belief: A new sociology of knowledge?* (pp. 196–233). London: Routledge & Kegan Paul.
Cannadine, D. (2001). *Ornamentalism: How the British saw their empire*. Oxford: Penguin.
Cohn, B. S. (1983). Representing authority in Victorian India. In E. Hobsbawm & T. Ranger (Eds.), *The invention of tradition* (pp. 165–209). Cambridge, MA: Cambridge University Press.
Cohn, B. S. (1996). *Colonialism and its forms of knowledge*. Princeton, NJ: Princeton University Press.
Cole, S. A. (2001). *Suspect identities*. Cambridge, MA: Harvard University Press.
Colley, L. (2003). *Captives: Britain, empire, and the world, 1600–1850*. New York: Anchor Books.
Collins, D., & Bromley, N. (2003). Private needs and public space: Politics, poverty, and anti-panhandling by-laws in Canadian cities. In Law Commission of Canada (Ed.), *New perspectives on the public-private divide* (pp. 40–67). Vancouver: UBC Press.
Conway, G., & Toenniessen, G. (2003). Science for African food security. *Science, 299*(5610), 1187–1188.
Drayton, R. (1993). *Imperial Science and a Scientific Empire: Kew Gardens and the Uses of Nature, 1772–1903*. Ph.D., dissertation, Yale University, Ann Arbor, MI.
Drayton, R. (2000). *Nature's government: Science, Imperial Britain, and the improvement of the world*. New Haven, CT: Yale University Press.
Edney, M. H. (1997). *Mapping an empire: The geographic construction of British India, 1765-1843*. Chicago: University of Chicago Press.
Ferguson, N. (2003). *Empire: How Britain made the modern world*. London: Penguin.
Ferguson, N. (2004). *Colossus: The price of America's empire*. New York: Penguin.
Foucault, M. (1978). *The history of sexuality; Vol. 1, An introduction*. New York: Pantheon.
Foucault, M. (1986). Governmentality. *Ideology and Consciousness, 6*, 5–21.
Friedman, T. L. (1999). *The Lexus and the olive tree: Understanding globalization*. New York: Farrar, Strauss, and Giroux.
Gieryn, T. F. (1999). *Cultural boundaries of science: Credibility on the line*. Chicago: University of Chicago Press.
Gottweis, H. (1998). *Governing molecules: The discursive politics of genetic engineering in Europe and the United States*. Cambridge, MA: MIT Press.
Grove, R. H. (1995). *Green imperialism: Colonial expansion, tropical island Edens, and the origins of environmentalism, 1600–1860*. Cambridge, MA: Cambridge University Press.
Hacking, I. (1999). *The social construction of what?* Cambridge, MA: Harvard University Press.
Hale, T., & Slaughter, A.,-M. (2005, May 26). Hardt and Negri's 'Multitude': The worst of both worlds. *Open democracy*. http://www.opendemocracy.net/globalization-vision_reflections/marx_2549.jsp. Accessed February 11, 2014.
Hall, C. (2002). *Civilising subjects: Colony and metropole in the english imagination*. Chicago: University of Chicago Press.
Hardt, M., & Negri, A. (2000). *Empire*. Cambridge, MA: Harvard University Press.
Hardt, M., & Negri, A. (2004). *Multitude*. New York: Campus Verlag.
Harvey, D. (2003). *The new imperialism*. Oxford: Oxford University Press.
Hazell, P. B. R., & Ramasamy, C. (1991). *The green revolution reconsidered*. Baltimore: Johns Hopkins University Press.
Hochschild, A. (1999). *King Leopold's ghost*. New York: Houghton Mifflin.
Hord, B. (2003, January 19). The road back: Prodigene and other biotech companies are moving ahead in an environment of increasing fear of crop contamination. *Omaha World Herald*.

Hughes, T. (1983). *Networks of power: Electrification in Western Society, 1880–1930*. Baltimore: Johns Hopkins University Press.
Ignatieff, M. (2003, January 5). The American Empire: The Burden. *New York Times Magazine*.
Jasanoff, S. (1990). *The fifth branch: Science advisers as policymakers*. Cambridge, MA: Harvard University Press.
Jasanoff, S. (2003). In a constitutional moment: Science and social order at the millennium. In B. Joerges & H. Nowotny (Eds.), *Social studies of science and technology: Looking back, ahead* (Yearbook of the sociology of the sciences, pp. 155–180). Dordrecht: Springer.
Jasanoff, S. (Ed.). (2004). *States of knowledge: The co-production of science and social order*. London: Routledge.
Jasanoff, M. (2005). *Edge of empire: Lives, culture, and conquest in the East, 1750–1850*. New York: Alfred A. Knopf.
Jasanoff, S. (2005a). *Designs on nature: Science and democracy in Europe and the United States*. Princeton, NJ: Princeton University Press.
Jasanoff, S. (2005b). In the democracies of DNA: Ontological uncertainty and political order in three states. *New Genetics and Society, 24*(3), 139–155.
Jasanoff, S. (2007). 'Let them eat cake': GM foods and the democratic imagination. In M. Leach, I. Scoones, & B. Wynne (Eds.), *Science and citizens: Globalization and the challenge of engagement* (pp. 183–198). New Delhi: Orient Longman.
Jasanoff, S. (2013). Epistemic subsidiarity: Coexistence, cosmopolitanism, constitutionalism. *European Journal of Risk Regulation, 2*, 133–141.
Johnson, C. (2000). *Blowback: The costs and consequences of American Empire*. New York: Henry Holt.
Judt, T. (2004, November 4). Dreams of empire. *New York Review of Books*, 38–41.
Kaczynski, T. (1995). *The Unabomber manifesto: Industrial society and its future*. Berkeley, CA: North Atlantic Books.
Kay, L. E. (1993). *The molecular vision of life: Caltech, the Rockefeller Foundation, and the rise of the new biology*. New York: Oxford University Press.
Khagram, S. (2004). *Dams and development: Transnational struggles for water and power*. Ithaca, NY: Cornell University Press.
Khalidi, R. (2004). *Resurrecting empire: Western footprints and America's perilous path in the Middle East*. New York: Beacon.
Latour, B. (1990). Drawing things together. In M. Lynch & S. Woolgar (Eds.), *Representation in scientific practice* (pp. 19–68). Cambridge, MA: MIT Press.
Latour, B. (1993). *We have never been modern*. Cambridge, MA: Harvard University Press.
MacKenzie, J. M. (Ed.). (1990). *Imperialism and the natural world*. Manchester: Manchester University Press.
Maier, C. S. (2009). *Among empires: American ascendancy and its predecessors*. Cambridge, MA: Harvard University Press.
Marris, C., Brian, W., Peter, S., & Sue, W. (2001). *Public perceptions of agricultural biotechnologies in Europe, final report of the PABE research project funded by the Commission of European Communities*. http://csec.lancs.ac.uk/archive/pabe/. Accessed February 11, 2014.
Mitchell, T. (2002). *Rule of experts: Egypt, technopolitics, modernity*. Berkeley, CA: University of California Press.
Morris, J. (1979). *Pax Britannica: The climax of an Empire*. London: Penguin.
New Scientist. (2004, May 11). GM wheat put on hold. *NewScientist.com* news service. http://www.newscientist.com/article/mg18224470.500-hold-the-wheat.html. Accessed February 11, 2014.
Nye, J. S. (2004). *Soft power: The means to success in World politics*. New York: Public Affairs.
Philip, K. (1995). Imperial science rescues a tree: Global botanic networks, local knowledge, and the transcontinental transplantation of cinchona. *Environment and History, 1*(2), 173–200.
Philip, K. (2004). *Civilizing natures: Race, resources, and modernity in colonial South India*. New Brunswick, NJ: Rutgers University Press.

Radcliffe, S. (2001). Imaging the state as space: Territoriality and the formation of the state in ecuador. In T. B. Hansen & F. Stepputat (Eds.), *States of imagination: Ethnographic explorations of the postcolonial states* (pp. 123–145). Durham, NC: Duke University Press.

Rajan, S. R. (Ed.). (1996). *Imperialism, ecology, and politics: Perspectives on the ecological legacy of imperialism*. Delhi: Oxford University Press.

Scott, J. C. (1998). *Seeing like state: How certain schemes to improve the human condition have failed*. New Haven, CT: Yale University Press.

Shiva, V. (1993). *Monocultures of the mind: Perspectives on biodiversity and biotechnology*. London: Palgrave MacMillan.

Shiva, V. (1997). *Biopiracy: The plunder of nature and knowledge*. Toronto, ON: Between the Lines.

Shiva, V. (2001). *Yoked to death: Globalisation and corporate control of agriculture*. New Delhi: Research Foundation for Science, Technology, and Ecology.

Slaughter, A.-M. (2004). *A new world order*. Princeton, NJ: Princeton University Press.

Slovic, P. (2000). *The perception of risk*. London: Earthscan Publications.

Stafford, N. (2004, May 17). Uproar over German GM Corn. *The Scientist*. http://www.thescientist.com/article/display/22179/. Accessed February 11, 2014.

Stoler, A. L. (1989). Making empire respectable: The politics of race and sexual morality in 20th-century colonial cultures. *American Ethnologist, 16*, 634–660.

Stoler, A. L. (2002). *Carnal knowledge and imperial power: Race and the intimate in colonial rule*. Berkeley, CA: University of California Press.

Storey, W. K. (1997). *Science and power in colonial Mauritius*. Rochester, NY: University of Rochester Press.

Storey, W. K. (2004). Plants, power, and development: Founding the Imperial Department of Agriculture for the West Indies, 1880–1914. In S. Jasanoff (Ed.), *States of knowledge* (pp. 109–130). London: Routledge.

Winickoff, D., et al. (2005). Adjudicating the GM food wars: Science, risk, and democracy in world trade law. *Yale Journal of International Law, 30*, 81–123.

Wynne, B. (1995). Public understanding of science. In S. Jasanoff, J. C. Petersen, T. Pinch, & G. E. Markle (Eds.), *The handbook of science and technology studies* (pp. 361–388). Thousand Oaks, CA: Sage.

Technological Collaboration and Nuclear Proliferation: A Transnational Approach

John Krige

Abstract Fifteen years ago Scott Sagan called upon scholars to enrich realism's balance-of-power model explaining a state's quest for survival in the cold war by opening the black box of national nuclear decision-making. This paper exhorts us to go further, and to break the walls of the national container that dominates realist models of security. Focusing on knowledge, it takes a transnational approach to the circulation of science and technology between states confronted by a common nuclear threat. In particular, it analyzes how the United States used its preponderance of scientific and technological knowledge/power after WWII as a political lever to shape the research trajectories of its Western European allies in line with its definition of the security needs of the 'free world,' so restricting the scope of their nuclear survival strategies.

Keywords Realism • Hegemony • Reciprocity • Positive disarmament • Technological sharing/Denial • Self-sufficiency

The plausibility and parsimony of a (neo)realist theory of international relations makes it an alluring framework for understanding the behavior of states in a competitive environment. By elevating survival to the primary motive of state behavior, realists bypass the foreign policy decisions that national actors make as they position themselves vis-à-vis their rivals (Mearsheimer 2001, 2006; Waltz 1979, 1993).[1] It is outputs that concern them, specifically those that secure, if only temporarily, a competitive advantage over an adversary and so shift the balance of

[1] I make no pretense at citing a comprehensive list of scholars who have adumbrated realist theory, but simply indicate a few works by two of its leading proponents.

J. Krige (✉)
School of History, Technology and Society, Georgia Institute of Technology, Atlanta, GA, USA
e-mail: john.krige@hts.gatech.edu

M. Mayer et al. (eds.), *The Global Politics of Science and Technology - Vol. 1*, Global Power Shift, DOI 10.1007/978-3-642-55007-2_11,

power in their favor. Science and technology enter this equation as resources mobilized to strengthen the nation and to protect it from external threats. This was particularly so in the cold war, characterized as a bipolar contest between two superpowers whose behavior, seen through the lens of outputs intended to maximize prospects for survival, emulated one another. A scientific and technologically-driven 'arms race' was an essential feature of such emulation: the quest for nuclear supremacy trumped domestic political priorities and ideological agendas as factors explaining international behavior.

In a classic paper published about 15 years ago Scott Sagan (1996) challenged this realist account. He proposed two alternative 'models' to the balance-of-power theory of nuclear decision making: states acquire nuclear weapons to satisfy the internal demand by a domestic coalition that successfully lobbies to have them and, secondly, they do so (or not) to symbolically express and legitimize their identity as modern states (what Sagan called a norms model). By thus cracking open the black-box of nuclear decision making Sagan took a fundamental step beyond balance-of-power realism. He did not deny its pertinence; he simply challenged its hegemony as *the* causal factor explaining why states acquire nuclear weapons. Such decisions, he insisted, were multi-causal, and engaged a complex cocktail of considerations—the quest for security, domestic demand, the symbolic expression of state power and identity among them.

The challenge Sagan posed to neorealism was immensely significant but constrained by its core assumption: that states act in "an anarchical international system and must therefore rely on self-help to protect their sovereignty and national security" (Sagan 1996: 57). In its emphases on national autonomy, self-sufficiency and self-help, this language eclipses relationships of interdependence between states who together confront a rival. In particular, it obscures the role of the transnational circulation of science and technology between allies as part of a collective response to a perceived nuclear threat. By sharing science and technology (or not) states can accelerate (or restrain) the nuclear weapons aspirations of their allies, so shaping the physiognomy of the changing nuclear landscape that they construct to secure their collective survival.

Scott Sagan called upon scholars to break open the black box of nuclear decision-making. This paper exhorts us to break the walls of the national container that dominates realist models of security. It takes a transnational approach to the circulation of knowledge between states confronted by a common nuclear threat, dissolving national boundaries, interrogating self-sufficiency, and diluting sovereignty. In particular, it analyzes how the United States used its preponderance of scientific and technological power after WWII to shape the research trajectories of its Western European allies in line with its definition of the security needs of the 'free world'. National research and development systems dedicated to consolidating the economic and military power of the modern state after 1945 have not only provided the vertical backbone needed to maximize national security in an anarchical international environment. They have also enhanced states' capacity to interact with each other (Herrera 2003), providing resources that could be used to build horizontal linkages that locked less powerful allies into the strategic agendas of more dominant states. This paper shows how technological collaboration was used as a political weapon to lever America's lead in nuclear and space science and technology, so shaping the strategic options of its European allies in the first decades of the cold war.

1 Surviving in the Nuclear Age

The danger of nuclear proliferation provided the major challenge to American power from the dawn of the nuclear age. The euphoria surrounding the early monopoly of the bomb was followed by the recognition that that monopoly was but temporary. True, General Groves, who had overall responsibility for the Manhattan Project, did originally hope to control the world's supply of uranium (Herken 1980). True, the McMahon Act of 1946 decreed that all knowledge pertinent to weapons development was 'born secret' so attempting to staunch its circulation outside the American weapons complex, even to a close ally like Britain (Morland 2005). True, the Baruch Plan sought to bring nuclear weapons under international control (Maddock 2010; Sokolski 2001). But neither nature, nor spies, nor other scientists, engineers and governments—including those in America's closest ally—obliged. By the early 1950s there were three nuclear powers; by the mid-1960s there were five and West Germany was increasingly reluctant to be excluded from the nuclear club. Watching the situation evolve in 1956 Secretary of State John Foster Dulles almost despaired of the situation. "It [was] appalling," he said, "to contemplate a multiplicity of uncontrolled national atomic developments leading to multiplying atomic weapons programs" among America's allies.[2] The legitimacy of the U.S. as the leader of the free world demanded that it respect first Britain's and then France's determination to become nuclear powers. It was equally important, in the interests of its own survival, that it 'control' those programs—and deny the same status to Germany—as long as the front line of the cold war was in Western Europe.

Faced with this situation the U.S. mobilized its scientific and technological leadership in strategic sectors to align the technological trajectories and political agendas of its allies along with its own interests in the European theater. As France's Defense Minister Pierre Mesmer put it "One is nuclear or one is negligible" (Mesmer 1963). No European ex-colonial power wanted to be reduced to irrelevance or to be humiliated by dependence, yet none had the resources—financial, industrial, intellectual—to build an independent nuclear capacity that could do more than tear the arm off the Russian bear (Gallois 1960). The challenge facing Washington was both to assist and to tether them, to integrate them into its strategic arsenal while ensuring that they did not jeopardize its survival. By regulating the flow of sensitive knowledge that European governments sought as

[2] *Memo of a Conversation with Franz Josef Strauss*, May 14, 1956, FRUS, 1955–57, IV: 438–441.

being crucial to their status as medium-sized world powers, the U.S. could control proliferation without subverting its legitimacy and undermining its hegemony. World order was stabilized by constructing horizontal bonds of technological interdependence in which allies traded a dilution of sovereignty for access to advanced scientific and technological knowledge that would give them a measure of control over their own survival.

2 Hegemony

The rank of great powers, writes Waltz, "depends on how they score on a combination of the following items: size of population and territory, resource endowment, economic capability, military strength, political stability and competence" (Waltz 1993: 50). Among these resources, scientific and technological capability was particularly prized after WWII. Everybody knows that there was an asymmetry in economic, political and military power between the United States and the rest of the world in 1945. But there was also an asymmetry in scientific and technological knowledge between America and its allies and enemies (Friedberg 2000; Krige 2006). The pursuit of preeminence in the name of leadership was being planned before the war ended; it became a priority as the cold war gained momentum. The quest for competitive scientific and technological advantage became embedded in domestic policies, and their supporting ideologies, that justified the construction and consolidation of the national security state. Federal support became an essential complement to the industrially sponsored research laboratory. The Federal government's R and D budget increased dramatically after the Korean War broke out, more than doubling to $1.3billion in fiscal year (FY) 1951 and more than doubling again to $3.1 billion in FY1953. It was given another enormous boost by the launch of Sputnik in 1957: by the early 1960s it had almost quadrupled to $15 billion (Kevles 1990). As Friedberg puts it,

> From the onset of the Cold War, top American decision makers tended to believe both that it was necessary for their country to seek a technological edge over the Soviet Union and its allies, and that such an edge could be achieved and maintained. These beliefs helped to keep technology at the forefront of American strategy and to sustain a massive four-decade flow of resources into research and development (Friedberg 2000: 297).

The pursuit of scientific and technological pre-eminence was driven, in the first instance, by the conviction that nothing less could protect America from an existential threat. But there was more to it than Friedberg says. The cold war was not simply a binary struggle for military superiority between superpowers. Scientific and technological leadership was also needed to enhance America's global reach. As Cristina Klein has noted, from an American perspective, the cold war was "as much about creating an economically, militarily and politically integrated 'free world,' as it was about waging a war of attrition against the Soviet Union" (Klein 2003:16). Washington sought to integrate Western Europe into its global agenda by encouraging it to play its part in the anti-communist struggle, while also striving to contain its ambitions within an American-led regime of world order. The challenge faced by U.S. policy makers in the 1950s and 1960s was not simply to combat Soviet communism; it was also to help rebuild Europe's scientific and technological strength, without unleashing 'runaway modernity' (Ninkovich 1999: 66) that would undermine its hegemony (Beeson and Higgott 2005).

As early as 1946 Truman was advised by White House Counsel Clark Clifford that America needed to do more than contain Soviet expansion: it needed to build a new 'world order'. According to Bright and Geyer, what distinguished this new regime from its predecessors, and above all from the imperial project of the European colonial powers, was that a transnational flow of knowledge enabled the United States to move "beyond the extension of power *over* others toward a direct and sustained organization *of* others, simultaneously, and in many parts of the world" (Bright and Geyer 2005: 205). American scientific, technical and intellectual pre-eminence, and the massive investment in education after the war that made that possible, were "as important as its economic and military power in making world order cohere and, more important, in developing and organizing the consent of subordinate participants" (Bright and Geyer 2005: 228). The postwar pursuit of an American-led regime of order was not a top-down project of command and obedience. It was an ongoing negotiated process in which science and technology were shared or denied in an asymmetric field of force defined by a knowledge-deficit between its partners and the United States. If this was hegemony, it was consensual not coercive (Krige 2006; Lundestad 1998).

American security depended on embedding European nuclear weapons and delivery systems in hegemonic structures that restricted the allies' autonomy of action and stopped them taking measures to ensure *their* survival that might drag Washington into a war that it did not want. Different circumstances called for different responses. Britain and France both sought to build independent nuclear deterrents, the first beginning in 1947, the latter in the mid-1950s, and with increasing emphasis after de Gaulle came to power in 1958. German Chancellor Konrad Adenauer pledged in 1954 that the country would never develop its own nuclear weapon. A decade later his successor Ludwig Erhard was far vaguer on the matter. Indeed by 1967 a large majority of the German political elite was vociferously against Germany signing the Nuclear Non-Proliferation Treaty that committed them to renounce nuclear weapons indefinitely. These affirmations were all driven by multiple considerations—the prestige associated with a nuclear capability, the capacity it provided to influence American strategic thinking in the European theater, and its value as a hedge against America reneging on its security guarantees—would the U.S. President be willing to risk New York to defend London, Bonn or Paris? While the U.S. feared that proliferation would undermine its national security, the Europeans realized that without their own 'autonomous' weapons systems they had no adequate control over their own defenses. Their security was impugned and their sovereignty crippled.

Washington could not, as leader of the free world, simply impose discipline by 'forcing' its allies to accept a status of permanent dependence. America's task was to strengthen Europe so that it could share in the burden of defense in Europe, without encouraging a "multiplicity of uncontrolled national atomic developments leading to multiplying atomic weapons programs" as Dulles put it (above). Scientific and technological collaboration squared the circle in the 1950s and 1960s by stimulating modernization without allowing it to spiral out of control. A regulated regime of transatlantic knowledge transfer was put in place by the State Department with help of the USAEC (Atomic Energy Commission) and of NASA (National Aeronautics and Space Administration), one for nuclear matters the other for missile delivery systems. These bonds of scientific and technological collaboration were mutually rewarding, locking the allies into America's strategic agenda and reinforcing the horizontal struts needed to stabilize world order.

3 Britain: Reciprocity

In 1955 the British government was presented with an analysis of the effects of a nuclear attack on the country. The report was prepared by a small cadre of experts chaired by a senior civil servant, William Strath. They imagined what would happen to the nation if it was attacked using just 10 hydrogen bombs targeted to make the best use of prevailing winds to spread radioactive fallout far and wide. The findings were sobering:

> half the population killed, major cities laid waste and the rest of the country and its people subject to enough radioactive fallout to contaminate huge areas and kill millions more. Agriculture and communications would be devastated and industry and the economy destroyed. Fire fighting, medical, transport, water and food supply services, and even the machinery of government itself, would likely collapse, and with them civil society. The immobilized and immiserated survivors would struggle alone against disease, starvation and the psychological effects of nuclear bombardment. Britain could not survive as a platform in a thermonuclear war, and its very survival as a nation would be in doubt (Hughes 2003: 258).

To meet this danger the British embarked on the construction of six new nuclear reactors in 1954 to fuel a thermonuclear (H-Bomb) weapons program. That program was also intended to secure closer collaboration with the U.S.

Britain's hopes of scientific and technological collaboration with America in nuclear matters were largely thwarted for the first decade after the war (Baylis 2002; Botti 1987; Melissen 1993). Much to the UK's chagrin, the McMahon Act of 1946 put a stop to any such collaboration "on pain of life imprisonment or even death" (Baylis 2002: 35). This was totally against the spirit of agreements reached between Churchill and Roosevelt on post-war nuclear sharing. This is partly why in January 1947 the British government embarked on its own independent nuclear deterrent. Such a program would not only secure Britain's status as a great power as the cold war gathered momentum. It was hoped that it would persuade the Americans that the UK were worthy partners for technological collaboration. These hopes were to be dashed time and again over the next 10 years, notwithstanding amendments to the Atomic Energy Act in 1954 and 1955 that facilitated a wider circulation of nuclear information, and strong support for closer collaboration from Eisenhower himself. The President ran into opposition from, and divisions between, the Congressional Joint Committee for Atomic Energy, the Atomic Energy Commission and other arms of the administration over a range of nuclear policy issues with the UK, e.g. sharing data on submarine nuclear propulsion, on the weight and dimensions of U.S. weapons carried by British bombers, and on controlling the launch of 60 nuclear tipped Thor missiles that were installed in silos around the country in the late 1950s. All involved sharing delicate nuclear secrets that the administration was most reluctant to divulge.

The Suez crisis of November 1956, the launch of Sputnik in October 1957, and a series of successful UK hydrogen weapons tests beginning in 1958 enabled the U.S. President to overrule his domestic opponents' determination to protect the U.S. nuclear monopoly at all cost. The desire to re-establish close ties after the Suez debacle, the advantage of building Britain more tightly into the defense of the West and, crucially for this argument, the recognition that Britain had important and useful knowledge to share on hydrogen bomb development, tilted the balance in the UK's favor. An amendment to the Atomic Energy Act signed at the end of June 1958 opened the way for far closer nuclear collaboration with countries that had made 'substantial progress' with their own weapons programs, a clause intended to include Britain but to exclude, specifically, France. Of course Britain was only told what the U.S. deemed it 'needed to know' rather than what its scientists, engineers and planners would have liked to know, the protection of U.S. security and competitive technological advantage *oblige*.

It is often said that U.S.-UK nuclear collaboration was driven by a "special relationship" informed by a shared history, culture and language. Against this, Baylis (Baylis 2002) emphasizes the significance of reciprocity in knowledge flows as fundamental to sustaining the bond between the two nations after 1958. The transatlantic circulation of knowledge (in both directions) saved the British a great deal of time and money and undoubtedly improved the technical aspects of their nuclear strike force. In return U.S. scientists and engineers learned about the paths their colleagues abroad had taken to solve nuclear weapons problems, confirming the wisdom of their choices or suggesting new ways of tackling similar bottlenecks. The balance was undoubtedly heavily in Britain's favor, but the very possibility of reciprocity sustained a two-way traffic, reinforced by personal ties of mutual respect, between scientists and engineers in both countries.

This first brief example challenges the core notion of (technological) self-sufficiency not simply for a declining world power like Britain—which is hardly surprising—but for a 'great power' like the United States. It cannot be said that the U.S. *depended* on Britain for technological inputs or that its global nuclear responses were shaped by scientific and technological knowledge acquired from British weapons programs in the 1960s. But nor should the fact that its scientists, engineers and industries were reassured and inspired by what they learnt from the British be dismissed in the name of 'self-sufficiency'. Realists concerned with American weapons policy today need to bear in mind that in 1999 the U.S.'s five major weapons laboratories, under assault for carelessly losing sensitive knowledge to potential foes, stressed that they could not but rely on the global pool of knowledge to consolidate the nation's defensive posture. As they put it, "The world is awash in scientific discoveries and technological innovations. If the United States is to remain the world's technological leader, it must remain deeply engaged in international dialogue [. . .] Indeed because DOE's [Department of Energy] laboratories conduct only 1–2 % of the world's research and development, the effectiveness of the laboratories depends substantially on their capacity to apply the 98–99 % of the work performed elsewhere" (Committee on Balancing Scientific Openness and National Security 1999: 11, 4). American 'self-sufficiency' in weapons development is a myth; U.S. weapons are hybrid objects, not purely national products (Krige 2012).

4 Germany: 'Positive Disarmament'

A few days before Christmas, 1965 German Chancellor Ludwig Erhard made a brief trip to Washington DC. After meeting with President Johnson during the day, he was treated to a state banquet on the evening of December 20th. The role of Germany in a nuclear strike force, and the 'offset' costs to Bonn of keeping the American military in the country topped the agenda of the formal discussions. Space exploration was highlighted at the banquet. Johnson's toast paid tribute to Erhard's faithful support of the war in Viet Nam and suggested that Germany participate with the U.S. in a major space science experiment, doing something "together [t]hat we cannot do so well alone." LBJ identified a probe to the sun and a probe to Jupiter as appropriate ventures that were both "very demanding," "quite complex," and as contributing "vastly to our mutual knowledge and our mutual skills."[3]

These two encounters, one during the day, the other at dinner are generally treated quite separately of one another; in fact they were not decoupled in the minds of the actors at the time, on the contrary. The American Ambassador in Bonn, George McGhee, advised Johnson before his December meeting with the chancellor that increased participation in the space program might provide an alternative outlet to nuclear sharing for German aspirations. Picking up the idea late in the afternoon Johnson mentioned to his guest that a cooperative venture in space "would do more to weld the two peoples together than all the conversations they could have on nuclear submarines and nuclear deterrence."[4] In short by the end of

[3] *The White House. Exchange of Toasts* between Johnson and Erhard, December 20, 1965, NSF, Country Files Europe and USSR, Germany, Box 192, Folder Germany. Erhard Visit [12/65], 12/19-21/65, LBJ Library.

[4] *Memorandum of Conversation. Part I of III. Subject: Space and the Great Society*, 20 December, 1965, NSF, Country Files Europe and USSR, Germany, Box 192, Folder Germany. Erhard Visit [12/65], 12/19-21/65, LBJ Library.

1965 the Johnson administration was beginning to see civilian space cooperation with Germany as a useful alternative to nuclear weapons to symbolically express (Sagan—above) its status as a modern power.

In his classic study of U.S.-German relationships in the nuclear domain Marc Trachtenberg has shown conclusively that, notwithstanding Adenauer's 1954 commitment to keep the FRG non-nuclear, both the first chancellor and his successor were, in fact, determined to acquire nuclear weapons if they could (Trachtenberg, 2012). Indeed in his meeting with the U.S. President on December 20, 1965, Erhard told Johnson that "it was impossible to assume that Germany would go forever without a nuclear deterrent." [5] The depth of this commitment is evident from the extremely hostile German opposition to the Nuclear Nonproliferation Treaty a few years later. Franz-Josef Strauss, the FRG's first minister for atomic affairs in the 1950s, and then the Minister for Defense under Adenauer, famously called the NPT a 'Versailles of cosmic proportions.' Henry Kissinger reported back to the State Department after a visit to Germany in November 1967 that not one of 50 or 60 top politicians in the country supported the NPT. [6]

The Johnson administration was under immense pressure, then, to provide Germany with some technological symbol of the prestige that derived from being a nuclear power without actually tolerating the acquisition by Bonn of an independent nuclear deterrent. LBJ had inherited from Kennedy the idea of integrating Germany into a multilateral nuclear force that would keep nuclear weapons within the FRG's reach but outside of its grasp, as Hal Brands puts it (Brands 2007: 397). The MLF envisioned a fleet of NATO nuclear surface ships, which would be manned by mixed-nationality crews. The project would supposedly satisfy German status anxieties by allowing the FRG partial control of nuclear weapons, while at the same time avoiding the dangers associated with an autonomous German capability. Bonn would remain tied firmly to NATO and the United States, preventing the emergence of a disruptive force in the center of Europe.

Political opposition at home and abroad apart, two main factors undermined this option in the mid-1960s. Firstly, the Chinese nuclear test in 1964 led many to believe that further proliferation was inevitable, and that it was only a matter of time before others tested a bomb. Secondly, the emergence of gas centrifuge enrichment as a technology for producing bomb material only amplified these fears as regards Nth country proliferation, and Germany in particular (Krige 2012). When they met in December 1965 Erhard explicitly told Johnson that "some nuclear scientists had told him that in the not too distant future nuclear weapons could be produced much more cheaply and technological aspects mastered much more easily, so that many smaller countries would be able to afford nuclear weapons and might actually find them cheaper to maintain than conventional forces."[7] With the FRG budget groaning under the weight of the offset payments,

[5] *Memorandum of Conversation. Nuclear Sharing*. December 20, 1965, FRUS 1964–68, XIII: 291.

[6] He was head of the State Department's Office of German Affairs at the time.

[7] *Memorandum of Conversation. Nuclear Sharing*. December 20, 1965, FRUS 1964–68, XIII: 291.

Erhard's message was clear: Germany might soon be able to contribute its own nuclear weapon to a multilateral strike force, exploiting the technology needed to enrich uranium without being beholden to American supplies from its Manhattan-era gas diffusion plants. This is the context in which Johnson's toast to Erhard proposed that they embark together on a major space science project.

As the space mission began to take shape, the technopolitical stakes were brought more sharply into focus. Johnson's suggestion to Erhard, along with developments elsewhere in Europe to be discussed shortly, precipitated a major discussion of the scope of U.S.-European scientific and technological collaboration in the space sector. It was discussed in a lively meeting of the National Aeronautics and Space Council on March 23, 1966. The meeting was attended by Vice President Hubert Humphrey, along with NASA Administrator James Webb, Defense Secretary Robert McNamara and some high-level State Department officials. The use of space collaboration as an instrument of non-proliferation was specifically touched upon in an exchange between Humphrey and McNamara. As an official report on the meeting put it, "In response to a question from the Vice President as to whether this was 'positive disarmament' Secretary McNamara replied that it was since at the present time many countries felt they should make technological progress through military hardware programs. A divergence to a space program would be wholesome for all concerned," McNamara added.[8] Webb picked this up in a letter to McNamara a month later where he again suggested that by stimulating foreign involvement in space technology the U.S. could divert its allies' energies away from the development of nuclear weapon systems.

Erhard returned to Washington for 2 days in September 1966. LBJ would not allow the offset arrangements that so pre-occupied his guest to dominate the encounter, however. He insisted on propelling space collaboration into prominence by accompanying Erhard down to Cape Kennedy during the Chancellor's brief visit. In an official address in the still incomplete giant Vehicle Assembly Building the President personally thanked all those who had come to the U.S. from Germany, including the 'Nazi' rocketeer Werner von Braun, for their significant contributions to the American space program. He also enumerated the many projects which NASA had engaged in with European partners, and re-iterated his desire to "vigorously pursue" international cooperation in space science, and to provide launchers for space efforts of mutual interest.[9] On the way back to Washington Webb took the opportunity to spend an hour with the German Chancellor. The "large on-going effort [at the Cape] made a deep impression" on Erhard, Webb wrote Secretary of State Dean Rusk. He went on: "it seems to me that Erhard had a different attitude when we left the Cape than when we arrived. In fact, he did say

[8] *Memorandum for the Record. International Space Cooperation.* March 23, 1966, FRUS 1964–68, XXXIV: 86.

[9] *The White House. Remarks by the President at Vehicle Assembly Building, Cape Kennedy, Florida.* 20 September 1966, NSF, Country File Europe & USSR, Germany 9/66, Erhard Visit, folder Papers, Cables, Memos [9/66], LBJ Library.

that it was impossible to learn from pictures, television, and documents the true scope and magnitude of what was being done and that he had a much better appreciation of its importance."[10]

It is most unlikely that these initiatives were deemed viable alternatives to a nuclear capability by Germany. They certainly stimulated space cooperation between the two nations. Johnson's toast led to a major bilateral project to send a probe to the sun (Helios). In the 1970s Germany took the lead in building Spacelab, an immensely expensive scientific laboratory that fitted in the Space Shuttle's cargo bay. As for nuclear weapons, the new chancellor Willie Brandt secured the ratification of the NPT in November 1969 after a tumultuous debate in the Bundestag.

5 France: Denial and Diversion

The United States was systematically hostile to France acquiring nuclear weapons and the delivery systems needed to use them effectively in war. Its first successful test of an A-bomb in February 1960 was followed in 1965 by the launch of France's first satellite with its own launcher derived from its ballistic missile program. As relations with France deteriorated the White House intervened to deny any technological sharing with French President de Gaulle in sensitive domains. The policy was enshrined in NSAM (National Security Action Memorandum) 294, signed by McGeorge Bundy and dated 20 April, 1964. NSAM 294 was a response to de Gaulle's determination to develop an independent nuclear deterrent, and the French President's dislike for international and supranational institutions that restricted France's sovereignty and autonomy of action. American preponderance in NATO (North Atlantic Treaty Organization) particularly irked de Gaulle: he refused to accept that the security of his country and the scope of retaliation by France to a nuclear-armed Soviet Union should be subject to restrictions imposed by the United States. "Given current French policy", NSAM 294 stated, "it continues to be in this government's interest not to contribute to or assist in the development of a French nuclear warhead capability or a French national strategic nuclear delivery capability." To that end the President directed that "effective controls be established immediately" to stop "exchanges of information and technology between the governments, sales of equipment, joint research and development activities, and exchanges between industrial and commercial organizations [...], which would be reasonably likely to facilitate these efforts by significantly affecting timing, quality or costs or would identify the U.S. as a major supplier or collaborator."[11]

---

[10] James E. Webb to Dean Rusk, 14 October 1966, record no.14465, International Cooperation and Foreign Countries, Foreign Countries, West Germany, folder Germany (West), 1956–1990, NASA Historical Reference Collection, Washington, DC.

[11] NSAM 294, U.S. Nuclear and Strategic Delivery System Assistance to France. http://www.lbjlib.utexas.edu/johnson/archives.hom/NSAMs/nsam294.asp. Accessed July 20, 2013.

Washington could not stop Paris developing an independent nuclear deterrent and delivery system. But it could retard its progress by refusing to collaborate and, by denying it cutting edge science and technology, make it less effective and destabilizing than it might otherwise have been.

De Gaulle's determination to go it alone posed something of a dilemma for the administration. On the one hand they were keen to strengthen European science and technology, notably in the space sector that was believed to have spill-over effects throughout the economy. On the other hand national security policy dictated that France was not to be helped in acquiring a 'strategic nuclear delivery capability.' How could one support the first without fostering the second? The European Launcher Development Organization (ELDO), established in the early 1960s was a potentially useful instrument to serve this dual role. To explain how that was possible, a short detour is called for.

ELDO brought together five of the six founding members of the European Economic Community (Belgium, the Federal Republic of Germany, France, Italy, and the Netherlands) plus Britain and Australia. In 1962 they agreed to build a three-stage satellite launcher called Europa. The first stage would be derived from Britain's intermediate range ballistic missile, Blue Streak, stripped of its military characteristics. France would build the second stage, Germany the third. The other participants would provide the test satellite, telecommunications and ground equipment, while the launching base would be in Woomera, South Australia.

NASA and the Department of State welcomed the formation of ELDO. To quote an early position paper on the issue, technological assistance to ELDO was coherent with "our objective of an economically and politically integrated European Community with increasingly close ties to this country within an Atlantic community."[12] Strengthening ELDO would also build up European intellectual and industrial capabilities that could be redirected towards defense if need be.

The strategy was not without its flaws, however. ELDO lacked a strong centralized system of project management and control; its Secretariat had little authority over the people and firms developing the separate stages in Britain, France and Germany. A report prepared by the CIA in May 1964 confirmed the danger: "the organization has no enforcement machinery to police compliance, and the possibility is raised that ELDO might contribute to the spread of ballistic missile technology."[13]

---

[12] *Draft U.S. Position on Cooperation with Europe in the Development and Production of Space Launch Vehicles*, attached to internal correspondence between Frutkin and Milton W. Rosen, 15 October and 30 October 1962. Record no. 14548, International Cooperation and Foreign Countries, Europe, folder US-Europe 1965–1972, NASA Historical Reference Collection, Washington, DC.

[13] Central Intelligence Agency, Office of Current Intelligence, *Special Report. Western European Space Programs*, May 1964, p.3, record no. 15707, International Cooperation and Foreign Countries, European Launcher Development Organization, folder ELDO, NASA Historical Reference Collection, Washington, DC.

How did NASA hope both to assist ELDO and to respect the injunctions of NSAM 294 under these circumstances? They suggested that it could be done by distinguishing between the *kind* of technology that could be shared. In the extremely sensitive domain of propulsion they drew a sharp line between solid propellants like powders, on the one hand, and non-storable liquid propellants like liquid hydrogen and liquid oxygen, on the other. France was heavily committed to using solid fuels in its military missiles. Non-storable and liquid fuels were unlikely to contribute much to the propulsion technologies France was developing for its strategic delivery objectives. In other words, by distinguishing between the military potential of solid (high security risk) and non-storable (low military interest) propellants, NASA and the State Department thought they could safely offer assistance to ELDO in a technologically crucial domain without significantly assisting France's missile program.

We see then that to strengthen European science and technology in the space sector, without encouraging the proliferation of missile programs, required drawing two interlocking distinctions, one institutional, the other technological. First a sharp distinction was drawn between national and multilateral programs—only the latter would be candidates for assistance. Then a distinction was drawn between the kind of technology that would be shared: help would only be given with propulsion systems that burnt non-storable, liquid fuels that posed a relatively minor proliferation danger. A paper drawn up by an interagency group in May 1966 summed up the argument with pristine clarity. In a framework like that provided by ELDO

> rocket programs tend to be more open, serve peaceful uses and are subject to international control and absorb manpower and financial resources that might otherwise be diverted to purely national programs. National rocket programs tend to concentrate on militarily significant solid and storable liquid fueled systems, are less open, and less responsive to international controls.[14]

The U.S. had to live with de Gaulle's developing nuclear weapons and ballistic missiles, though it was determined that Germany should not follow suit. By encouraging a major investment in space science and technology in Germany, by refusing to share any sensitive technology with France, and by enthusiastically supporting ELDO, the administration sought to 'divert' limited resources in both countries away from nuclear weapons systems into civilian space activities. In so doing they aimed to restrict their war-fighting capability and to force them into dependence on Washington's security guarantees, so stabilizing the region under American leadership.

[14] *Meeting No. 1, Committee on Expanded International Cooperation in Space Activities. Subject: Cooperation Involving Launchers and Launching Technology*, T.H.E. Nesbitt 17 May 1966, folder Cooperation in Space—Working Group on Expanded International Cooperation in Space ELDO # 1 [2 of 2], Charles Johnson File, National Security Files, box 14, LBJ Library.

6 Concluding Remark

In deciding whether or not to embark on a nuclear weapons program (and a missile delivery system) a state has to consider whether it has the resources needed—major investments of money, but also trained manpower, and industrial capability to name just a few. That investment can be reduced by scientific and technological collaboration, a collaboration that is all the more efficient if conducted with powers that have already made major strides in acquiring nuclear weapons and missiles. This was the trump card that the U.S. could play in the first two or three decades after the cold war. By sharing or denying sensitive nuclear and space technology they could have a major impact on the security choices their allies made. America's ceaseless quest for scientific and technological superiority after WWII was not only needed to match its communist rival. It also provided it with a political weapon that it could deploy to manage proliferation among its allies by shaping the contours of indigenous nuclear and space programs. By reciprocally sharing nuclear technology with Britain Washington could lock their allies' advanced weapons systems tightly into U.S. strategic planning, tying us "hand and foot to them" in nuclear matters, as Labour Minister Tony Benn commented (Benn 1998: 127). By exploiting domestic opposition to a nuclear Germany, and denying the Federal Republic any technological support for sophisticated weapons systems, they could keep Germany non-nuclear. And by refusing to collaborate technologically with France in the 1960s they could seriously hamper the offensive capability of de Gaulle's 'independent' nuclear deterrent (Ullman 1989).

Scott Sagan has identified two of the reasons, in addition to security concerns, that help explain why states get nuclear weapons. This paper insists that we complicate the realist model even more by factoring in the scientific and technological capability of the country concerned, including the opportunities and constraints provided by technological collaboration in transnational networks through which sensitive nuclear and space knowledge circulates. Scientific and technological capability is a core factor shaping a state's decision to go nuclear, and regulating the circulation of nuclear and missile materials is a crucial policy instrument to curb proliferation.

References

Baylis, J. (2002). Exchanging nuclear secrets: Laying the foundations of the Anglo-American nuclear partnership. *Diplomatic History, 25*(1), 33–61.

Beeson, M., & Higgott, R. (2005). Hegemony, institutionalism and US foreign policy: Theory and practice in comparative historical perspective. *Third World Quarterly, 26*(7), 1173–1188.

Benn, A. W. (1998). *Office without power. Diaries 1968–1972*. London: Hutchison.

Botti, T. J. (1987). *The long wait. The forging of the Anglo-American nuclear alliance*. New York: Greenwood Press.

Brands, H. (2007). Non-proliferation and the dynamics of the middle cold war: The superpowers, the MLF and the NPT. *Cold War History, 7*(3), 389–423.

Bright, C., & Geyer, M. (2005). Regimes of world order. Global integration and the production of difference in twentieth-century world history. In J. H. Bentley, R. Bridenthal, & A. A. Young (Eds.), *Interactions. Transregional perspectives on world history* (pp. 202–238). Honolulu, HI: University of Hawai'i Press.
Committee on Balancing Scientific Openness and National Security. (1999). *Balancing scientific openness and national security controls at the nuclear weapons laboratories*. Washington, DC: National Academies Press.
Friedberg, A. (2000). *In the shadow of the garrison state: America's anti-statism and its cold war grand strategy*. Princeton, NJ: Princeton University Press.
Gallois, P. (1960). *Stratégie de l'âge nucléaire*. Paris: Calmann-Lévy.
Herken, G. (1980). 'A most deadly illusion'. The atomic secret and American nuclear weapons policy, 1945-1950. *Pacific Historical Review, 49*(1), 51–76.
Herrera, G. I. (2003). Technology and international systems. *Millenium, 31*(3), 559–593.
Hughes, J. (2003). The Strath report: Britain confronts the H-Bomb, 1954-1955. *History and Technology, 19*(3), 257–275.
Kevles, D. (1990). Cold war and hot physics: Science, security and the American State, 1945-56. *Historical Studies in the Physical Sciences, 20*(2), 239–264.
Klein, C. (2003). *Cold war orientalism. Asia in the middlebrow imagination, 1945-1961*. Berkeley: University of California Press.
Krige, J. (2006). *American hegemony and the postwar reconstruction of science in Europe*. Cambridge, MA: MIT Press.
Krige, J. (2012). Hybrid knowledge: The transnational co-production of the gas centrifuge for uranium enrichment in the 1960s. *British Journal for the History of Science, 45*, 337–357.
Lundestad, G. (1998). *'Empire' by integration. The United States and European integration, 1945-1997*. New York: Oxford University Press.
Maddock, S. J. (2010). *Nuclear apartheid. The quest for American atomic supremacy from World War II to the present*. Chapel Hill, NC: University of North Carolina Press.
Mearsheimer, J. J. (2001). *The tragedy of great power politics*. New York: Norton.
Mearsheimer, J. J. (2006). Structural realism. http://mearsheimer.uchicago.edu/pdfs/StructuralRealism.pdf. Accessed February 7, 2014.
Melissen, J. (1993). *The struggle for nuclear partnership. Britain, the United States and the making of an ambiguous alliance*. Groningen: Styx.
Mesmer, P. (1963, January 24). *Journal Officiel, 1613*.
Morland, H. (2005). Born secret. *Cardozo Law Review, 26*(4), 1401–1408.
Ninkovich, F. (1999). *The Wilsonian century. U.S. foreign policy since 1900*. Chicago: University of Chicago Press.
Sagan, S. D. (1996). Why do states build nuclear weapons? Three models in search of a bomb. *International Security, 21*(3), 54–86.
Sokolski, H. D. (2001). *Best of intentions: America's campaign against strategic weapons proliferation*. Westport, CT: Praeger.
Trachtenberg, M. (2012). *The cold war and after. History, theory, and the logic of international politics*. Princeton, NJ: Princeton University Press.
Ullman, R. H. (1989). The covert French connection. *Foreign Policy, 75*, 3–33.
Waltz, K. N. (1979). *Theory of international politics*. Reading, MA: Addison-Wesley.
Waltz, K. (1993). The emerging structure of international politics. *International Security, 18*(2), 44–79.

Part III
Interlude: Exploring a Multi-Disciplinary Field

Specialist Knowledge, Carbon Democracy, and Local Technicalities as Basis for the 'International': An Interview with Timothy Mitchell

Peer Schouten

Timothy Mitchell moves across disciplinary borders to study entanglements of technology, governance, and fossil energy. Tracing the global through tiny details and mundane processes at the local level, Mitchell suggests replacing technological determinism with infra-theory to understand the limits of government, the role of technical experts, and the change of political systems.[1]

[1] An extended and modified version of this interview, including additional references, can be found on http://www.theory-talks.org/

P. Schouten (✉)
School of Global Studies, University of Gothenburg, Gothenburg, Sweden
e-mail: peer.schouten@globalstudies.gu.se

M. Mayer et al. (eds.), *The Global Politics of Science and Technology - Vol. 1*, Global Power Shift, DOI 10.1007/978-3-642-55007-2_12,

How is your understanding of technology different from Marxist understandings of technology, which often stand accused of determinism?

If you read Marx, there is an extraordinary level of interest in the technical—that is, in both the technical aspects of political economy as a field of knowledge in the nineteenth century, and the factory as a technical space. So, conventional political economy to him was not just an ideological mask that had to be torn away so that you could reveal the true workings of capitalism. Political economy had produced a set of concepts—notions of value, of exchange, of labor—that actually formed part of the technical workings of capitalism. The factory was organized at a technical level that had very specific consequences. Critically speaking, the trouble with a significant part of Marx's theories is that he stopped doing that kind of technical work and Marxism froze itself with a set of categories that may or may not have been relevant to a moment of nineteenth century capitalism. Nevertheless, I find it really interesting and profitable to read some of the contemporary Italian Marxist theory, as well as some of the work in Marxist geography that continues to be very productive.

To turn to the questions about oil and democracy that have been the focus of my recent work, to me much of the literature that already existed—including earlier research on coal, industrialization and democracy—suffered from a kind of technical determinism because it actually did not go into the technical. It was determinist because such accounts had actually jumped over the technical side much too fast. In the case of the resource curse literature, it was only interested in the oil *once it had already become money*. And once it was money, then it of course corrupts: you buy people off, or you do not have to seek their votes. The whole question of how oil *becomes* money and how you put together the technical system that turns oil or coal into forms of political power, does not get opened up.

Even though there is not much of the technical in them, arguments like that are technically very determinist. Technology is just something that is kept external to International Relation (IR) theory's explanatory models but accounts for everything else that the model cannot explain. That ends up being a terrible kind of technical determinism. But as soon as you start opening up the technical side of the discussion, you realize that there are many ways things can go and many different ways things can get built. Energy networks for example can be built in different ways and there can be different mixes of energy. Of course most of the differences are technical differences, but they are also human differences. It is precisely by being attentive to the technical aspects of politics that one finds the only way to get away from a certain kind of technical determinism that otherwise rules us. In the study of the economics of growth, for instance, there is this great externality of technological change that drives every sort of grand historical explanation.

What Is the Nexus Between Expertise, Technologies, and Governance?

Parallel to the theoretical aspects I have just mentioned, some aspects of my work draw on Foucault and address the topics of expertise and modes of power. Let me suggest a couple of thoughts in this regard. One is that modern government has unfolded through the development of new forms of expertise, which among other things define problems and issues upon which government can operate. This can concern many things, whether it is problems of public health in the nineteenth or twentieth century; or problems of economic development in the twentieth century; or problems of energy, climate change and the environment today. Again and again government itself operates—as Foucault has taught us—through simultaneously constituting fields of knowledge and power. The objects brought into being in this way—defined in important ways through the development of expert knowledge—become in themselves modes through which political power operates. Thanks to Foucault and many others, that is a way of thinking or field of research that has been widely developed, even though there are vast amounts of work still to be done.

But I think there is another relationship between modes of government and expertise, and this goes back to things I have been thinking about ever since I wrote an article about the theory of the state—*The Limits of the State*. The point I made then is that it is interesting to observe how one of the central aspects of modern modes of power is the way that the distinction between what is the state and what is not the state; between what is public and what is private, is constantly elaborated and redefined. So politics itself is happening not so much by some agency called 'state' or 'government' imposing its will on some other preformed object—the social, the population, the people—but rather that it concerns a series of modern techniques that create what I have called the *effect* of a state: the very distinction between what appears as a sort of structure or apparatus of power, and the objects on which that power works.

Why Do You Focus on the "Local" in Order to Study Global Issues?

One of the biggest challenges is to learn to develop ways of seeing even what seem like the most global of issues, as things that are very local. Part of the problem with fields such as 'global studies', the term 'globalization', and other terms of that sort, is that they tend to define their objects of study in opposition to the local, in opposition to even national-level modes of analysis. By consequence, they assume that the actors or the forces that they are going to study must themselves be in some sense global, because that is the premise of the field. So whether it is nation states acting as world powers; whether it is capitalism understood as a global system—they have to exist on this plane of the global, on some sort of universal level, to be topics of IR and global studies, for instance. And yet, on close inspection, most of the concerns or actors central to those modes of inquiry tend to operate on quite local levels; they tend to be made up of very small agents, very particular arrangements that somehow have managed to put themselves together in ways that allow them take on this appearance and sometimes this effectiveness of things that are global. I am very interested in taking things apart that are local, on a particular level, to understand what it is that enables such small things, such local and particular agents, to act in a way that creates the appearance of the global or the international.

Now this relates to substantive concerns that we face as a global community. When I began writing *Carbon Democracy* there was a lot of attention for the problem of 'creating a more democratic Middle East', as it was understood at the time of the Iraq war. It struck me that when debating this problem—of oil and democracy, of energy and democracy—we saw it as somehow specific to these countries and to the part of the world where many countries were very large-scale energy producers. We were not thinking about the fact that we are all in a sense caught up in this problem that I call carbon democracy, and that there are issues—whether it is in terms of the increasing difficulty of extracting energy from the earth, or the consequences of having extracted the carbon and put it up in the atmosphere—that we, as democracies, are very challenged by. Those issues—and I think in particular the concerns around climate change—when you look at them from the perspective of U.S. politics, and the inability of the U.S. even to take the relatively minor steps towards ending the use of fossil fuels that several other industrialized democracies have taken: this inability to act suggests a larger problem of oil and democracy that needs explaining and understanding and working on and organizing about. I also think there is a whole range of contemporary issues related to energy production and consumption that revolve around the building of more egalitarian and more socially just worlds. These problems of what I call carbon democracy present themselves very powerfully as limits to democracy in American politics, whereas the limits are experienced in other ways in other parts of the world.

What Way of Theorizing, in Terms of Disciplinary Boundaries, Does Your Work Suggest?

I was always drawn into the local and the particular and the specific, and I was never very good at thinking at the level of large-scale grand theory. The first thing that shaped my scholarship was a constant shifting of fields and disciplines, which was not to me a problem—it was rather that there was a kind of intellectual curiosity that drove me from academic field to field. I actually began as a student of law and then moved into history while I was still an undergraduate, but then I became interested in political theory; and decided that I liked it better than political science. However, by the time I arrived in political science to study for a PhD, I had become interested in politics of the Middle East and I found myself in this field of comparative politics. I can say, if there was one thing that helped me arrive at where I am, it was this constant moving outside of the boundaries of one subfield or discipline and trespassing on the next one—trying to do it for long enough that scholars from different disciplines started to accept me as someone who they could debate with. And I think all along that has been important to the kind of scholarship I do.

Traveling across different contexts, I have not developed a set of theories that I take with me. Rather, I would say I have developed a way of seeing—I would not necessarily call it 'meta', I see it as much more as sort of 'infra': much more mundane and everyday. While I have this intellectual history of moving across disciplines and social sciences in an academic way, there is another sort of moving across fields, another sensibility, and that sensibility provides me with a sense of rootedness or grounding. That is a more traditional way of moving across fields, because whether when one is writing about contemporary politics or more historically about politics, one is dealing constantly with areas of technical concern of one sort or another, with specialist knowledge. Engaging with that expert knowledge has always provided both a political grounding in specific concerns and with a kind of concern with local, real-world, struggles on the ground. So that might have been things like the transformation of irrigation in nineteenth-century Egypt, or the remaking of the system of law; or it might be the history of malaria epidemics in the twentieth century, or the relationship between those epidemics and transformations taking place in the crops that were grown; or, more recently—and more obviously—of oil and the history of energy, and the way different forms of energy are brought out of the ground. And I should mention beside those areas of technical expertise already mentioned, economics as well: a discipline I was never trained in, but that I realized I had to understand just as much as I had to understand agricultural hydraulics or something of petroleum geology, as a form of technical expertise that is shaping the common world. In sum, what keeps me grounded is the idea that to really make sense of the politics of any of those fields, one has got to do one's best to enter and explore the more technical level—with the closest attention that one can muster to the technical and the material dimensions of what is involved—whether it is in agricultural irrigation, building dams or combating disease. And entering this level of issues does not only mean interviewing experts but arriving at the level of understanding the disease, the parasite, the modes of its movement, the hydraulics of the river, the properties of different kinds of oil... So as you can see it is not really 'meta', it really is 'infra' in the anthropological way of staying close to the ground, staying close to processes, things and materials.

How Are Fossil Energies and the Emergence of Mass-Democratic Politics Related?

Let me take an example from my book *Carbon Democracy*. I was interested in what appeared to be the way in which the rise of coal—the dominant source of energy in the nineteenth century and the emergence of modern industrialized states—seemed to be very strongly associated with the emergence of mass democracy, whereas the rise of oil in the twentieth century seemed to have if anything the opposite set of consequences for states that were highly dependent on the production of oil.

I wanted to examine these relations between forms of energy and democratic politics in a way that was not simply some kind of technical- or energy determinism, because it is very easy to point to many cases that simply do not fit that pattern—and, besides, it simply would not be very interesting to begin with. But it did seem to me, that at a particular moment in the history of the emergence of industrialized countries—particularly in the late nineteenth century—it became possible for the first time in history and really only for a brief period, to take advantage of certain kinds of vulnerabilities and possibilities offered by the dependence on coal to organize a new kind of political agency and forms of mass politics, which successfully struggled for much more representative and egalitarian forms of democracy, roughly between the 1880s and the mid-twentieth century. In general terms, that story is known; but it had been told without thinking in particular about the energy itself. The energy was just present in these stories as that which made possible industrialization; industrialization made possible urbanization; therefore you had lots of workers and their consciousness must somehow have changed and made them democratic or something.

That story did not make sense to me, and that prompted me to research the question in detail, drawing on the work of others who had looked even more in detail at the history of struggles for a whole set of democratic rights. The accounts of people at the time were clear: what was distinctive was this peculiar ability to shut down an economy because of a specific vulnerability of the supply of energy. Very briefly, when I switched to telling the story in the middle of the century with oil, it is different: partly just because oil was a supplementary source of energy, countries and people now had a choice between different energy sources, but also because oil did not create the same points of vulnerability. There were fewer workers involved compared to coal; it is a liquid, so it can be routed along different channels more easily; there is a whole set of technical properties of oil and its production that are different. That does not mean to say that the energy is determining the outcome of history or of political struggles, and I am careful in the book to introduce examples that do not work easily one way or the other: the history of oil industry in Baku, which is much more similar to the history of coal, or the oil industry in California for example. But you can pay attention to the technical dimensions in a certain way, and to the new political possibilities that arise with this enormous development of sources of energy, which reflects both an exponential increase in the amount of energy but also an unprecedented concentration of the sites at which energy is available and through which it flows. You can tell a new story about democratic politics and about that moment in the history of industrialized countries, but also the subsequent history in oil-producing countries in a different way. That would be an example of how attention to technical expertise and technical processes translates into a different understanding of the politics of oil.

The Triple Helix, the Complexity of Technological Innovations, and the Decomposition of National Innovation Systems: An Interview with Loet Leydesdorff

Ruth Knoblich

Loet Leydesdorff suggests a view on technological innovations as a complex process in which he disaggregates national innovation systems. The theoretical model of the "Triple Helix" puts emphasis on transnational and potentially uncoupled dimensions of innovation dynamics. Leydesdorff shows the limits of governmental policies and proposes communication processes as a crucial lens to explore global, national, and regional levels of the knowledge economy.[1]

[1] An extended and modified version of this interview, including additional references, can be found on http://www.theory-talks.org/

R. Knoblich (✉)
IEE/Ruhr-University Bochum, Bochum, Germany

IPWS/Bonn University, Bonn, Germany
e-mail: ruth.knoblich@rub.de

M. Mayer et al. (eds.), *The Global Politics of Science and Technology - Vol. 1*, Global Power Shift, DOI 10.1007/978-3-642-55007-2_13,

How Are States and Technological Innovations Connected?

If we take a long-term perspective, then one can say that the nation state—the national or political economies in Europe—were shaped in the nineteenth century, somewhat later for Germany (after 1871), but for most countries it has been the first half of the nineteenth century. This was after the French Revolution and in relation to industrialization. These countries were able to develop an institutional framework for organizing the market as the wealth-generating mechanism. The institutional framework permits to retain wealth, to regulate market forces, and also to steer them to a certain extent. However, the market is not only a local dynamics; it is also a global phenomenon.

Nowadays, another global dynamics is involved: science and technology add a different dynamics than the market. The market is an equilibrium-seeking dynamics at each moment of time. According to traditional perspectives on economy, markets are inherently equilibrium-seeking. The evolutionary dynamics of science and technology now add a non-equilibrium-seeking dynamics on top of that, and this puts the nation state in a very different position. Thus, one obtains a complex adaptive dynamics, or an eco-dynamics or however you want to call it—these are different words for approximately the same thing. For the nation state, the question arises how it relates to the market dynamics on the one side, and to the knowledge dynamics on the other. Thus, the nation state has to combine two tasks. Let me illustrate this model of three subdynamics with a figure from my *The Knowledge-Based Economy* (2006: 24) (Fig. 1).

The figure shows that first-order interactions generate a knowledge-based economy as a next-order or global regime on top of the localized trajectories of nation states and innovative firms. These complex dynamics have to be specified and then to be analyzed empirically. For example, knowledge dynamics change the relation between government and economy; they also change the position of the state in relation to the wealth-retaining mechanisms. How can the nation state be organized in such a way as to retain wealth from knowledge locally, while knowledge (like capital) tends to travel beyond boundaries? One can envisage the complex system dynamics as a kind of *cloud*—a cloud that touches the ground in certain places, as Harald Bathelt illustrates.

How can national governments shape conditions for the cloud to touch and remain on the ground? One tends to think of university-industry-government relations first as neo-corporatist arrangements between these institutional partners. However, I am interested in the ecosystem shaped through the tri- and bilateral relationships. This ecosystem can be shaped at different levels. It can be a regional ecosystem or a national ecosystem for instance. One can ask whether there is a synergy between these three (sub-)dynamics and where that synergy can generate wealth, knowledge, and control; in which places and along trajectories for which periods of time—that is, the same synergy as meant by "touching the ground". A complex system can be expected to generate niches with synergy, but as unintended consequences. To what extent and for which time span can these effects be anticipated and facilitated?

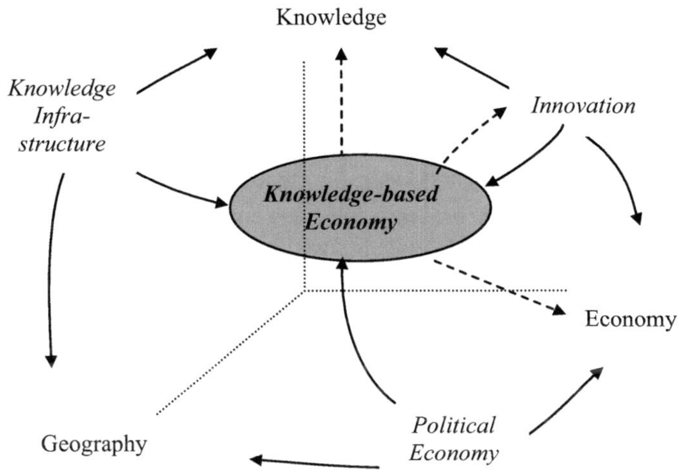

Fig. 1 The first-order interactions generate a knowledge-based economy as a next-order system. (Source: Leydesdorff, Loet (2010). The Knowledge-Based Economy and the Triple Helix Model. *Annual Review of Information Science and Technology, 44*, 367–417, at p. 379.)

At this point, Luhmann's theory comes in because he has this notion of different codifications of communication, which then, at a next-order level, begin to self-organize when symbolically generalized. Codes are constructed bottom-up, but what is constructed bottom-up then may begin to control top-down. Thus, one should reflexively articulate what is constructed from the bottom up and why. What are the selection mechanisms? Observable relations (such as university-industry relations) are not neutral, but mean different things for the economy and for the state for example. And this meaning of relations can only be evaluated in terms of codes of communication.

Against Luhmann's model, I argue that codes of communication can be translated into one another while interhuman communications are not operationally closed, as in biology. One needs other social-scientific perspectives on these fluidities, such as proposed by Michel Callon for example. In evolutionary economics, one distinguishes between market and non-market selection environments. But not among selection environments that are differently codified. Here, Niklas Luhmann's theory offers a heuristic: The complex system of communication tends to differentiate in terms of the symbolic generalizations of codes of communication because this allows the system to process more complexity and thus to be more innovative. The selection environments, however, have to be historically specified because they are not given but rather constructed over long periods of time.

What Is the Most Relevant Aspect of the Dynamics of Innovation for the Discipline of International Relations?

The main challenge is to endogenize the notions of technological progress and technological development into theorizing about political economies and nation states. The endogenization of technological innovation and technological development was placed on the research agenda of economics by evolutionary economists like Nelson and Winter in the late 1970s and early 1980s. In this context, the question was how to endogenize the dynamics of organized knowledge, science and technology into economic theorizing. However, one can just as well formulate the problem of how to reflect on this globalizing subdynamic of knowledge in political theory and International Relations.

Knowledge dynamics provide the system with another degree of freedom. In the study of political economies, the focus has rested on two subdynamics: economic wealth generation and political control. In a knowledge-based economy, however, a third subdynamic—organized knowledge production—has to be endogenized into the model. The Triple Helix of University-Industry-Government Relations can be considered as an eco-system of bi- and tri-lateral relations. The three institutions and their interrelations can be expected to form a system carrying the three functions of (1) novelty production, (2) wealth generation, and (3) normative control.

Since political dynamics are always involved, this has implications for International Relations as a field of study. The dynamic analysis is different from comparative statistics. The knowledge dynamics can travel and be "footloose" to use the words of Vernon, whereas it leaves footprints behind. Grasping "wealth from knowledge" (locally or regionally) requires taking a systems perspective. However, the system is not "given"; the system remains under reconstruction and can thus be articulated only as hypothesis. In the social sciences, however, one can also use the concept of a system heuristically.

For example, which level of analysis would lead to better explanations: scrutinizing the German knowledge-based economy at the country level (e.g., in terms of the East-West or North-South divide) or at the level of its Federal States? How can one provide political decision-making with the required variety to operate as a control mechanism on the complex dynamics of these eco-systems? For example, when studying Piemont as a region in Northern Italy, it is questionable whether the synergy in university-industry-government relations is optimal at this regional level or should better be examined from a regional perspective including Lombardy. On the one hand, the administrative borders of nations and regions result from the construction of political economies in the nineteenth century, but, on the other hand, the niches of synergy that can be expected in a knowledge-based economy are bordered also; for example, in terms of metropolitan regions (e.g., Milan–Turin–Genua).

In evolutionary economics, one distinguishes furthermore between technological trajectories and regimes. Trajectories can result from "mutual shaping" between two selection environments; for example, markets and technologies. Nations and firms follow trajectories in a landscape. Regimes are global and require the specification of three selection environments. When three (or more) dynamics interact, symmetry can be broken and one can expect feed-forward and feedback loops. Such a system can begin to flourish auto-catalytically when the configuration is optimal. From such considerations, a confluence of the neo-institutional program (of Henry Etzkowitz) and my neo-evolutionary view, the Triple Helix model emerged: how do institutions and functions interrelate?

What Does the "Triple Helix" Reveal About the Fragmentation of "National Innovation Systems"?

Colleagues from the Economics and Management Studies at the Erasmus-University in Rotterdam offered firm data from the Netherlands with these three dimensions: the economic, the geographical and the technological dimension; based on data from more than a million Dutch firms. I presented the results at the Schumpeter Society and asked whether someone in the audience had similar data for other countries. I expected Swedish or Israeli, but someone from Germany stepped in, Michael Fritsch, and so we did the analysis for Germany. These studies were first published in *Research Policy*. Thereafter, we did studies on Hungary, Norway, Sweden, and recently also China and Russia.

Several conclusions arise from these studies. The data can be decomposed along different dimensions. One can decompose national systems geographically into regions, but one can also decompose them in terms of the technologies involved. We were mainly relying on national data. And of course, there are limitations to the data collections. Actually, we now have international data, but this is commercial data, and therefore more difficult to use than governmental statistics.

For the Netherlands, we obtained the picture that can more or less be expected: Amsterdam, Rotterdam, and Eindhoven are the most knowledge-intensive and knowledge-based regions. This is not surprising although there was a bit of surprise: We know that in terms of knowledge bases, Amsterdam is connected to Utrecht and then the geography goes a bit to the east in the direction of Wageningen. What we did not know was that the niche also spreads to the north in the direction of Zwolle. The highways to Amsterdam Airport (Schiphol) are probably most important.

In the case of Germany, when we first analyzed the data at the level of the "Laender" (Federal States), we could see the East-West divide still prevailing, but when we repeated the analysis at the lower level of the "Regierungsbezirke" we did not find the East-West divide as dominating (using 2004 data). So, Dresden for example was more synergetic in Triple-Helix terms than Saarbruecken. And that was nice to see considering my idea that the knowledge-based economy increasingly prevails after the fall of the Berlin Wall and the demise of the Soviet Union. The discussion about two different models for organizing the political economy—communism or liberal democracy—has become obsolete after 1990.

After studying Germany, I worked with Balázs Lengyel on Hungarian data. Originally, we could not find any regularity in the Hungarian data, but then the idea arose to analyze the Hungarian data as three innovation systems: one around Budapest, which is a metropolitan innovation system; one in the west of the country, which has been incorporated into Western Europe; and one in the east of the country, which remains the old innovation system that is state-led and also dependent on subsidies. For the western part, one could say that Hungary has been "europeanized" by Austria and Germany; it has become part of a European system.

When Hungary came into the position to create a national innovation system, free from Russia and the Comecon, it was too late, as Europeanization had already stepped in and national boundaries were no longer as dominant. Accordingly, and this was a very nice result, assessing this synergy indicator on Hungary as a nation, we did not find additional synergy at the national level. While we clearly found synergy at the national level for the Netherlands and also found it in Germany, mainly at the Federal States' level, we could not find synergy at a national level for Hungary. Hungary has probably developed too late to have a nationally controlled transformation.

A similar phenomenon appeared when we studied Norway: my Norwegian colleague (Øivind Strand) did most of our analysis there. To our surprise, the knowledge-based economy was not generated where the universities are (Oslo and Trondheim), but at the West Coast, where the off-shore, marine and maritime industries are most dominant. FDI in the maritime and marine industries constitutes regional development on the West Shore of Norway. Norway is still a national system, but the Norwegian universities like Trondheim or Oslo are not much involved in entrepreneurial networks. These are traditional universities, which tend to keep their hands off the economy.

Actually, when we had discussions about these two cases, Norway and Hungary, which both show that internationalization had become a major factor, either in the form of Europeanization in the Hungarian case, or in the form of foreign-driven investments (off-shore industry and oil companies) in the Norwegian case, I became uncertain and I thought: Do we not believe in our own artifacts? Therefore, I proposed to Øivind to study Sweden, given the well-organized data available on this national system. We expected to find synergy concentrated in the three regional systems of Stockholm, Gothenburg, and Malmø/Lund. Indeed, 48.5 % of the Swedish synergy is created in these three regions. This is even more than one expects on the basis of the literature. Some colleagues were upset, because they had already started trying to work on new developments of the Triple Helix, in Linskøping for example. But the Swedish economy is organized and centralized in this respect. Perhaps, that is why they need to talk so much about "regionalization" in policy documents. Sweden is very much a national innovation system, with additional synergy between the regions.

Can Governments Alter Historical Trajectories of National, Regional or Local Innovation Systems?

Let me mention the empirical results for China in order to illustrate the implications of empirical conclusions for policy options. We had no Chinese data set, but we got access to the database Orbis of the Bureau van Dijk (an international company, which is Wall Street oriented, assembling data about companies) that contains industry indicators such as names, addresses, NACE-codes, types of technology, the size of the enterprise, etc. However, this data is very incomplete. Using this incomplete data for China, we said that we were just going to show how one could do it if one had full data. We know that the National Bureau of Statistics of China has complete data. I did the analysis with Ping Zhou at the Zhejiang University.

We analyzed China first at the provincial level, and as expected, the East Coast emerged as much more knowledge intense than the rest of the country. After that, we also looked at the next-lower level, that is the 339 prefectures of China. And then, four of them popped up as far more synergetic than the other ones. These four municipalities were: Beijing, Shanghai, Tianjin, and Chongqing. These four municipalities became clearly visible as an order of magnitude more synergetic than other regions. The special characteristic about them is that—different from the others—these four municipalities are the ones administered by the central government. Actually, it came out of my data and I did not understand it; but my Chinese colleague said that this result was very nice!

The Chinese case thus illustrates that government control matters. It shows—and that is not surprising as China runs a different model—that the government is able to organize the four municipalities in such a way as to increase synergy. Of course, I don't know what is happening on the ground. We know that the Chinese system is more complex than these three dimensions suggest. I guess the government agencies consider the option of extending the success of this development model, to Guangdong for example or to other parts of China. Isn't it worrisome that all the other and less controlled districts have not been as successful in generating synergy?

Is There a Connection Between Infrastructures and the Success of Innovation Processes?

Throughout all advanced economies, knowledge-intensive services (KIS) are not synergetic because they can be disconnected—uncoupled—from the local economy. For example, if one has a knowledge-intensive service in Munich and receives a phone-call from Hamburg, the next step is to take a plane to Hamburg, or to catch a train inside Germany perhaps. Thus, it does not matter whether one is located in Munich or in Hamburg as knowledge-intensive services uncouple from the local economy. The main point simply is proximity to an airport or to a train station.

This is also the case for high-tech knowledge-based manufacturing. But it is different for medium-tech manufacturing, because in this case the dynamics are more embedded in the traditional parts of the economy. If one looks at Russia, the knowledge-intensive services operate differently from the Western European model, where the phenomenon of uncoupling takes place. In Russia, KIS contribute to coupling, as knowledge-intensive services are related to state apparatuses. In the Russian case, the whole knowledge-based economy is heavily concentrated in Moscow and St. Petersburg. So, if one aims—like the Russian government—at creating not "wealth from knowledge" but "knowledge from wealth"—that is, oil revenues—an advice might be to uncouple the knowledge-intensive services from the state apparatuses. Of course, that is not easy to do in the Russian model because traditionally, the center (Moscow) has never done this. Uncoupling knowledge-intensive services, however, might give them the degree of freedom necessary to move around, from Tomsk to Minsk or *vice versa*, steered by economic forces more than they currently are (via Moscow).

Gaia Theory, Ecovillages, and IR Taking Material Limits Seriously: An Interview with Karen Litfin

Maximilian Mayer

Karen Litfin argues that IR needs to seriously engage with the idea of limited energy resources and the essential embeddedness of human society in the earth system. Stressing the multiple purposes and often unexpected applications of technology, Litfin talks about interdisciplinary research as an important way of overcoming a mechanistic worldview and the example of ecovillages as a sustainable form of living.[1]

[1] An extended and modified version of this interview, including additional references, can be found on http://www.theory-talks.org/

M. Mayer (✉)
Center for Global Studies, Bonn University, Bonn, Germany
e-mail: maximilian.mayer@uni-bonn.de

M. Mayer et al. (eds.), *The Global Politics of Science and Technology - Vol. 1*, Global Power Shift, DOI 10.1007/978-3-642-55007-2_14,

What Is the Conceptual Place of Technologies Within International Relations Theory?

I think there is a huge gap: International relations (IR) does not pay nearly enough attention to technological systems—and when they do, it's generally from an uncritical and mechanical perspective. Even though much of the constructivist critique of liberal institutionalism is that the latter is overly materialistic, it actually is not. Institutionalists talk about economics as if that were material reality, but it's not: economics is a secondary human system overlaid on material systems. I think that IR needs to get really serious about understanding the actual material basis for politics. Climate change will probably be the issue that drives such a move. So what kinds of technologies and institutions are we going to have to facilitate a global civilization? This is a worthwhile question. We have a more or less globalized economy, but we do not have a global polis; we do not have the institutions that are commensurate to the economy that we have got. So the question is: can we sustain current civilization on the energy budget that is available to us and not wreck the climate?

Having achieved something of a global economy, the central problem now is to formulate the political institutions that are commensurate to our globalized economic institutions. We have far to go on that project. It also means doing so within the carrying capacity of the earth. The challenge, in terms of actual politics, is to find those institutions. The challenge for the discipline of international relations is to do the necessary thinking to facilitate that institutional transition. But few IR scholars even acknowledge that political institutions must attend to the carrying capacity of the earth. In general, the discipline of international relations, political science and even most of social science behave as if there are no natural constraints to our behavior. Yet our freedom to even be able to theorize about the international system is completely dependent upon a vast web of life, other people growing our food, and a whole technological infrastructure that we had nothing to do with creating. International Relation theory talks a lot about interdependence, but do we really take it seriously?

Technological systems are driven by energy; energy is the master resource. Some energy analysts say that in order to have a global civilization, we need to have energy return on investments something like 5 to 1. Right now petroleum is getting, depending on where you find it and how it's getting to you, somewhere between 15 and 25 to 1. It used to be 100 to 1 at the beginning of the nineteenth century. Currently, we are getting, say, 20:1. I have seen analyses of tar sands that put that energy source at somewhere between 3 and 5 to 1. Solar panels, if they work well, are maybe getting 5:1. The trend is worsening and we are starting to push that envelope of 5:1 energy return on investment. And if we exploit some of the new unconventional hydrocarbons—like fracking and, worse, methane hydrates—to their maximum potential, we'll fry the planet. My question is how can we leverage existing resources, technological, economic, financial and political resources to sustain a global civilization.

I wish more people were putting their attention on that question. The assumption underlying where most people are putting their attention, is that business as usual can continue. Stephen Quilley, an environmental sociologist, coined the term "low energy cosmopolitanism." A huge challenge lies before us: If it is possible to have a global civilization on the energy budget that we have available, it is going to be some form of a low energy cosmopolitanism, where we make some very conscious choices about what we are going to globalize. For instance, Germany probably would not be importing grapes from Africa and none of us would be going on luxury vacations We would be making a lot of conscious choices, but if we want to have a global civilization we have to be globalizing something, so what is it that we are globalizing?

How Do You Explain the Non-linear and Unpredictable Pathways of Technologies?

I could address that question in a very specific way with respect of specific technologies. Satellites, for instance, do have certain properties embedded in them. Initially, my feminist theoretical critique of earth observing satellites argued that this kind of gaze from space actually does downplay or preclude certain perspectives. But as I thought about it more deeply, I saw very concretely that a lot of people are using those technologies to do what *they* want—not what the centralized political and scientific institutions that gave rise to the satellites wanted. So I would say the wildcard here is consciousness and human inventiveness, because that's what will shape how people deploy the technologies once there are on the ground. For example, satellites were devised for spying and are certainly still being used for spying, but they are being used for so much else, such as Google Maps. I think some people might have been able to foresee that kind of development, but most of us did not expect that this sort of thing could come about. Or that you could have indigenous people mapping their traditional lands in order to make land rights claims. So the wildcard really is human consciousness and that is why nothing really is deterministic.

The greater the complexity in a living system, the more surprising are its emergent properties. Seven billion human brains linked together in global technological and ecological systems are bound to yield surprises! But if you just look at human systems as living systems there can be a kind of materialistic reductionism there. We should not be materialistic reductionists. There is this wildcard of human consciousness. The fact of the matter is, we can assemble all the data we want but we do not know where we are going. But what we do know is that we have created a tremendously complex and complicated world that nobody can actually understand.

Does This Mean That IR Scholars Should Become Energy Engineers?

I do not do the technical analysis; I just read the conclusions and assume if it is passing peer review in the scientific journals, it is probably a valid conclusion. On the basis of that I do my thinking. Energy return on investment is a huge concept and hardly anybody is paying attention to it. Thomas Howard-Dixon, in his 2006 book *The Upside of Down*, paid attention to it, but he is a rare exception among IR scholars. When I started teaching international relations back in the early 1990s, I started realizing that petroleum holds the whole thing together. The whole global system was held together by petroleum. You could also say fossil fuels, but coal and natural gas do not power that much transnationally; it is petroleum. Yet hardly anybody in IR talks seriously about petroleum—or energy or biodiversity or soil or the atmosphere. That's what I mean about getting to the material basis.

I have always been interested in science and technology. As an undergraduate, I studied physics and astronomy, but I did not finish those majors because I realized, that if I graduated with those degrees I would most likely be working indirectly or directly for the military. I got politicized and I began to see that the political agenda drives the scientific agenda. This was in the 1970s and it was possible at that time that we were going to have an all-out nuclear war. I did not want to be a part of that. I began to see that there is a dialectical relationship between science and politics. Because science facilitates the technological changes, which make the basic backdrop for politics, it's very important. Science also provides metaphors through which we understand politics. I did my M.A. thesis on the mechanistic worldview and the devitalization of nature in the seventeenth century. I also traced it back to the ancient Greek philosophy. A reductionist and mechanistic worldview underlies a lot of IR theory, as well most of our political institutions. We need to really start questioning that. I'm interested in biological metaphors because I think the leaning edge of human thought understands human systems as living systems. From this vantage point, we can begin to reshape our institutions in ways that mimic, sustain, and regenerate living systems. There is a long history of natural law and I do not exactly put myself in that camp, but I think there are ways that we need to understand ourselves as thoroughly embedded in natural systems and then move consciously from that place.

Having said that, I think how we interact with the material basis is a reflection of our consciousness. This does not imply a material reductionist stance. Rather, I am looking for a wholeness that understands our approach to material reality as being a reflection of our consciousness. One thing that I think IR can do, as well as political science, the social sciences and every discipline, is to really look at system theory. If we think this world system that we've created of a globalized economy and rudimentary international law is not a part of a living system, we are living in a big delusion. So to actually understand how living systems function, we need the literature on system theory that of course has been used in biology and ecology, but has also been applied a lot in the business world and organizational development. And I think it is also making its way into IR.

How Is "Gaia Theory" Related to Your Research?

Looking through the lens of Gaia Theory, we would first understand that we exist within certain spheres such as biosphere, atmosphere and hydrosphere. We have taken geological time and inserted it into human time by digging up fossil fuels. As a consequence, we have kind of checkmated ourselves and are now forced into having to think in geological terms. We have to start thinking in geological time scales, which was never the case before. If we are going to find a way of inhabiting this planet sustainably, particularly if we are going to have anything approaching a global civilization, we have to understand that we live within a living system and then go about the rather daunting but exciting project of developing international law and institutions that reflect that reality. There is a whole subfield of earth system governance in which Earth system scientists, IR theorists and international legal experts are coming together to think through these questions. The literature on earth system governance starts from the premise that the Earth is a living system and draws heavily on earth system science, which draws heavily from Gaia theory. You cannot separate atmosphere, oceans, lithosphere, biosphere: they are all intertwined as one big living system—and now humanity is functioning as a geophysical force on a planetary scale.

That's the meaning of the Anthropocene, and it will require an entirely new way of going about politics and economics. The idea that we are coming up against real limits is a very frightening idea from the perspective of a certain idea of freedom rooted in liberal politics. We really need to rethink the meaning of freedom in an era of limits. My own feeling is that human beings are kind of hard-wired towards unlimitedness—but the world is now pressing us to interrogate this impulse. We don't do well with limits. But the fact of the matter is, we are not evolutionarily adapted to abundance, we don't even know what to do with abundance. We are squandering resources in the most absurd ways. So we really need to rethink what freedom is in a world of limits.

It is not all together a bad thing that we are facing these limits. Those of us who have at least the privilege of being well fed and reasonably comfortable, can actually turn our attention to this question of consciousness—because the question of "what is freedom" is a problem of human consciousness. Rather than turning our desire towards mastery—I think as human beings we have an innate desire towards mastery—rather than turning that desire onto the external world, we have pretty well mastered it; except turns out that we live in it so it is coming back to bite us and we are facing huge climate change most likely.

Do "Ecovillages" Offer Alternative Solutions for Global Problems Such as Climate Change, Water Scarcity, or Desertification?

I visited a number of ecovillages and ecovillage networks in both developing countries and affluent countries. It is different in the developing countries and in the affluent countries, and I think it is important to clarify that at the outset. In the latter, there is a greater possibility for what I consider 'post-individualist' that both transcends and includes individualism. A very simple 'post-individualistic' approach to property rights, for instance, would be co-housing, where the land is owned in common and people own their own homes. But their private homes would be a lot smaller because so many amenities are shared. The common house would have a community kitchen, so that, depending upon how much people are willing to share, private kitchens can be very small. If there is a collectively owned guest space, then you do not need a guest room in your house. And if you do a lot of your socializing together, then you can do that in the common house. So your own house could be quite small but you would still have access to all the comforts of a private existence and more. The more people are willing to share, the more will be collectively owned. And that really does require trust, because it is a big problem if the relationships blow up and you have your finances entangled with those people! This is just one example of how property rights can coexist with the softening of boundaries between individuals.

The flipside of this is occurring in developing countries, where the post-individualistic arrangement that I have been making does not really apply. And this is important because that's where most people in the world live. There you have cultures where people already have much more of a collective orientation. So we really need to pay attention to what is happening there. Actually, in many cases, their developmental task is to become more individuals. And the question is: how do they become more highly individualized rather than being subsumed by traditional moral codes—how do they that without over-consuming. In the west, we had a fossil fuel subsidy that enabled us to become highly individualized, as I said before, the only reason we can be having this interview is because somebody else is growing our food.

Going to ecovillages, I asked myself who is living in ways that can actually work for the long run. I wanted to see collective efforts and particularly larger communities that were generally at least a hundred people, because you can do a lot more collectively, than you can on your own. Some of these communities are reducing their ecological footprint radically. In some cases, we are talking about per capita reductions in material consumption and waste production of 80–90 % as compared to their home country averages. This is very big news—especially given that these communities are still tied to the larger system. They are not tiny isolated enclaves. For instance, they're still using the mass transit of the larger society; most of them have Wi-Fi and high-speed Internet. They are not living in caves and many of them are very much globally engaged. On a material level, they're much closer to living within the Earth's carrying capacity. So in that way, I was very interested in just seeing what are their physical systems. But I began to see that their physical systems were only made possible because of the degree of trust and reciprocity that they have created.

On a practical level, one of the ways that they reduce their footprint is by withdrawing to some extent from the global economy. Having very low consumption and being fairly energy efficient and self-reliant, reliance on food self-sufficiency, but withdrawing from global society. To me, they are answering the question I raised earlier: What would a low-energy cosmopolitanism look like? And they are doing this not just because they consume less and live more simply but because by and large ecovillagers actually have a cosmopolitan identity. They might be growing their own food and composting their garbage, but they are also tied into the global system. They are actively engaged in the Internet, sometimes attending global conferences and many of them are politically active on issues such as genetically modified organisms and nuclear waste disposal and human rights. They are little nodes of positive examples, but they are very small and hardly anybody lives in an ecovillage. So, the question becomes how these principles could be scaled up to the level of cities, regions, national government and international norms.

The Role of Science and Technology in the Dynamics of Global Change and the Significance of International Knowledge Cooperation in the Post-Western World: An Interview with Dirk Messner

Ruth Knoblich

Based on three waves of global change specifically challenging the management of global commons, Dirk Messner points to the role of science and technology particularly in managing the climate system. Illustrating that global power shifts lead to new patterns of international cooperation, he argues that investment in joint knowledge creation and knowledge exchange is vital for managing the earth system. Messner emphasizes the significance of interdisciplinary research for global development, mentions initiatives taken by development institutions to set up open access to research pools, discusses opportunities held by big data and refers to the present redefinition of the discipline of development research itself.[1]

[1] An extended and modified version of this interview, including additional references, can be found on http://www.theory-talks.org/

R. Knoblich (✉)
IEE/Ruhr-University Bochum, Bochum, Germany

IPWS/Bonn University, Bonn, Germany
e-mail: ruth.knoblich@rub.de

M. Mayer et al. (eds.), *The Global Politics of Science and Technology - Vol. 1*, Global Power Shift, DOI 10.1007/978-3-642-55007-2_15,

What Role Do Science and Technology Play in the Dynamics of Global Change?

There are multiple important dimensions, but I would like to focus on three waves of global change that make it specifically challenging to manage the global commons. When I say global commons I have in mind the atmosphere, the climate system, and other parts of the earth system, but also the international financial markets and global infrastructures, such as the Internet—the stability of global commons such as these is a much required public good. The first wave of global change is the globalization wave; the economic globalization, cross-border dynamics, global value chains. It becomes evident that in many areas and especially when it comes to the global commons, regulation exceeds the capacity of individual nation states. The international community is required to institutionalize multilateralism and efficient global governance mechanisms in order to properly address issues arising from global dynamics. The second big global change is the shift from a Western to a post-Western world order. Global power shifts remaking the international system impede governing global commons. The third wave of global change is related to climate change, which adds a new dimension of global dynamics; human beings now have to learn how to steer, to stabilize, and how to govern the earth system as such. We are not only a species living on this planet, depending from resources and ecosystems of the earth systems. With the acceleration of economic globalization during the 1990s and the emergence of new, non-Western economic drivers of change, like China, humankind impacts now significantly the physical structures of the earth system. This trend is new. For the first 4.6 billion years of the existence of the earth system it was driven by the laws of physics, the dynamics of biology and bio-chemical processes. Homo sapiens appeared 220,000 years ago, and the impacts of our species on the earth system have been marginal until the industrial revolution started 250 years ago. During the last decades human mankind became a major driver of change at a planetary scale.

Technology is driving economic globalization, the first wave of global change. So we need to understand the dynamics of new technologies, especially the impact of ICTs, in order to understand the dynamics of economic globalization. The World Wide Web and social communication media are restructuring industrialization processes and global value chains. ICT infrastructure is also displaying a big potential for less developed regions. In Africa, for example, we saw many African countries jumping from the old telephone technologies to smartphones within less than a decade, because the old, maintenance and capital intensive communication infrastructure was no longer needed. Many African people now have access to smartphones, thus to communication- and information networks, and begin to reshape prize constellations and the global economy. Because of its restructuring effects, the impact of ICTs is relevant in all areas of the global economy. The global trend towards urbanization is similarly related to ICTs. Currently, we approach the global economy via data on national economies. But this might be about to change, as global mega-cities develop into global knowledge and financial hubs, building their own networks. In 2030, 80 % of the global production, global GDP, global consumption, global exchange might be concentrated in 70–80 global cities or city regions.

Technology is also linked to the second wave of global change—the tectonic global power shift—in the way that investment in technology and knowledge in emerging economies are growing rapidly. We are not only facing economic and political power shifts, but also a remaking of the global science and research system itself. From my perspective, international cooperation in the field of science and technology research between "old powers" and "new powers", between Western countries and non-Western countries is extremely important for two reasons: First, we need to pool know-how in order to solve core global challenges and to develop patterns for managing the global commons. Interaction and cooperation in the field of science and technology is especially important for the creation of knowledge that is "better" in any way. For instance, in the field of adaptation policies to the impacts of climate change, most of the knowledge on how societies and local communities actually work or respond under these conditions exists in non-Western societies. The generation of knowledge is context dependent. We need to interact with colleagues from the respective countries for mutual learning and common knowledge improvement. My second argument is that, as an effect of the global power shift, traditional development cooperation is losing legitimacy. Many of these societies, from China to Peru, from Kenya to Vietnam, are no longer interested in our usual business, in our "aid-packages", our money, our experts or our concepts. What they are more interested in is true and reciprocal knowledge exchange and joint knowledge creation. Therefore, investments in respective forms and institutions of knowledge exchange and creation will be a central pillar of/for future oriented development cooperation or international cooperation and beneficial for all partners involved.

The role of technologies concerning the implications of climate change is crucial and multifaceted. In the German Advisory Council on Global Change we put forth suggestions concerning the transformation towards a low-carbon global economy. We are relatively optimistic in a technological sense. This statement is partly based on the Global Energy Assessment (GEA) research, which has been driven by Nebojsa Nakicenovic, one of our colleagues, who is working on energy modeling. The perspective there is that we know which kind of technologies we need for the transformation into a low-carbon or even zero-carbon economy. We can even calculate the investment costs and structures of different countries and regions. But we do know relatively little about the transformation processes of entire societies, economies and, eventually, the international system. The transformation towards a low-carbon society is a "great transformation". In the entire history of mankind there might be only two examples for such a profound change: the industrial revolution 250 years ago and the Neolithic revolution 10,000 years ago, which induced the practices of agriculture. Today, we thus witness the third great transformation: the decoupling from fossil resources, from high-carbon to zero-carbon. To achieve the 2° Celsius goal, a complete decarbonization of the basic infrastructures of the global economy (the energy systems, the urban infrastructures and systems, the land use systems) is required. Comprehensive knowledge is key to achieve this. Let me emphasize once more the significance of international cooperation in the field of science and technology research, particularly in the IPCC context. I am sure that politicians from China, India, or Brazil only accept what the IPCC is presenting as objective knowledge, as the state-of-the-art knowledge, because their national scientists are deeply involved. If this were a classical western-based knowledge project it would have resulted in a lack of legitimacy. In the case of global climate policy, it is obvious that investment in joint knowledge creation is also about creating legitimacy for joint action.

In What Ways Has the Role of Scientific Knowledge and Technical Expertise Changed Within Development Institutions?

Exemplarily, we could have a look at the World Bank. The current reorientation of the World Bank as a Knowledge Bank originates from the assumption that knowledge is just as important as money for global development. The second point is that more and more of their partners in non OECD-countries, classical developing and emerging economies, are more and more interested in the knowledge pools of the World Bank and less in their experts. And: dynamic developing countries and emerging economies are even more interested in investments in their own knowledge systems and joint knowledge creation with the World Bank. The old North-South knowledge transfer model is eroding. You might say that currently there are two contradictory global trends: on the one hand via social media and the Internet knowledge is being widely distributed—broader than ever before and actually, theoretically accessible at any point in the world—, on the other hand the proliferation of knowledge is accompanied with access restriction and control, and the growing privatization of knowledge. Aiming to play a constructive role in collaborative knowledge generation, the World Bank invests a lot in building up freely accessible data bases and open research tools, including the provision of governance or development indicators of any kind. However, this is a difficult process that is developing slowly.

What Are the Opportunities of Big Data Analysis for Global Development?

The access to any kind of data is important for any kind of knowledge creation. It has been very limited for many developing countries for a very long time. So, thinking about how to assure access to serious data is significant. This would be my first point. My second point is that, when it comes to big data and the question of managing large amounts of indicators on, for example, cross-country or cross-sector modeling, I think the new technologies are opening up new research possibilities and opportunities. Big data provides the opportunity to identify patterns.

Looking for similar dynamics in very different systems is a very interesting exercise, because you get deeper insights into the basic dynamics of systems. This is what I have learned from my colleague Nakicenovic, whom I have mentioned before, and who is working on the Global Energy Assessment, or from Juergen Kurths, from the Potsdam Institute for Climate Impact Research, who is studying basic structures and dynamics of very different systems like air traffic networks, global infrastructures and social media networks. Managing big data allows you to see patterns which cannot be seen if you only work with case studies. However, to understand the dynamics of countries and sectors, new actor constellations or communities, you need to go into details and in this specific moment, big data is of no use since you need qualified and serious data on the ground. Big data and qualified, specific data: they complement each other.

For sure, an important aspect of big data is that for the most part, it is gathered and stored by private businesses. We started this interview talking about global commons and we actually just defined a global commons: data on development should be global commons, and we need standards and rules of managing those. Private actors could play a role, but within a set of rules defined by societies and policies, and not the private business sector.

What Is the Value of Interdisciplinary Research in the Context of Global Challenges Such as Environmental Change?

We talk a lot about interdisciplinarity, however, we do not have career paths that systematically build interdisciplinary teams. Looking particularly at global environmental changes and the future of the earth system, at the end of the day, social scientists and natural scientists need to learn how to work together and to understand each other. The future of the oceans, for example, is not a question that can be understood by ocean biologists only. They are the people studying how this element of the earth system is actually working, its dynamics, its drivers—focusing on physical, chemical, and biochemical processes. But when we look at the oceans towards 2100 from the perspective of global change, the most important drivers are us human beings, our economies, and our consumption patterns, our greenhouse gas emissions and their impacts on the oceans. And this implies that in order to understand the dynamics of global change, we need to analyze the interactions, interdependences and feedback loops between three systems: the ecological system(s); social systems (our economies and societies) driven by humans; the technical systems and infrastructures; therefore natural scientists, social scientists, and engineers need to interact very closely. In the German Advisory Council we call this approach: Transformation Research. Currently, we do not possess the appropriate university structures to adequately address this sort of problem. This is an immense institutional challenge. If I were a young scholar I would move into this direction, crossing disciplinary boundaries as much as possible.

How Far Is the Discipline of Development Research Affected by Global Change?

This is a complex question, to which I do not have a definite answer. The whole field of development research is currently about to get redefined. In the past, the concept of development was clear: On the one side, there was the developed world, the OECD-world, consisting of 35–40 countries and on the other side, the "underdeveloped" part of the world, all other countries. Understanding the differences between developed and developing, along with thinking about the basic drivers of modernization and wealth creation in less developed countries was at the core of development research for a long period. How can poor countries become rich and as developed as OECD countries already are?

Today, it is highly questionable if even the broader categories of "development research" still serve to analyze the new realities. Do we currently still need "development economists", and how would they differ from classical "economists" doing research in those European countries suffering most from the debt crisis, high unemployment and weak institutions? Situations in many OECD-countries nowadays look like what one would expect from a still developing or emerging economy, and the other way around. So, what makes development research specific? This is a real question. Studying non-OECD countries, do we still need governance theories, democratization theories—thus, theories that are systematically different from those we apply in our research on OECD-countries? The discipline of development research is under immense pressure. This debate is linked to the second wave of global change we talked about: the post-Western world order, emerging economies catching up, convergence trends in the global economy.

If you look at the role of international technology transfer, the same scenario arises: the North-South, donor-recipient categories have dissolved. Technology transfer has lost its distinct direction, and it is much more reciprocal and diffuse than it used to be. There are several studies currently pointing to the fact that investment rates in R&D and in technology creation are growing fast in several regions around the globe, whereas in many OECD-countries, investment is stagnating, or even decreasing. The whole map of knowledge, if you like to say so, is about to undergo deep changes. This implies that the common assumption that knowledge is based in OECD-countries and transferred to the South via development cooperation is just not working any longer. We need new patterns of cooperation between different countries in this area. And we need research on global development dynamics which will be different from classical development research which has been based on the basic assumption of a systemic North-South divide.

Why Is the Low-Carbon Transformation so Difficult to Achieve?

The first two great transformations have been evolutionary processes. No one "planned" the industrial revolution, not to mention the Neolithic revolution. These have been evolutionary processes. The sustainability transformation process instead needs to be a governed process right from the beginning. In our institute, we looked at different transformation dynamics, not only the really big ones, the Neolithic, industrial, and the current sustainability transformation. We also examined structural adjustment programs in Latin America and Africa, the collapse of communism at the end of the 80s, the abolition of slavery, and similar other transformations of human societies. Based on this historical perspective, we have identified four main drivers of transformation: The first one is crisis, this is the most important one. Confronted with strong crisis, society and probably also individuals re-act and change direction. The second important driver is very often technology and scientific revolution. The third driver is vision: if you are confronted with a problem but you do not know where to go to, transformation becomes very difficult. The European Union is the product of a fresh vision among elites after World War II; the United Nations is a result of the disasters of the first half of the twentieth century. Advancing a vision is an essential means to move or to transform in a goal-oriented manner. Sustainability, of course, is also a vision. The fourth and last driver of transformation is "knowledge": you know that you have a certain problem constellation, and though the crisis is still not there, you react based on your knowledge in a preventive way.

For the low-carbon transformation, the fourth driver currently is absolutely key. We are able to address problems which would otherwise become much worse in the future, although the climate crisis is latent still—in contrast to, for example, the financial crisis, which is more visible in its effects. The impacts of a global warming of 4 or 5° are still not visible. This makes for a huge difference. In fact, humans are not very good at acting and transforming significantly based on knowledge only. In combination with visible, tangible crises, knowledge is a strong driver of change, but without crisis, it is merely sufficient. Transformations based on knowledge and preventive action only are rare. The ozone hole is one positive an example; solving the problem was possible, because it required less complex technological change, affecting few industries only. Human beings are risk-averse in a sense, we are conservative, we do not like to change rapidly; we are path-dependent. John Maynard Keynes once said: "It is easy to develop new concepts and ideas. The difficult thing is to forget the old ones". Therefore, scientific tools are needed in order to sketch out future scenarios. Based on scientific knowledge, we need to convince our societies, our political decision-makers that it is necessary and possible to transform societies and economies towards sustainability. Pushing towards sustainability at a point where the crisis has not yet materialized implies a specific and new role for science in managing global dynamics. Organizing a deep transformation towards sustainability, avoiding significant crisis driven by Earth system changes, would be a cultural learning process—a civilizational shift.

Nuclear Ontologies, Technopolitics in Postcolonial Spaces, and the Cold War as Transnational History: An interview with Gabrielle Hecht

Maximilian Mayer

Gabrielle Hecht argues that nuclear exceptionalism is a form of technopolitical claims-making. Tracing the historical process by which something comes to count as a "nuclear" thing not only reveals the ontological insecurity of the nuclear, but also shifts the research focus from high politics to the (post)colonial spaces that play a key role in nuclear weapons programs. The study of the diversity of global technopolitics requires a close collaboration between science and technology studies and IR.[1]

[1] An extended and modified version of this interview, including additional references, can be found on http://www.theory-talks.org/

M. Mayer (✉)
Center for Global Studies, Bonn University, Bonn, Germany
e-mail: maximilian.mayer@uni-bonn.de

M. Mayer et al. (eds.), *The Global Politics of Science and Technology - Vol. 1*, Global Power Shift, DOI 10.1007/978-3-642-55007-2_16,

How Could International Relations Theory Conceptualize Science and Technology?

It does not make sense to think about international relations without thinking about the technologies, systems, and infrastructures that make any kind of global movement possible. The flows of people, of products, of culture, political exchanges—these are all mediated through and practiced in the technological systems that permeate our globe. I draw attention to the specific political practices that are enacted through technological systems with the notion of technopolitics. I initially used this concept in my work on nuclear power in France to capture the ways in which hybrid forms of power are enacted in technological artifacts, systems and practices. There I used the term in a rather narrow sense to talk about the *strategic* practices of designing technologies to enact political goals. My paramount example was that of the French atomic weapons program. In the early 1950s, France's political leaders insisted that France would never build atomic weapons. But engineers and other leaders in the nascent nuclear program were designing reactors in a way that optimized the production of weapons-grade plutonium rather than electricity. When politicians finally signed on, the technology was ready to go. This example problematizes the very notion of a "political decision." Instead of a single, discursive decision, we see a complex process whereby political choices are inscribed into technologies, which subsequently favor certain political outcomes over others.

In this example, both engineers and politicians *consciously* engaged in technopolitics. By contrast, Timothy Mitchell has used the hyphenated term "techno-politics" to emphasize the unpredictable and unintended effects of technological assemblages. Over the last 15 years, I have also developed a broader notion of the term, particularly in its adjectival form, "technopolitical." I find this to be a useful shorthand for describing both how politics can be strategically enacted through technological systems, and also how technological systems can be re-appropriated for political ends in ways that were unintended by their designers. The point, really, is to highlight the myriad politics of materiality.

Does the Lens of "High Politics" Capture the Particular Characteristics of Nuclear Weapon Programs?

It is not that there are no high politics, of course there are. But they cannot offer a sufficient or straightforward explanation for how or why any particular country develops a nuclear program. A focus on high politics implies a focus on decision makers and moments. But that is really misleading. In pretty much every case, the apparent "moment" of decision is in fact a long process involving a tremendous amount of technopolitical, cultural, and institutional work, rife with conflicts and contingencies of all kinds. I think a more productive approach is to try to understand nuclear capacity-building. Itty Abraham has done some fantastic work on India's nuclear program, which helps us think about other cases as well. For example, he analyzes the symbolic importance of the nuclear test, noting that IR uses "the test" as kind of "aha!" moment, the *moment* in which one knows that a country has nuclear weapons. Instead, Abraham sees the test as a process for the cultural production of meaning: a process in which certain meanings get fixed, but by no means the most important moment for understanding the actual technology and politics behind the production of nuclear weapons.

As You Stress the Ontological Insecurity of Nuclear Materials, Do You Mean the "Nuclear" Is Purely a Matter of Social Construction?

No, definitely not. But to explain what I mean by all this we should take a few steps back and start with what I like to call *nuclear exceptionalism*. This is a technopolitical claim—emerging immediately after the end of World War II—that there was something radically unique about nuclear things. From 1945 onward, both cold warriors and their activist opponents cultivated this nuclear exceptionalism. Atomic weapons were portrayed as fundamentally different from any other human creation. The bomb was the ultimate geopolitical trump card, and it was imagined as replacing empire in one fell swoop. You see nuclear scientists and engineers gaining prestige, power, and funding far beyond their colleagues in conventional research. In the meantime, anti-nuclear groups make their own claims to exceptionalism by talking about the unprecedented dangers posed by nuclear things. Everywhere you see nuclearity and morality intertwined. Nuclear things either represent salvation or moral depravity... or the apocalyptic end of mankind. But regardless of where you stood politically, this notion of nuclear exceptionalism rested on the sense that the difference between nuclear and non-nuclear things was transparent—ultimately a clear-cut, physical matter of radioactivity.

The nuclear thus emerged not just as a category, but also as a universal and universalizing ontology, one that seemed to apply in the same way all over the globe. And frankly, historians, political scientists, anthropologists, and sociologists have reproduced that nuclear exceptionalism. I did it myself in the *The Radiance of France*.

All of which has made it hard to see is what I call nuclearity—the process by which something comes to count as a "nuclear" thing—has a history, a politics, and a geography. Things that count as nuclear in one time and place might not count as nuclear at another. Rendering something as nuclear and exceptional is a form of technopolitical claims-making. It follows that insisting that certain things are not especially nuclear, or that they are banal, is also a form of technopolitical claims-making.

You can see this in the response of the nuclear industry to activist opponents. In the late 1960s and over the course of the 1970s, the nuclear industry began to represent nuclear power not as a life-saving technology for the human race, but as simply another way to boil water. Radiation was just another industrial risk. Such representations sought to banalize nuclear things.

Nuclearity could thus get made, unmade and remade. My favorite example comes from a 1995 US government report on nuclear proliferation. The appendix has a table that summarizes the nuclear activities of 172 nations. Neither Gabon, nor Niger, nor Namibia are listed as having any nuclear activities, despite the fact that those nations together, during that very year, produced something like 25 % of the world's uranium. So when does uranium count as a nuclear thing? When does it lose its nuclearity? And what does "Africa" have to do with it?

The argument is not that radioactivity does not have to do anything with nuclearity, or that nuclearity has nothing to do with the technologies and physical processes we typically associate with the word. Rather, I argue that nuclearity is one thing, and radioactivity and fission are another; sometimes they overlap, but not always and not necessarily. Understanding where and why they *don't* map onto each other is politically revealing.

Let's Apply the Notion of Nuclear Exceptionalism to the Case of Iran: What Alternative Questions Would This Raise About the Issue?

Nuclear technology has played an important role in shaping modern Iranian national identity. This began in the 1970s under the Shah, who—with the support of the US—developed a grandiose plan to build a fleet of nuclear reactors. Things took a different turn after the 1979 Iranian revolution. For a while, the new regime sidelined the nuclear program as an unwelcome manifestation of western corruption. But after a few years leaders reappropriated nuclear development and sought to invest it with Iranian-ness. The dynamics of nuclear exceptionalism have operated in Iran much the same way they did in France and in South Africa. Nuclear exceptionalism has served to give material form to national identity. And materialized national identity is most emphatically *not* something that you can negotiate away in the P5 + 1 talks.

What Do You Expect from the Growing Academic Exchange Between Science and Technology Studies and IR?

Science and technology studies (STS) is really good at exploring practice, and especially at calling attention to the differences between principles and practice—for example, between regulation on the one hand, and the actual practices that regulations are meant to control (without ever entirely succeeding) on the other hand. STS can bring to IR an understanding of how the intimate details of practice matter politically—of how everyday technopolitical and techno-scientific exchanges can be more important loci for politics than treaties, diplomacy, and other forms of what might be called high politics.

I can also answer this question wearing my historian's hat. The IR scholarship on nuclear weapons that I am familiar with seems to be quite focused on models—on using history to produce predictive models that will in turn serve to shape international policy on nuclear weapons regulation. But if history tells us one thing, it is that models are basically useless for understanding how countries develop nuclear weapons. Instead, history and STS both teach us about which questions to ask (in this instance, about nuclear development). Identifying the important questions—rather than prescribing the applicable model—leaves open the list of possible answers. It also leaves open solutions and policies, letting us being more attentive to the specificities and uniqueness of individual cases.

Why Are Science and Technology Hardly Ever Studied in the Postcolonial World from a STS Perspective?

I think there are a number of reasons why STS has paid relatively little attention to the postcolonial world. One is that in STS—like many disciplines—the prestige of the subject matter maps onto the prestige of the researcher. So STS researchers who study cutting-edge science or large-scale technological systems seem somehow to be getting at "harder" topics, those that focus on active creation. Engineering and other acts of creation appear more prestigious than acts of maintenance, or acts of dismantling. Even studying small-scale creation seems to confer more prestige than studying mundane practices. This brings us back to the theme of rupture vs. continuity: studying or proclaiming rupture seems somehow sexier—and certainly more radical—than studying continuity. Another, more trivial answer is just that most STS researchers so far have come from Europe and North America, and they tend not to be trained in area studies.

In Which Ways Does Exploring the Cold War as Transnational History Change Our Understanding of Global Politics?

Understanding the history of the Cold War—even its most iconic technology, nuclear weapons—as a form of transnational history really calls attention to spaces that have previously been considered marginal, even perhaps not fully nuclear. Ultimately, it should provoke us to problematize the "the Cold War" as a frame for global or transnational history (and social science). In *Entangled Geographies*, my colleagues and I build on the work of Odd Arne Westad, whose book *The Global Cold War* was an argument for understanding the non-superpower, non--European dimensions of the Cold War. We propose a technopolitical spin, which offers a de-provincializing of the Cold War that is complementary to Westad's. By focusing on places like Saudi Arabia, or Zimbabwe, or Brazil, or South Africa, we show how even the central struggles of the Cold War were intimately bound up in "northern" relationships to colonial and post-colonial worlds, and in the imaginaries that characterized those relationships.

In *Being Nuclear* I focus on uranium from Africa—more specifically South Africa, Namibia, Gabon, Madagascar, and Niger. Uranium from Africa has long been a major source of fuel for nuclear power and atomic weapons, including the bomb dropped on Hiroshima, but it has been almost completely absent from accounts of the nuclear age, whether scholarly or popular. This changed in 2002, when the US and British governments claimed that Iraqi leader Saddam Hussein "sought significant quantities of uranium from Africa" (later specified as the infamous "yellowcake from Niger"). Africa suddenly became notorious as a source of uranium. But that did not admit Niger, or any of Africa's other uranium-producing countries, to the select society of nuclear states. Nor did it mean that uranium itself counted as a nuclear thing. The books explores what it means for something — a state, an object, an industry, a workplace — to be "nuclear." I show that such questions lie at the heart of today's global order and the relationships between "developing nations" and "nuclear powers."

Being Nuclear argues that "nuclearity" is not a straightforward scientific classification but a contested technopolitical one. In the first part I follow uranium's path out of Africa and analyze the invention of the global uranium market. In the second part I enter African nuclear worlds, focusing on miners and the occupational hazard of radiation exposure. In both parts, I show that nuclearity requires instruments and data, technological systems and infrastructures, national agencies and international organizations, experts and conferences, and journals and media exposure. When (and where) nuclearity is densely distributed among these elements, it can offer a means of claiming expertise, compensation, or citizenship. It can serve as a framework for making sense of history, experience, and memory. When (and where) network elements are absent, weak, or poorly connected, nuclearity falters, fades, or disappears altogether, failing to provide a resource for people claiming remediation or treatment. Nuclearity in one register does not easily transpose to another: geopolitical nuclearity does not automatically translate into occupational nuclearity. Yet these domains remain connected. African uranium miners depend on the transnational movement of nuclear things, but that movement also depends on African miners. Ultimately, I conclude, nuclear security must be considered in tandem with other forms of human security—food and health and environmental and political security. By placing Africa in the nuclear world, and the nuclear world in Africa, the book seeks to remake our understanding of the nuclear age.

I should note that it is not only uranium production that connects colonial and postcolonial spaces with nuclear things. African countries were not the only places where uranium was produced. Much of the rest of the world's uranium came from the Navajo Nation in the United States, Aboriginal territories in Australia, First Nation territories in Canada, colonized spaces in the Soviet Empire, etc. French nuclear weapons were tested in the Algerian desert and French Polynesia; the United States tested its weapons on the Bikini Islands; Britain tested its weapons in Maralinga, in Aboriginal Australia; the Soviet Union tested its weapons on the planes of Kazakhstan.

In Which Ways is the Picture You Have Chosen Representative for the Global Politics of Science and Technology?

I chose the picture that also is on the cover of *Being Nuclear* in part to disrupt facile distinctions between the "global" and the "local." The image portrays a uranium miner touching a big pile of rocks in Niger. He is wearing two pieces of protective gear: an industrial mask, designed to prevent inhalation of radioactive dust, and a *tagelmust*, a headwrap that Tuaregs and other Saharans use to prevent inhalation of desert dust. This juxtaposition disrupts the expectations of many North American and European viewers, who do not associate nuclear things with nomadic people... or even, to be blunt, with black people. So on a really basic level, the picture invokes the dependence of nuclear technology on the labor of men like this one, living in places like Niger (which is now one of the world's largest uranium producers).

Once you read the book, the picture acquires additional dimensions. For example, I examine how levels of radon (a radioactive gas produced by uranium) were—

or were not—measured in African mines, and how health data about African uranium miners did not get included in the production of "global" science about low-level radiation exposures, in part because much of their work never became nuclear. So the picture presents us with an irony: the mask is an icon of global occupational health, but the more you learn, the more you realize that this mask offers too little too late. As you continue to ponder the picture, it may occur to you that the *tagelmust* is necessary protection against high desert winds... winds which blow the dust from the uranium tailings piles over a wide swath of land (some of which is inhabited). Finally, if you look really closely you may notice that part of the man's index finger is missing—a reminder of the ordinary dangers of mine work.

GPSR Compliance

The European Union's (EU) General Product Safety Regulation (GPSR) is a set of rules that requires consumer products to be safe and our obligations to ensure this.

If you have any concerns about our products, you can contact us on ProductSafety@springernature.com

In case Publisher is established outside the EU, the EU authorized representative is:

Springer Nature Customer Service Center GmbH
Europaplatz 3
69115 Heidelberg, Germany

Batch number: 10739625

Printed by Printforce, the Netherlands